D1380568

*Now available in a lower priced paperback edition in the Wiley Classics Library.

Continued on back end papers

Nonsampling Error in Surveys

Nonsampling Error in Surveys

JUDITH T. LESSLER

Research Triangle Institute
Research Triangle Park, North Carolina

WILLIAM D. KALSBEEK

University of North Carolina
Chapel Hill, North Carolina

A Wiley-Interscience Publication
JOHN WILEY & SONS, INC.
New York • Chichester • Brisbane • Toronto • Singapore

A portion of this book was prepared with the support of NSF Grant SOC-7804597. However, any opinions, findings, conclusions and/or recommendations herein are those of the authors and do not necessarily reflect the views of NSF.

In recognition of the importance of preserving what has been written, it is a policy of John Wiley & Sons, Inc., to have books of enduring value published in the United States printed on acid-free paper, and we exert our best efforts to that end.

Library of Congress Cataloging in Publication Data:

Lessler, Judith T.
    Nonsampling error in surveys / Judith T. Lessler, William D. Kalsbeek.
        p.   cm.—(Wiley series in probability and mathematical statistics. Applied probability and statistics)

    "A Wiley-Interscience publication."
    Includes bibliographical references and index.
    ISBN 0-471-86908-2
    1. Error analysis (Mathematics) 2. Surveys. I. Kalsbeek, William D., 1946–   . II. Title. III. Series.

QA275.L44 1992
511′.43—dc20
                                                    91-31950
                                                    CIP

Printed and bound in the United States of America by Braun-Brumfield, Inc.

10 9 8 7 6 5 4 3 2 1

To Dan Horvitz
for His Long and Continuing Commitment
to High Quality Surveys.

Contents

Preface

Current survey practice has evolved through efforts by a broad segment of the research community. Practitioners and methodologists trained in various disciplines have played integral but distinctive roles in this dynamic process. The practitioner's operation has generally provided the source of and testing ground for new ideas, while the problems that inevitably accompany the implementation of new ideas have often been defined and solved in the methodologist's laboratory. Both orientations have contributed to a better understanding of survey error, thus enabling survey research to evolve steadily from its crude beginnings centuries ago. Moreover, broad interest in survey error research since the 1940s has led to a voluminous literature on the subject.

Nonsampling Error in Surveys evolved from a National Science Foundation study (SOC-7804597) whose goal was to investigate the feasibility of creating a unifying taxonomy of sampling and nonsampling survey errors, along with a mechanism for archiving empirical data on these errors in the interest of improving future survey design. One important outcome of that study was the sobering realization that research on nonsampling error is much too extensive and diverse to allow one to meet these goals without first attempting to consolidate and summarize existing research. This book was born out of that realization and is intended to serve as a general reference for students and practitioners of the survey method.

Although nonsampling survey error can reasonably be classified in several ways, we presume for the sake of presentation that it originates from either of the following three general sources: problems with the sampling frame (frame error), nonresponse in obtaining data from sample members (nonresponse error), and inadequacies in the process of obtaining survey measures from respondents (measurement error). We chose this classification because these sources of survey error are distinguishable in the survey operation and because most of the literature on nonsampling error divides quite neatly into these three categories.

We begin (Chapter 1) with a discussion and some history of survey error in general terms, followed thereafter (Chapter 2) by an examination of where, in the course of a survey, problems leading to error in estimates may occur. Sampling and nonsampling errors are included in both discussions. Except for Chapter 12, where two general error models are suggested, the remaining chapters of the book can be categorized according to survey error arising from the three types of error just described: frame error (Chapters 3 through 5), nonresponse error (Chapters 6 through 8), and measurement error (Chapters 9 through 11). A Compendium of Nonsampling Error Terminology, which contains referenced verbatim definitions for many of the key terms in the text, is presented following Chapter 12.

The presentation format for the three types of nonsampling error is similar. Each set of three chapters devotes some attention to defining key terms, formulating known effects, and examining suggested remedies. Terminology related to each source of error is reviewed while historically tracing the major contributions to the literature. Emphasis is given to those statistical models that have been posed to facilitate our understanding of the effects of each error source. We have also tried to provide a reasonably comprehensive treatment of the many ways that have been devised to deal with these effects.

The emphasis given to definitions, effects, and remedies varies somewhat among types of nonsampling error. This variation largely reflects differences in the amount of existing knowledge and attention in the literature. For example, the chapters on error due to problems with the sampling frame and nonresponse tend to give relatively higher priority to statistical remedies, while the chapters on measurement error focus more heavily on the statistical models that have been suggested.

Our relative preoccupation with terminology was motivated by an important barrier to understanding that we discovered in some parts of the literature. Either the same name was used for different concepts, or different names were used for the same concept. This revelation led us to believe that it would be important to share with the reader some of the actual words and phrases we found for key concepts in the literature as well as a sampling of word-for-word definitions that some authors have suggested.

Use of the book may vary somewhat, depending on the reader's background and interest. For instance, new students of the survey method who are interested in a largely nontechnical introduction to nonsampling errors may wish to focus their attention on Chapters 1–4, 6, 7, 9, and 10. These chapters pinpoint the origins of various specific errors and emphasize the effects associated with specific errors. Comparative discussions of terminology are presented for each source of error. More advanced students may be more interested in Chapters 5, 8, and 11, which emphasize remedies for specific error sources, and in Chapter 12, which suggests two ways to isolate marginal and joint effects of survey error.

For the practitioner requiring a reference guide to assist in finding a solution to a design or analysis problem tied to nonsampling error, selected material in Chapters 4, 5, 7, 8, 10, and 11 may prove to be useful. Unfortunately, the breadth of the literature in this field precludes an in-depth treatment of most topics. Thus our presentation specifically aimed at the practitioner must be limited to a general description of the statistical manifestations of the different forms of nonsampling error and to a relatively limited exposition of the various published strategies for dealing with these manifestations. References are intermixed in this discussion to satisfy the reader who seeks a more detailed treatment of the presented topics. We hope that this format will be useful to the full range of practitioners, from the novice in a low-budget survey, who wishes to better understand and thereby anticipate the common pitfalls of survey research, to the experienced specialist in a large complex study, who must find creative solutions to the many problems that survey error can cause.

No book can be written without significant time and effort by many beside the authors. Although the ultimate responsibility for content is of course ours, we wish to extend a sincere note of thanks to those who have contributed to the final product. We are, first of all, indebted to Daniel Horvitz for his encouragement and continued support for our efforts, but perhaps most important, because it was his dream of a mechanized information system to improve survey design that first motivated the work which led to this book. Ralph Folsom also made many important contributions to our effort as a valued colleague in the NSF grant, which laid the groundwork for the book. We also wish to acknowledge gratefully his many helpful insights and sound advice during the planning and preparation of the manuscript. We wish to thank Frank Andrews, Barbara Bailar, Joseph Duncan, Kent Marquis, and Donald Rubin, who served in an advisory capacity for the NSF grant. This advisory group was used as a sounding board for several ideas as well as a source of numerous helpful suggestions in the course of our work. We are also deeply indebted to J. Michael Bowling, Robert Casady, Brenda Cox, Graham Kalton, Roderick Little, Colm O'Muircheartaigh, and Michael Weeks for their critical reviews and helpful comments on various partial drafts of the manuscript. We wish to recognize the excellent editorial help provided by Lynn Igoe and the superior typing skills displayed by Ernestine Bland, Jennie Caparella, Cynthia Coates, Laurine Johnson, Pat Parker, Pat Penland, and Brenda Porter, all of whom made significant contributions in preparing various drafts of the manuscript. Last, but by no means least, we express our sincere thanks to Beatrice Shube and Kate Roach at Wiley for their constant encouragement and seemingly infinite patience with us during the book's preparation.

JUDITH T. LESSLER
WILLIAM D. KALSBEEK

Nonsampling Error in Surveys

CHAPTER 1

Introduction

In this book we review methods developed to measure and reduce nonsampling errors in survey statistics. The contents are delimited by our definition of *survey*. We reviewed a number of books before developing our definition. Few of the sources reviewed provided a short, succinct definition of a survey, and it was most common to see a fairly lengthy discussion of the nature of surveys illustrated by examples. In these discussions, surveys were distinguished by their purpose, subject matter, and type of methods used in gathering data (Moser and Kalton, 1972). Usually, a distinction was made between scientific studies based on observation data and those based on experimental data (Murthy, 1967; Jessen, 1978; Dalenius, 1983b). Surveys were considered to be a form of observational study. Textbooks by social scientists often described a survey as being restricted to a study of people or human populations (Babbie, 1973; Warwick and Lininger, 1975; Hoinville et al., 1978; Backstrom and Hursh-César, 1981). Others recognized the role of surveys in studying institutions, agricultural or industrial production, forestry, business, inventories, and so on (Hansen et al., 1953a; Deming, 1960; Murthy, 1967; Dalenius, 1974).

Based on this review, we have developed the following definition of a survey: A *survey* is a scientific study of an existing population of units typified by persons, institutions, or physical objects. A survey attempts to acquire knowledge by observing the population as it naturally exists and making quantitative statements about aggregate population characteristics.

We will consider a survey to include censuses in which attempts are made to study all members of the population and sample surveys in which a scientific sample of the population is studied. Thus a study that asked all U.S. automobile manufacturers to report on their expenditures for employee health care would be considered a survey, as would a study in which a scientific sample of auto workers is asked to report on their personal health care expenditures. Although one may argue that nonprobability samples are in some instances scientific, we choose to exclude from consideration studies using nonprobability samples.

1

By stating that a survey is the study of a population as it naturally exists, we mean to exclude experimental studies in which the material to be studied is manipulated by the researcher and the results observed. However, observational studies and epidemiologic studies in which the aim is to make analytic comparisons between certain subgroups of the population are considered to be surveys. Thus a study comparing the mathematics achievement levels of 9-year-olds in rural and urban schools would be considered a survey, whereas a study measuring achievement levels of children randomly assigned to methods of mathematics teaching would not. A clinical trial on the efficacy of a drug is not a survey; a study of adverse reactions to anesthesia by examining a sample of hospital records is.

We also exclude from consideration case studies in which only a few specific members of the population are included and no quantitative inference to the aggregate population is attempted. Thus our definition of a survey excludes an anthropological study of patterns of drug use among teenagers in Dade County, Florida; however, it includes a study that selects a sample of in-school 16- and 17-year-olds and administers a questionnaire on drug use.

Methods used in surveys are used in other areas of scientific study, and there is no universally accepted definition for a survey. In some studies, the distinctions above are not maintained and experimental treatments and survey techniques are combined. In addition, the techniques presented in this book for studying survey errors are appropriate for other types of endeavors. However, we believe the definition will encompass most of what is commonly considered to be a scientific survey.

1.1 HISTORICAL DEVELOPMENT

Studies of the history of surveys note that this history is characterized by the increasing use of surveys, the development of probability sampling methods, a growing awareness of the errors that can be present in survey results, and an increasing sophistication of methods used for gathering information and controlling for errors.

Short histories of the use of surveys can be found in Moser and Kalton (1972) and Warwick and Lininger (1975). Surveys were conducted during the ancient Roman and Egyptian empires principally to obtain information for military conscription, taxation, and other governmental matters. During the period 1700–1900 use of surveys increased with a focus on studying social problems. Motivated by the economic deprivation that plagued much of society during those times, numerous studies of the poor, the imprisoned, and the outcast were conducted. Survey investigators also became aware of the need for improved survey techniques. Notable among those aware of the importance of using sound survey practices are Charles Booth, who studied London's poor (Moser and Kalton, 1972), and Fredric LePlay, who investigated the economic plight of European families (Warwick and Lininger, 1975). Concern with methodology established their surveys as among the most scientifically sound prior to the twentieth century.

Usefulness of sample estimates relative to complete enumeration was also recognized during this period. Kruskal and Mosteller (1980) trace the development of the idea of sampling during a review of use of the term *representative sample*. At the 1895 meeting of the International Statistical Institute (ISI), Anders Kiaer advocated the use of representative sampling rather than complete censuses for social investigations. In later articles, Kiaer stressed his preference for "a small number of careful observations carried out with great care to a large number of superficial observations made superficially on a large scale" (Kruskal and Mosteller, 1980:177). Today, this preference is still cited as the principal advantage of surveys relative to censuses.

Kiaer did not advocate use of probability sampling but rather suggested that the sample be spread widely over the population to achieve a miniature of the larger population. Subsequent to the 1895 ISI meeting, the concept of using random samples was developed, along with methods for using probability models to assess the accuracy of the sample estimates.

The idea of using samples instead of censuses generated considerable controversy. During the 1895 ISI meeting George von Mayer criticized the use of samples, commenting that there should be "no calculation when observations can be made" (p. 175). By the time of the 1925 ISI meeting, the idea of sample investigations had been accepted; however, there was continued controversy about the relative usefulness of purposive versus random selection. Neyman's landmark 1934 paper served to strengthen greatly the case for probability methods that "make possible an estimate of the accuracy of the results obtained … irrespectively of the unknown properties of the population studied" (Neyman, 1934:585).

Once the use of samples was accepted, methods for probability sampling and control of sampling error rapidly developed. A theory of sampling was developed in which size of the sampling error depends on the variability in the entire population, type of sample design, sample size, and type of estimation procedure used. Although this theory has spawned a wealth of practical procedures, there continues to be some controversy as to the basic theoretical underpinnings of survey statistics. Reviews of the development of sampling methodology may be found in Stephan (1948), Dalenius (1957, 1962), and Hansen and Madow (1976). The article by Kruskal and Mosteller (1980) contains an extensive bibliography of very early work. For a discussion of the continuing theoretical controversies, see Smith (1976), who reviewed work on the foundations of survey sampling theory and an article by Hansen et al. (1983), followed by comments from conflicting points of view.

In addition to adopting sampling methods and theory, survey practitioners have been concerned with improving the quality of other aspects of survey research. In 1915, Bowley, who made very important contributions to the development of sampling methods, reported on a survey of employment and poverty. He stated that there were four possible sources of uncertainty or error in an investigation as to social conditions:

> The information obtained may be incorrect; the definitions and standards used [for determining poverty] may be loose, unsuitable, or wrongly conceived; the households actually visited may not contain a fair sample of the whole population; and the calculable possibilities of error arising from the process of estimating the whole by measuring a part (Bowley and Burnett-Hurst, 1915:174).

At the 1926 ISI meeting Bowley further stressed the need to control multiple sources of errors; he pointed out that it was necessary to define the population in question exactly, to have adequately defined attributes and variables, and to make sure that every person or thing selected is observed. He states that "any breach of these conditions, however slight, introduces an unknown element of error in the result, and destroys the relevance of the [sampling] formula" (Bowley, 1926).

There was continuing concern with survey errors, and in 1944 Deming attempted a classification of "factors which affect the ultimate usefulness of a survey" which was much broader than these Bowley cited nearly 20 years before. In addition to sampling errors and biases, Deming listed the following:

- Variability in response
- Differences between different types and degrees of canvas
- Bias and variation arising from the interviewer
- Bias of the auspices
- Imperfections in the questionnaire design and tabulation plans
- Changes that take place in the population before the tabulations are available
- Bias arising from nonresponse
- Bias arising from late reports
- Bias arising from unrepresentative data for the survey or the period covered
- Bias arising from unrepresentative selection of the respondents
- Errors in interpretation

Deming's classification is neither complete nor are its categories mutually exclusive; however, it illustrates well the range of factors that must be considered when attempting to assess and control the errors in surveys.

Despite this long recognition of the need for the control of nonsampling errors, progress in the development of theories and methods for controlling them has been much less satisfactory than progress in the understanding and control of sampling errors. This is because of the complexity of the problem.

In some cases defining error is difficult. Also, most surveys involve a complex sequence of procedures carried out by many different people, so that it is difficult to control the process. For sampling we have a theory that allows us to calculate the error that results from a conscious choice to use a certain sample design. We make many other choices of methods; however, we do not have a comprehensive theory that allows us to calculate the error that results from these choices.

1.2 THE NATURE OF SURVEY ERROR

The results of a survey are used to make quantitative statements about the population studied. These may be descriptive statements about the aggregate population, analytic statements about the relationship among subgroups of the population, or interpretative statements about the nature of social or economic processes. A survey error occurs when there is a discrepancy between the statements and reality.

Survey errors are generally divided into two major types, sampling errors and nonsampling errors. *Sampling errors* are present by design and result from the conscious choice to study a subset rather than the whole population. Efforts to control sampling error are grounded in a well-developed theory, so that in designing a sample a researcher focuses on development of estimation formulas and random selection techniques suitable to the particular problem and falling within the context of their theory. Sampling errors are not the result of mistakes per se, although mistakes in judgment when designing a sample may cause larger errors than necessary.

Nonsampling errors encompass all other things that contribute to survey error. Nonsampling errors are often thought of as being due entirely to mistakes and deficiencies during the development and execution of the survey procedures. Nonsampling errors are said to arise from wrongly conceived definitions, imperfections in the tabulation plans, failure to obtain response from all sample members, and so on. Thus a perfect design perfectly implemented would be free of nonsampling errors.

In many cases it is helpful to think of nonsampling errors as arising from avoidable deficiencies and mistakes. However, in other instances it might be better to conceive of the errors as resulting from the conscious choice to use a certain method during the survey, with full knowledge that the method has some survey error associated with it. The problem then becomes one of defining, measuring, and ultimately controlling the errors associated with the chosen methods.

That there is no comprehensive theory for assessing the impact of nonsampling errors is not surprising when one considers the complex nature of surveys and the multiple opportunities for error. Often, survey information is wanted as a basis for a policy decision. The decision to do a survey may itself be an error because even the best conducted survey could not provide the information needed. There may be problems in defining the measurements. Correctly defined measurements may be incorrectly made. Measurements may be missing for members of the sample. Tabulation methods may be incorrect. The wrong population may be studied. Many, many factors can cause a disagreement between survey results and true population values. As one might expect, even the notion of true population values is controversial. For some survey error researchers, the notion of an absolute standard for comparison is a fundamental element of their conceptual framework. Other researchers are satisfied with a purely operational view of reality where measurements are simply defined as a product of a specified data collection process. An absolute standard of truth plays no role in this purely operational framework.

1.3 TOTAL SURVEY DESIGN

The attempt to control the total error of estimates considering all sources of error has come to be called *total survey design*. During the design phase of the survey, the practice of total survey design involves assessing the level of error associated with alternative procedures and choosing that combination of sampling design, measurement procedure, and analysis method which will minimize the total error of the final estimates within the resources available for the survey. For example, the sampling error in a survey estimate is reduced by increasing the sample size. Thus, since telephone interviews are cheaper than personal interviews, a larger sample can be used if telephone interviewing is chosen for a survey. However, the researcher may still decide to use personal interviewing because the other errors associated with telephone interviewing may offset any reduction in error through increased sample sizes. This situation may be because the subject matter of the survey dictates a lengthy, complex interview that respondents have difficulty answering by telephone. Or it may be that the goals of the survey require information from people without telephones.

The success of total survey design methods at the planning stages depends on obtaining good information on costs and errors of alternative procedures and on the availability of total error and total cost models that can be used for choosing an optimum design. While the survey is being carried out, the practice of total survey design involves the use of quality control procedures that monitor progress of the data collection and data processing. The goal of the quality control procedures is to detect errors when they occur or soon after so that the survey work can be repeated if necessary. Response rates, item completion rates, edit failure rates, consistency checks, resurveying, and recoding are methods used to detect errors in the ongoing survey process.

Applying total survey design at the analysis and reporting phase of the survey entails attempting to calculate and report on the total error of the survey estimates. Use of suitably designed probability sampling procedures allows one to calculate accurate measures of the sampling error. Rigorous measurement of the effects of other sources of error also depends on introducing experimental procedures into the survey process that allow determination of the magnitude of the effect of a particular error source on the total error of estimates. An example is the use of record check studies in which a subsample of respondents is selected and more accurate record data obtained to verify respondent-reported data. Information in a properly designed record check study can be used to estimate the bias caused by respondent misreporting. The information can also be used to adjust the results to correct for this bias before publication.

Ideally, the practice of total survey design operates in a comprehensive and integrated fashion. Surveys are carefully planned with consideration given to all known sources of error. Resources available for conducting the survey are then directed toward minimizing total error and not any single error component. Quality control procedures are included which alert researchers in a timely manner to mistakes in the data collection and processing. During analysis an attempt is made to determine the likely total error of the estimates, and an assessment of probable total error is included in the survey reports. Finally, in anticipation of future surveys on similar topics, estimates of the components of the total error are made so that they may be used in the planning phase of surveys that follow.

Clearly, the preceding paragraphs describe an idealized standard of performance. The present state of the art falls considerably short of this goal. Attempts to employ total survey design in surveys in the design, execution, or analysis phases have been hampered by several problems. Additional complexities are introduced into the survey. Extra data must be collected or experimental methods must be included to permit estimation of the impact of certain sources of nonsampling errors. Once data are available, effort and money must be expended to arrive at the estimates. Quality control procedures also take time and money that could be devoted to the primary activities of sampling and data collection.

In many cases researchers are simply reluctant to face the problems that may be present in the survey. An "ignorance is bliss" attitude seems to prevail and gratuitous assumptions are made about the quality of the data (the 7000 nonrespondents are adequately represented by the 3000 respondents). For a one-time survey, conventional wisdom often dictates using methods that are believed to give good results for the funds available without adequate investigation of alternatives. Selections of methods are based on expert opinion reinforced by flimsy empirical justifications.

Lack of data on costs of survey methods and their associated errors seriously impedes the use of total survey design methods. Most surveys are directed at gathering information about a certain population and readers of the reports about them are interested in the substantive results. Although general users may be interested in the total error of the estimates, they probably have little interest in the detailed technical aspects of the design, such as components of the variance, cost of training interviewers, amount of time and number of miles spent traveling between sample clusters, cost of calculating the standard errors, and so on. This information is exactly the detail needed for designing future surveys. If this information is available at all, it is usually in the internal records of the survey research organization conducting the survey; it is rarely published.

Even within survey research organizations, the information needed to design the next survey may not be available. Cost accounting methods are not designed to give cost components related directly to the variable components of the survey design. Thus, whereas we may know the total mileage and time interviewers spent traveling, we may not know how much of that was between primary sampling units, how much was for callbacks to not-at-home households, and so on.

This unavailability of information led to the idea of a survey design information system suggested by Daniel G. Horvitz (1978) at the Second Symposium on Survey Sampling. As conceived by Horvitz, this computerized system "would store information on the magnitude of each of the components of error identified for a specific social variable measured in a specific manner with a specific population group along with associated cost components." Users would employ the data when designing new surveys and would report information on cost and error components for their current surveys.

Horvitz cited several advantages of such a system: (1) promoting the use of standard terminology, (2) integrating knowledge about the sizes of survey errors associated with the measurement of particular characteristics, (3) facilitating the use of total survey design methods by permitting retrieval of information that could be used in the design phases of a survey, (4) providing standards to compare survey estimates, and (5) encouraging the execution of methodological studies.

To apply total survey design methods successfully, one needs a mathematical model for the impact of the various sources of error present in surveys. Kish (1965) presents a very general model for survey error that decomposes the total error into fixed biases and variable errors. The mean square error (MSE) of a sample estimate, $\bar{y}$, of the population, $\bar{Y}$, is expressed as

$$\text{MSE}(\bar{y}) = \sum_r \frac{S_r^2}{m_r} + \left(\sum_r B_r \right)^2, \tag{1.1}$$

where there are r sources of error. The first term on the right is the sum of all variance terms arising from the multiple sources of error, and the second term is the square of the net biases.

One feature of the model is that it contrasts bias with variances, illustrating that biases are not generally reduced by increasing sample sizes, whereas variances can be reduced by increasing the number m_r of the units. Kish notes that this general model can be adjusted to accommodate covariances due to clustering, covariances between errors in measurement of a single characteristic, covariances between the measurements for several characteristics, and covariances between individual true values and measurement errors.

The model can be used to illustrate the practice of total survey design during the planning phases of a survey. An associated linear cost model is developed and expressed as

$$C = \sum_r c_r m_r. \tag{1.2}$$

The design of the survey generally involves placing constraints on either overall cost or overall MSE and solving for the minimum value of the other in terms of the m_r, the number of units of various kinds to be used in the survey. Successful application is, of course, dependent on having good estimates of S_r^2, c_r, and B_r.

Most of the models used to evaluate the errors in survey statistics are mean-square-error models of the general type above. The principal statistical motivation for using variances and mean square error to gauge the accuracy of biased statistics is the Chebyshev inequality, which states that

$$\Pr\left\{|\hat{\theta} - \theta| > K\left[\text{MSE}(\hat{\theta})\right]^{1/2}\right\} \le \frac{1}{K^2}, \tag{1.3}$$

where $\hat{\theta}$ is the biased estimate and θ is the population value. Furthermore, if $\hat{\theta}$ is asymptotically normal, then

$$\Pr\left\{|\hat{\theta} - \theta| > 2\left[\text{MSE}(\hat{\theta})\right]^{1/2}\right\} \le 0.05. \tag{1.4}$$

Specific comprehensive models for multiple sources of error have not been developed. Most frequently, one encounters models that simultaneously consider sampling errors and one source of nonsampling error. We review several models of this type in subsequent chapters.

1.4 CLASSIFICATION OF SURVEY ERROR

Our purpose is to review and classify methods used in the study of nonsampling survey errors. This book is seen as a first step in the development of a survey design information system as conceived of by Horvitz. The subsequent discussions of concepts, mathematical models, and remedial procedures associated with survey error provide an initial structure for gathering information on sizes of errors in surveys. We hope that this may one day give rise to a survey design information system; however, our immediate goal is to provide information that will assist other researchers in their attempts to study and control survey errors.

To organize the discussion, we need a method of classifying errors. Many methods have been used to classify errors. A major distinction made in this introductory chapter is between sampling and nonsampling errors. Errors have been classified according to where they occur in the various stages of the survey process (Murthy, 1967; Zarkovich, 1966), according to the type of contribution they make to a mean square error model (e.g., the Kish classification discussed above), and in terms of the information needed to detect them. Deming (1960) uses the latter method in which he classifies errors into those that could be detected by repetition of the survey process and those that could not.

Our classification has some features of each of these approaches. We divided the techniques that have been used into four areas that correspond roughly to the types of activities in a survey: (1) constructing a sampling frame, (2) designing and selecting a sample, (3) locating sample members and soliciting their participation in the survey, and (4) collecting data and converting it into machine-readable form. We call the survey errors associated with these four basic activities *frame errors*, *sampling errors*, *nonresponse errors*, and *measurement errors*. We consider frame errors, nonresponse errors, and measurement error as components of nonsampling errors. This division into three major components forms the first dimension of classification in the text.

Within each of these groups, we have considered the concepts and terminology used and mathematical models for the impact of the error type. These models are mainly mean-square-error models; and thus, within each error type, we consider contributions to variance and bias. Error rates or measures of the extent to which an error is occurring are also included. We review the ways in which sizes of errors can be estimated in the context of the mathematical models for the extent or impact of the error. Finally, we consider the survey methods that can be used to control or reduce the impact of a particular error source on the survey estimates. The following discussion clarifies the types of errors included in the three major sections.

1.4.1 Frame Errors

Some method must be developed for accessing and identifying members of the study population. The mechanism that provides access to the population is called the *sampling frame*. Suppose that we wish to study the health status of workers in the United States. We could try to reach these workers by visiting their workplaces, visiting their homes, or telephoning them at home. If only specific illnesses that required medical treatment were of interest, we might try to gain access to these people through physicians' offices and hospitals. Thus candidates for a sampling frame could include lists of areas, telephone numbers, business establishments, physicians, or hospitals. The frame chosen will affect the quality of the survey results. Some of the frames might not cover all of the population of interest; others might include large numbers of persons who are not of interest; it might be difficult to distinguish the members and nonmembers of the population in other frames. In addition, the type of frame chosen will influence the type of sample design that can be used and the efficiency of potential designs.

An area frame would cover all workers but would also include large numbers of nonworkers. A telephone frame would not cover workers without telephones and would also include large numbers of nonworkers. Business establishments would contain large concentrations of workers; however, it might be extremely difficult to construct a complete list of establishments. If we were interested only in specific illnesses that required medical care, each of these frames would include a very large proportion of persons who were ineligible for the survey and hospitals and/or physicians might be a better choice. However, if medical records were used, it might be easy to identify persons who had the disease but difficult to determine if they were among the workers of interest. Thus, when deciding to adopt a particular frame, one would need to consider the errors that would be introduced as a result of this choice. We consider methods for studying frame errors and for conducting surveys in the presence of imperfect frames in Chapters 3 through 5.

1.4.2 Nonresponse Errors

We consider nonresponse errors to be all errors arising from failure to include a designated sampling unit, population element, or data item in the survey. Often, multistage samples are used in surveys. If we decided to use a frame of business establishments to survey the health of workers in the United States, we might first select a sample of business establishments and then select a sample of workers within each establishment. Data might be collected by interviewing each worker about personal health status. The survey would be subject to errors of nonresponse if we failed to obtain data for an establishment, a worker, or a particular item in the questionnaire.

There could be several reasons for failure to obtain a response. The establishment might refuse to participate in the survey, with the result that none of its workers could be included. Alternatively, we might have difficulty in locating a particular establishment because of an incorrect address. Similarly, worker data may be missing because the worker refused to be interviewed, was ill, or was away for the entire survey period; or that person's questionnaire might be lost after the interview. Thus there are a variety of causes for nonresponse. Chapters 6 through 8 are devoted to the impact of nonresponse and to methods for estimating and adjusting for nonresponse in survey data.

1.4.3 Measurement Errors

Measurement error occurs when an incorrect value is associated with a population element, such as errors caused by problems of conceptualizing what measurements are needed, or in defining a method of making the measurement, errors occurring during the implementation of the methods, coding errors, and so on. Numerous survey participants can contribute to these errors. For example, in the worker health survey, individual workers may have difficulty recalling episodes of illness and thus make incorrect reports to the interviewers. The interviewers may make errors when asking questions and recording answers. A series of questions aimed at investigating attitudes toward the health care system may have been poorly designed and fail to measure what was intended. We discuss methods for studying the effect of measurement errors in Chapters 9 through 11.

The distinctions made in the discussion above are not always precise. For example, missing data items are often imputed, resulting in measurement errors; in a sample survey we may develop a listing of business establishments to use for sampling, and failure to include some business would be considered a frame coverage error; however, if our goal was to do a complete census of business establishments, missing an establishment could be considered a measurement error. Despite this lack of uniqueness, we believe that most types of nonsampling errors can be comfortably included in one of the categories above and that the categories are helpful for clarifying the impact that a particular source of error can have on survey estimates.

In addition to the sections of the book devoted to these three major areas, we have included a chapter on sources of survey error (Chapter 2) in which we discuss where in the survey process various types of errors can occur. The final chapter of the book is devoted to reviewing various attempts to develop total error models.

CHAPTER 2

Sources of Survey Error

Having proposed a working definition and broad classification of survey errors, we now consider where in the course of doing a survey these errors will appear. We refer to those things about surveys that may contribute to survey error as *sources of survey error*. Our goal in this chapter is to point out where frame, sampling, nonresponse, and measurement errors may appear in survey practice. Our discussion follows the major operational stages as they appear in most surveys, noting along the way where problems leading to error are likely to appear. Examples used to illustrate some of the sources will be taken from three hypothetical surveys.

Identifying all of the many things that can contribute to survey error is a difficult, if not impossible task. Survey research is simply too broad in scope. Among those things that contribute to error are the topic of the survey, the survey's objectives, the amount of money available to do the survey, the population being surveyed, the expertise of those conducting the survey, the nature of the list or lists used to select the sample, the approach used to collect the data, and procedures used in processing the data. Since we cannot possibly consider all of the settings in which surveys are conducted, it would be naive for us to suggest that all sources of survey error will be mentioned. Instead, we mention those sources that are particularly relevant to survey research and seem to appear time and again in surveys. Nonspecific human errors such as programming or computational mistakes are ignored in our discussion, although certainly they contribute to the total amount of error in a survey. Similarly, we do not mention many of the less typical sources of error. For example, the impact of using "key informants" to profile community attitudes is not mentioned as a source of measurement error since this approach to data collection is seldom used.

Before moving ahead we note two things about survey errors. First, the effect of a particular source of error in the overall survey error is determined by both its inherent potential for causing error and by the degree to which the potential can be controlled by good survey practice. A source with great potential for causing error will not cause major problems if means to control the potential are used. For example, the use of diaries may reduce recall

13

errors. Thus we can judge the relative importance of different sources of error only after we consider the ability and willingness of the investigator to control the problems the source may cause. Second, although surveys are planned and conducted by people and may involve obtaining measurements from people, it is inappropriate to equate all errors in a survey with fault or frailty on the part of people. It is true, of course, that many errors can be traced to shortcomings on the part of participants. For example, an interviewer may misrecord the respondent's age or the respondent may be unable to remember how many times he visited his dentist. However many forms of survey errors occur despite the best efforts of the participants. The best available lists for sample selection may have numerous uncorrectable problems; some people will refuse even after our best efforts to solicit participation, and estimates will differ from the value we wish to estimate regardless of how efficiently the sample is designed.

Finally, note that our primary intent is to describe what and where things can go wrong in a survey. We are less interested in describing in detail how one goes about doing a survey. Step-by-step instructions are available in any of several excellent books; see, for example, Parten (1966), Moser and Kalton (1972), Babbie (1973), Warwick and Lininger (1975), and Dillman (1978).

We will discuss sources of survey error in the following way. Roughly following the order in which they occur, we will briefly describe each major survey task and identify specific sources of error. Those sources of error thought to be self-evident are listed and not discussed, while sources thought to need added comment are illustrated using examples produced from three imaginary but representative surveys, described below.

2.1   THREE HYPOTHETICAL SURVEYS

2.1.1   Mail Survey

High school athletic directors in the United States are surveyed to assess the need for athletic trainers in varsity sports. Using a self-administered questionnaire, the investigator asks the respondent for personal opinions on the need for trainers and the capabilities they should have. A stratified simple random sample of 1000 athletic directors is selected from the membership list of a national high school coaches' association. Disproportionate sampling rates are chosen to yield roughly equal sample sizes in each geographic region of the country. Prior to keypunching and verifying the data, questionnaires are manually edited by the survey staff.

2.1.2   Telephone Interview Survey

A survey is conducted by telephone to learn how often adults in the local coverage area of a large university hospital visit the hospital and how much they know about the services it has to offer. Three hundred adult respondents, selected from several neighboring communities in the vicinity of the hospital, complete a 10-minute interview in which they are asked factual questions about service of the hospital and opinion questions about health care in general. The respondents are identified by stratifying telephone numbers according to community, selecting an optimally allocated but approximately proportionate stratified sample of residential telephone numbers, and then choosing one adult at random in each selected household. Survey staff enter data using key-to-disk processing equipment in which certain simple edits are performed automatically as the data are keyed. Final editing of the data is done by programming the computer to perform range checks for individual responses to questions and checks of consistency among responses to different questions.

2.1.3 Personal Interview Survey

Personal interviews are done in a sample of about 5000 households to obtain information about health care received by persons in the noninstitutionalized population of a state. Military personnel are excluded except for those living in off-base housing. Sixty interviewers are trained in one of five training sessions conducted simultaneously around the state. Respondents participate in a 45-minute interview dealing with health care services they have recently received. They are asked how often they have used the services, how accessible they are, how they feel about their quality, what they were treated for, and what kind of health insurance coverage they have.

The sample is selected in four stages with stratification by region of the state and socioeconomic level in the first stage. Minor civil divisions are used as sampling units in the first stage with enumeration districts or block groups, area segments of 50 to 100 dwellings, and dwellings serving as sampling units in the second through fourth stages, respectively (see U.S. Bureau of the Census for definitions of the area units above). Each primary sampling unit (PSU) is selected with probability proportional to its 1980 population. The questionnaire is manually edited just prior to data entry, which, as with the telephone survey, is done key-to-disk with later machine editing.

2.2 SURVEY PLANNING

In the planning phase of the survey the groundwork for the project is laid and the direction for subsequent activities is set. The investigator at this point determines which questions are to be addressed, and having concluded that the survey is the best research strategy to answer the questions, organizes available human resources and funds with the intent of finding answers to the research questions.

2.2.1 Establishing the Scope of Work

What Is Done?

The first step of the survey is to delineate the scope of work. The investigator begins this task by listing specific research objectives, thereby identifying which research issues are to be covered in the study. The population to be studied is also identified. The investigator must also decide if a survey is the most appropriate way to meet the objectives. In some instances it is not. A review of the literature may also reveal that prior studies have already investigated some of the stated research objectives. Experience from these studies may enable the investigator to avoid procedural pitfalls and to reduce the scope of the study if some objectives have been adequately addressed by previous efforts. Some of the working definitions for the survey are also created at this time, with preliminary plans being made for analysis as well. The investigator might, for example, construct some "table shells" (entire tables formatted for presentation but containing no numbers) for an enumerative or descriptive study. Finally, a provisional operational plan for the survey might be made, including such things as a list of survey tasks, a timetable of events, a list of potential problems, early thoughts about sampling, and a summary of personnel needs.

What May Cause Problems?

- The investigator may misunderstand the sponsor's intended objectives and chart a misguided course for the study.
- "Competing" objectives may call for different design strategies. For example, to produce the most precise estimates of knowledge about the hospital in the telephone survey may require a substantially different allocation of the sample among strata than would estimates of hospital use. Some kind of compromise allocation would then be required.
- The population intended for study (target population) does not correspond to the population actually studied (survey population). For example, in our hypothetical telephone survey, some adults in the surrounding communities may not have telephone numbers listed in the local directories. Those with unlisted numbers or no telephones will be missed, thus causing target and survey populations to differ.
- A survey may not be the best way to meet the objectives. If one of the stated objectives of the personal interview survey were to determine which of two treatment regimens for hypertension is superior, a survey would be less desirable than a prospective clinical study in which the two regimens are randomly allocated to patients, who are followed through time to observe the results of their treatment.
- By not using the experience gained in prior similar surveys, avoidable mistakes might be made. Consider, for example, questions asked about

high school athletic trainers in the mail survey. Suppose that an earlier study about the use of high school athletic trainers had suggested that *trainer* be defined in the questionnaire. By failing to heed this suggestion, the investigator may be causing unnecessary respondent confusion and increased measurement errors.

- The investigator may misinterpret the sponsor's goals and create working definitions that do not meet the sponsor's needs. In the personal interview survey, the investigator may incorrectly assume that "accessibility to health care" refers to the distance traveled to the health care provider, while the sponsor may have intended for it to refer to the mode of transportation used.
- The investigator may fail to anticipate the steps taken and problems faced in doing the survey and choose unqualified staff, thus reducing the quality of measurement. For example, developing medically correct and understandable questions on the occurrence of chronic disease in the personal interview survey may be impossible without the assistance of a physician with survey experience.

2.2.2 Budget Consideration

What Is Done?
Because surveys are seldom done with unlimited budgets, it is often important to determine during the planning phase if there is enough money to carry out the survey as planned. Sometimes there is not enough to do everything one hopes to do, in which case the scope of the study is narrowed or additional funds are sought. Then, when sufficient funds are found, the total budget may be divided among various components of the survey, such as administration, sampling, data collection, data processing, and analysis.

What May Cause Problems?

- Failure to request enough money may force the investigator to tighten the survey's budget and seek compromises, which in turn reduce the quality of the data.
- Misappropriation of funds to the survey components may lead to budgetary shortfalls and reduced survey quality. In the telephone survey, the investigator might allocate an unusually large portion of the budget to sampling, with the idea of creating a current listing of residential telephone numbers. At the same time, he may underbudget for training. Assuming that appropriation of the remainder of the budget was done properly, the investigator might eventually be forced to settle for less costly but less effective training methods.

2.2.3 Communication Among Members of the Survey Team

What Is Done?

Done selectively and in moderation, good communication among those conducting the survey contributes to improving the survey operation. This communication may take several forms. The investigator communicates to his or her staff by preparing task assignments for them (thereby clarifying responsibilities) and by telling them how decisions will be made. On the other hand, survey staff members communicate among themselves and to the investigator by presenting periodic verbal and written reports. These reports generally cover one or more of the following: what has been done in completing each task assignment, what problems have been encountered, and what has been done to solve the problems.

What May Cause Problems?

- The investigator fails to establish or to clarify the rules for making decisions, thus creating staff confusion and ultimately survey error. This situation could arise in the telephone survey if the person responsible for designing the questionnaire decides to delete an item asking the respondent how many adults currently live in the household. This particular item, however, is important to the analyst, who does not hear about the deletion until it is too late. Because the investigator has failed to make it clear that changes in the questionnaire are not to be made without the analyst's knowledge, it will be more difficult for the analyst to produce sound statistical estimates.
- Members of the survey team do not share important information about their work. In the personal interview survey the person in charge of developing manual editing specifications for clerical staff may fail to include a check on birth date information, assuming that the person planning the computer edits will include this check. Unfortunately, the person planning the computer edits excludes the check also, thinking that it will have already been done manually. Failure to communicate these decisions leaves undetected data collection errors on birth date information.

2.3 SAMPLING

Another major activity early in the survey is choosing the sample to be studied. Designs for most present-day surveys use probability sampling methods. In probability sampling, randomization is used to select the sample with the selection probability for each member of the population being nonzero and theoretically known. Since some who are selected fail to respond, the probability of being selected and responding will be less than or equal to the selection probability. In most cases, the response probability (given selection) is impossible to determine precisely and can only be estimated.

2.3.1   Preliminary Activities

What Is Done?
Survey sampling often requires information that can only be obtained external to the survey itself. The kind of information varies depending on the type of survey and how it is to be used in the survey. Several examples from the three hypothetical surveys illustrate this point.

First, the frame used to select the sample of athletic directors in the mail survey would be obtained from the membership list of an association of high school coaches. Second, maps or aerial photographs obtained from local public or private organizations would be used to construct area segments for sampling in the third stage of selection in the personal interview survey. Third, information from the 1980 census would be used to construct region-by-socioeconomic-level strata and population size measures for selecting minor civil divisions in the first sampling stage of the personal interview survey. Finally, in the personal interview survey, census-based population projections obtained from a state agency could be used for poststratification by age, race, and sex. In poststratification one ratio-adjusts the data so that the adjusted distribution of the sample by age, race, and sex corresponds to the figures from the state agency (see Chapter 11 for a more complete discussion of poststratification).

In personal interview surveys where area units are sampled, another preliminary activity is listing dwellings. In this step, a trained field worker (often a data collection supervisor) is sent to each area unit selected. Following a set of instructions, the field worker prepares a list of all dwellings in the area unit. The resulting list becomes the sampling frame for choosing sample households and ultimately, for identifying survey respondents. Each entry on the list must have enough information so that any dwelling selected can easily be found.

What May Cause Problems?

- The criteria used for stratification may not produce estimates with the smallest possible sampling errors. For example, a composite measure of education level, rather than region of the state, may be more strongly related to important survey measurements and thus be a better criterion for stratifying minor civil divisions in the personal interview survey. This is especially true when considering estimates related to people's health care attitudes, which are known to be strongly associated with their educational level.
- The list used as a sampling frame may be imperfect, thus giving some members of the survey population no chance or several chances of

selection and thus create bias in survey estimates. In the mail survey, not all members of the coaches' association may be athletic directors (commission error), and some athletic directors may not be members of the coaches' association (omission error). In the telephone interview survey some households may have more than one telephone number. Numerous additional frame problems can occur when listing dwellings in personal interview surveys. For one, instructions for locating the dwellings may be inadequate because definitions used by the listers are vague or because listing procedures are not clearly spelled out. Second, the listers may not record enough accurate information to enable the interviewers to find each dwelling listed. A dwelling identified only as having a "black asphalt shingled roof" in a neighborhood of dwellings with similar roofs will be hard to find. A misrecorded house number may send the interviewer to the wrong dwelling. Third, the wrong area may be listed. Differences may be because of misinterpretation of boundaries or the lister's negligence.

2.3.2 Sample Design

What Is Done?
A sample design is the strategy one devises for selecting those members of the population from whom one attempts to obtain survey data. Normally, samples are designed after the analytic objectives of the survey have been identified. Most designs can be distinguished by the number of stages of selection, the type of units chosen at each stage, how units are stratified before selection, and the method and number of sampling units chosen in each stage. Because the nature of the design dictates the statistical properties of estimators one might formulate for certain population values, the sample design must be considered in choosing appropriate survey estimators.

Choosing stratum sample sizes and deciding on the number of units to be selected at each sampling stage are seldom arbitrary decisions. Using information on the unit cost and the proportion of total variance attributed to each stage or stratum, the sample can be allocated in a cost-effective manner. Often, the cost of doing a survey is the most important factor influencing the sample design. One usually wishes to choose that allocation which will maximize the precision of estimates, given the amount of money available to do the survey. However, overall precision of survey estimates is sometimes the most important factor, in which case one uses that allocation that will produce for the least amount of money the desired precision level.

What May Cause Problems?

- The number of units chosen at each stage of sampling fails to produce the most cost-effective estimates from the survey data. For example, suppose that for practical reasons the 50 sample PSUs used in the personal interview survey are the same as those used in an earlier survey. Suppose further that the most cost-effective number of sample PSUs is 200. Estimates obtained from a sample containing 50 PSUs and an average of 100 households per PSU will have larger sampling errors than estimates obtained from comparable sample units with 200 PSUs and an average of 25 households per PSU.

- Nonsampling errors might be increased because of a decision to increase the sample size. Increasing the sample size of the telephone survey from 300 to 500 might leave the investigator with no choice but to spend less on those things that contribute to less measurement error (e.g., interviewer supervision, quality control for data entry, interviewer training, etc.).

- The overall sample size may produce acceptable precision for estimates that apply to the total population but may yield unacceptably imprecise estimates for small but important domains (population subgroups). This problem typically occurs when the precision of total population estimates is considered in determining sample size but the precision of domain estimates is not.

2.3.3 Sample Selection

What Is Done?
Once the sample design has been developed, the next step is to select a sample following a selection protocol dictated by the design. Data collection assignments can then be produced. In some surveys the entire selection process can be completed in the central office (the place which, except for any interviewers, houses the survey staff). In personal interview surveys, a portion of this work, such as selection of dwellings, may be done in the field. Finally, in some surveys where complex sampling procedures are required or where special frames must be constructed (e.g., from commercial mailing lists), the entire sample selection process may occur away from the central office.

What May Cause Problems?

- Subjective judgement may enter into sample selection even though probability sampling is intended. If, for example, field interviewers select the sample of dwellings, there may be a strong temptation to favor those dwellings where interviews will be easier to complete (e.g., those without barking dogs or with easy access from the road).

- Data collection assignments, listing the dwellings slated for an interview, may not provide interviewers in personal interview surveys with enough accurate information to be able to locate each dwelling. Those who prepare the assignment sheets may fail to provide enough information to identify uniquely each selected dwelling. The problem could also arise if the assignment sheets have transcription errors.

2.4 DEVELOPMENT OF QUESTIONNAIRES AND FORMS

Doing a survey implies that at some point measurements must be taken from certain members of the survey population. A survey questionnaire for recording measurements is developed along with administrative forms that guide the measurement process. Questionnaires typically contain questions related to the survey's objectives; numbers that uniquely identify each respondent; numbers that indicate what part of the sample the respondent came from; and questions that gather descriptive information about the respondent, such as age, race, and marital status. Often, the questionnaire contains a place to record the attempts to survey respondents, although separate survey forms may be used. Special administrative forms generally serve to facilitate survey operations and are used specifically to assist in such things as paying interviewers, logging the movement of questionnaires from the field or through data processing steps, keeping track of corrections resulting from edit checks, and other administrative matters.

2.4.1 Designing Measurement Devices

What Is Done?
Designing the survey questionnaire requires that one determine which, how, and in what order questions will be asked of the respondent. In like manner, other survey forms are designed by considering what information is to be recorded and in what format it would best appear. The questionnaire is developed largely with the respondent in mind, although some items may be added to assist interviewers or data processors. Most administrative forms, on the other hand, are developed to assist the interviewers, supervisors, editors, and others involved in survey operations.

What May Cause Problems?

- Questionnaire items may be put in the wrong place. In the telephone interview survey, for instance, questions on which area hospital the respondent prefers should be asked before questions mentioning the names of specific area hospitals. Otherwise, responses to the preference questions would be biased in favor of those area hospitals whose names have been mentioned.
- The respondent might not understand the specific intent of a particular question. For example, in the personal interview survey a question on

satisfaction with received health care might fail to make it clear to the respondent which aspect of health care is being addressed: accessibility, cost, or quality. Respondents may also not understand questions that use technical jargon or unfamiliar words (e.g., insurance eligibility criteria, dental caries, etc.). Finally, some questions might ask more than one question simultaneously (e.g., "Do you plan to quit your job and find another one next year?").

- The respondent may feel compelled to answer a question rather than to admit ignorance. In the telephone interview survey, the respondent may know very little about the hospital's reputation as a health care facility. Yet when confronted by a question about its reputation, he or she may state an impromptu opinion rather than appear uninformed.

- Certain words or phrases in the questions can influence a respondent's answer. For example, opinion questions in the mail survey that begin with the phrase "Wouldn't you agree that" could influence some respondents to answer in the affirmative. Other characteristics of the questions may taint the respondent's answers: the use of inflammatory words (e.g., "*radical* members of the coaching staff"), links to the status quo (e.g., "most coaches think that"): and suggesting hypothetical circumstances for opinion questions (e.g., "How much would you pay an athletic trainer if there were no limit on how much money your school could pay?").

- If the questionnaire is too long, the respondent may lose interest and end participation prematurely.

- One may construct a questionnaire that fails to build rapport with the respondent. Starting with complicated or sensitive questions that are difficult to answer may cause the respondent to feel inadequate or uncomfortable. Starting with simple and innocuous but interesting questions, on the other hand, tends to put the respondent at ease and engenders harmonious feelings toward the survey topic.

- If the researcher uses open-ended questions requiring long and complicated answers, the interviewers will have difficulties recording responses and computer coding will be difficult.

2.4.2 Instructions for Data Collection

What Is Done?
After the questionnaires and forms have been prepared, instructions for using them must be devised. In most telephone and personal interview surveys, standard instructions are written for all data collectors in a training manual. The manual serves as the teaching guide during training sessions and as a reference guide during data collection. Most manuals include the following: survey background information, general data collection instructions, specific instructions for the current survey, and reference information for each individual question in the questionnaire.

What May Cause Problems?

- The training manual may fail to provide data collectors with helpful suggestions for soliciting respondent participation. For example, interviewers who are not given a set of answers to anticipated questions about the survey might fail to convert skeptical potential respondents.
- The instructions included in a self-administered questionnaire may be unclear and therefore misinterpreted by the respondent. For example, this problem could occur in the mail survey if the procedure for ranking the importance of athletic trainer qualification is not clearly explained.
- The training manual may fail to motivate the data collectors to do good work.

2.4.3 Small-Scale Testing of Survey Forms and Procedures

What Is Done?
Once working drafts of survey questionnaires and other recording forms have been developed and a preliminary plan for collecting survey data has been worked out, it often behooves the investigator to test these forms and this plan under surveylike conditions to be more certain that data collection will run smoothly. When the usefulness of forms and procedures is clear, a small dress rehearsal of the data collection plan on some typical respondents may be sufficient. However, when important procedural questions must be answered, a more extensive field test may be needed to provide a rational basis for making final decisions.

What May Cause Problems?

- Unwarranted decisions may be made on issues related to the plan for data collection. For example, there may be some uncertainty as to whether nurses or experienced household interviewers would be better data collectors in the personal interview survey. Nurses would understand the medical terminology better but would lack the survey experience needed to get people to respond. Conversely, experienced household interviewers would have the survey experience but lack the health orientation. Without an appropriately designed field experiment, one could only speculate on which type of data collector would be most effective.
- Failing to conduct a dress rehearsal may result in a survey where seemingly operable forms and procedures do not produce the desired results. Questionnaire wording may be confusing or the forms may be difficult for an interviewer to administer. Procedures may be too complicated and extensive for interviewers to complete on schedule.

2.5 DATA COLLECTION

In the data collection phase of the survey, measurements needed to meet the study's objectives are made on selected respondents. Recording these measurements is usually done by specially trained data collectors in telephone and personal interview surveys and by the respondent in surveys using self-administered questionnaires.

2.5.1 Preparation

What Is Done?
Preparing for data collection often begins early in the survey. Endorsements from relevant organizations and professional associations may be requested to bolster the legitimacy of the survey, thus making the reluctant potential respondent more likely to participate. In addition, one may begin recruiting of data collectors as soon as the final plan for data collection is completed. In interview surveys, data collector training sessions are planned. These sessions typically include a step-by-step review of the written instructions data collectors must follow, work periods in which procedures and the use of survey forms are demonstrated and practiced, question-and-answer periods for resolving problems trainees may be having with the training material, and a concluding period in which trainees are evaluated on their understanding of the material and in which data collection assignments are made.

What May Cause Problems?

- Endorsements obtained for the study may be ineffective and fail to contribute to higher response rates. In the mail survey, which considers the role of athletic trainers in high school sports, selected high school athletic directors might be less impressed by a supporting letter from the American Medical Association than they would be by a similar letter from the National High School Athletic Coaches Association.
- Unsuitable data collectors may be recruited. Interviewers in the personal interview survey should be bright, pleasant, responsible, experienced, and willing to follow instructions faithfully. When persons of these qualities cannot be found, the likelihood of problems during data collection increases. Interviewers with abrasive personalities will alienate some potential respondents, irresponsible interviewers may fabricate responses, and so on.
- Data collectors may be inadequately trained, thereby causing them to make mistakes while implementing planned procedures or to improvise with their own procedures when they do not understand what they are supposed to do. The following training-related problems could arise in the personal interview survey: Interviewer instructions may be poorly written or nonexistent; trainers in the five simultaneously conducted

training sessions might emphasize different things, thus causing a discordance in the interpretation of procedures; the sessions may not allow enough time to cover each topic thoroughly or they may be too long and drawn out to the point of boredom; and finally, trainers may not be knowledgeable enough to teach or enthusiastic enough to stimulate the interviewers' interest.

2.5.2 Operations

What Is Done?

Several things happen during the operations phase of data collection: Participation is sought from those selected for study; questionnaires are completed by or with the assistance of survey respondents; and completed questionnaires are sent to be processed. Measurement at this point in the survey is out of the purview of the investigator. In most surveys, either a hired interviewer or the respondents themselves record responses with the investigator absent. The investigator must thereby establish some degree of control of the measurement process.

In interview surveys, for example, data collection supervisors might be hired to assure that interviewers are following instructions. This is especially true in a data collection operation involving dozens of interviewers and hundreds or thousands of respondents; this operation requires that the investigator create methods of coordination so that things proceed according to plan. For example, a system for monitoring the status of data collection activities might be used to provide the investigator with information on where efforts are lagging behind schedule.

One further ingredient of this phase of the survey is quality control, which consists of a series of planned checks of the operation whose purpose is to improve the overall quality of the work. Quality control is best inserted into any part of the survey process where instructions are to be followed by a trained staff (e.g., data collection, editing, coding, and data entry). For instance, it would be advantageous for the personal interview survey to include some effort to recontact a representative subset of selected dwellings to check the following: the outcome of call attempts made by the interviewer, the attitude of the interviewer toward members of the household, and answers to a number of factual questions asked in the original interview (verifying first that it has occurred). Quality control measures in the data collection operation might also include having interviewing supervisors check each interviewer's work or possibly having all interviewers double-check their own work before sending it in to be processed.

What May Cause Problems?

- Supervision of the data collection operation from the central office alone might be inadequate. For example, this might be a problem in the personal interview survey, where 60 interviewers would be conducting

5000 interviews across an entire state. Without an interviewing supervisor to oversee the work in each section of the state, control and the overall quality of the data collection operation may be reduced. No one would be available to monitor the progress of interviewing, watch for problem interviewers, or help out if and when interviewing lags behind in difficult areas.

- Failure to coordinate response solicitation efforts adequately may result in some segments of the sample having a greater chance for participation than others. In the telephone survey, for example, one might intend for interviewers to make up to six call attempts at certain times of the day and week. Without a way to record the history of call attempts, and supervision to assure that all allowable efforts are expended, the sample of 300 households may be overloaded with those residents who are more likely to be at home, such as the elderly and the unemployed.

2.6 MANUAL EDITING AND CODING

Once respondents have answered the questions, the next job is to edit the questionnaires to determine if they were completed correctly. Some answers may have been recorded as a written response in narrative form and need to be coded. Coding is the process in which a number or letter is used to represent the substance of a written response. The code can then be processed and analyzed by computer. Editing and coding are often done together, hence our later reference to editing/coding or editor/coder. In addition, some of the editing/coding work is done manually and some by machine (e.g., computer, key-to-disk data-entry equipment, etc.). Manual editing often includes the following checks: that responses fall within an accepted range of values, are consistent with other responses, and are legible, and that the right questions were answered, and in the proper sequence.

2.6.1 Preparation

What Is Done?
Prior to the operation phase of manual editing/coding, several things usually happen: Workers are recruited, a set of written instructions is prepared, and a training session is conducted. An editor/coder should be intelligent, thorough, and resilient. Instructional materials typically include a description of the edits to be performed, codes to be assigned to particular responses, and procedures to be followed when problems are detected during editing.

What May Cause Problems?

- Unresponsive edits and codes may be used. In the mail survey, for example, coding categories used for a question requiring a narrative response to describe the duties of an athletic trainer may not correspond to the categories that would be most useful to the study's sponsor. Analysis might, thereby, be unresponsive to the sponsor's needs.

- Edit checks will not detect all errors. For example, editing instructions in the personal interview survey may call for age to be flagged if it exceeds 130, which is a reasonable upper endpoint for longevity. However, if the interviewer had accidentally recorded "124" for a 24-year-old respondent, the error could pass undetected.

- Coding categories may not be unique and lead to coding problems. To illustrate, consider the personal interview survey in which the respondents are asked to discuss their last doctor visit. Suppose that two of the coding categories for the length of time since the respondent's last doctor visit are "less than six months ago" and "less than a year ago." The problem here is that a response of three months could be assigned to either category.

- Unsuitable editors/coders may be recruited. Those lacking thoroughness may fail to detect some obvious mistakes; those lacking resilience will allow boredom eventually to become negligence; and so on.

- Training of editors/coders may be inadequate. One might spend a great deal of time teaching editors/coders to find mistakes and too little time on how to handle the mistakes they discover. Confusion and mishandling of error corrections could result.

2.6.2   Operations

What Is Done?
If manual editing/coding is done as the first step in data processing, questionnaires will often be logged and batched prior to being edited and coded. Logging a questionnaire in from the field consists of checking the identification number of the questionnaire's respondent against a list of known respondents. The list of respondents is often compiled from status reports produced during data collection or from a listing of the original sample. Then, questionnaires may be batched into groups of 5 to 20. The questionnaires in each batch are physically joined by some means (e.g., rubber band, insertion into a storage envelope) and remain attached throughout data processing. This batching is done to reduce the chances of losing individual questionnaires during processing, because it is usually much more difficult to misplace a group of questionnaires than a single questionnaire.

Once the questionnaires have been logged and batched, manual editing/coding can begin. As with the data collection operation, status monitoring, supervision, and quality control may be part of the plan. Measures like the number of questionnaires edited/coded per hour, the percentage of questionnaires where edits have identified potential problems, and the cumulative count of questionnaires edited/coded may be collected to enable the investigator to observe the status of the manual editing/coding operation. Supervisors might also be needed to resolve errors or inconsistencies discovered during the edits, to help out when work loads are heavy, and to answer questions editors/coders may have. Finally, quality control may be built into the operation. One way is to have the supervisor recheck the first few questionnaires completed by each editor/coder and then to do a periodic spot-check of the work thereafter.

What May Cause Problems?

- The likelihood of losing track of questionnaires, and thus increasing nonresponse, will increase if they are not logged and batched. For an example of where logging may help, consider the personal interview survey, where completed questionnaires must be passed from interviewers in the field to the central office for processing. In this case a list of questionnaires expected in the central office might best be developed from interviewer status reports so that questionnaires actually reviewed could be checked against the list. Then, if at some later point a question arises as to whether a particular questionnaire was ever sent to the central office from the field to be processed, the answer can readily be found.
- Since manual editing/coding is often a relatively laborious task, an unsuitable work environment (e.g., a work area that is too warm or too cold, no work breaks or rotation in work assignments, etc.) is likely to contribute to low-quality work. Uncomfortable workers may become sloppy workers.
- Supervision of the editors/coders may be inadequate. For example, supervisors lacking a clear understanding of individual edit checks may misinform those asking questions about these checks.
- Procedures for checking quality of the work done by editors/coders may be either lacking or nonexistent, thus increasing the chances that undetected mistakes will be made.

2.7 DATA ENTRY

Editing and coding activities are usually the final step prior to data entry, the phase of the survey in which specially trained clerks possessing keyboard skills convert information recorded on the questionnaire and other survey forms into a format interpretable by a computer. Punched cards, magnetic tape, and computer disks are three of the most common types of machine-readable storage.

2.7.1 Preparation

What Is Done?
Preparatory activities for data entry bear strong resemblance to comparable activities of data collection and manual editing/coding. As before, recruiting staff, developing operating instructions, and conducting training sessions are the major ingredients. Regarding recruitment, one usually prefers data-entry clerks who can do fast and accurate work on a keyboard device. In large surveys with large and more complex questionnaires, written data-entry instructions might include a brief description of the survey and what has happened prior to data entry, a description of the questionnaire and other survey forms being keyed, and a discussion of quality control procedures being used. The training of data-entry clerks might then include a review of the written instructions and a session to enable them to practice before the questionnaires begin to arrive from manual editing/coding.

What May Cause Problems?

- Procedures developed for data entry may fail to produce data that are useful for analysis. In the mail survey, for example, the identification number keypunched for each responding athletic director may not indicate the sampling stratum from which the respondent was chosen, thus making a proper analysis of the data difficult, if not impossible.
- Unsuitable data-entry clerks may be recruited. Effective clerks do their work with speed and accuracy: ineffective clerks do theirs slowly and make many mistakes.
- Training of data-entry clerks may be inadequate. They might, for example, not be given enough time to practice on a difficult questionnaire before beginning their work. In this instance, procedures would be misunderstood and mistakes would be likely in the early stages of data entry.

2.7.2 Operations

What Is Done?
The operational phase of data entry is often done in a confined area where several clerks, working on some type of data-entry equipment, can be properly monitored and the quality of their output adequately controlled. Questionnaires will often be assigned to data-entry clerks in batches, with the final outcome of their work eventually being collated together into a computer-readable file containing the data that had been recorded on all survey questionnaires. We call this file the *raw data file*.

As in the previous discussion of data collection and manual editing/coding before, status monitoring, supervision, and quality control are important in this phase of the survey. Periodic figures indicating production status and the work efficiency of the clerical staff are often produced while the questionnaires are being keyed. One or more supervisors may be named to make work assignments, to answer questions, or to assist with keying when necessary. Finally, some method of detecting and solving quality problems may be devised. For example, the data on a sample of all questionnaires may be rekeyed independently and checked against the data as keyed originally. Remedial action may then be needed if keystroke error rates are unacceptably high.

What May Cause Problems?

- Data-entry equipment may fail. If key-to-disk processing equipment were used in the personal interview survey, for example, a power failure may cause the data for the questionnaire being keyed at the time of the outage to be lost and never recovered, thus increasing the amount of nonresponse.
- Supervision during data entry may be inadequate. Supervisors may, for example, fail to notice that production is falling behind and then be forced to compromise quality to meet study deadlines. Their inability to make efficient work assignments and to solve problems or to answer questions by their staff may all contribute to this problem.
- Quality control measures may be inadequate or nonexistent, thus contributing to unacceptable keystroke error rates. Suppose, for example, that quality control was limited to occasional over-the-shoulder check of clerks with slow production rates, the rationale being that slow clerks have limited keyboard skills and are therefore prone to making mistakes. The weakness of this approach is that clerks doing fast but sloppy work may be contributing even more data-entry errors than may slow, deliberate clerks.

2.8 FINAL DATA PROCESSING

Only rarely does one expect that analyses can be performed immediately on raw data. In most cases, supplementary editing is needed to "clean" the data further, and additional codes, which are best assigned by computer, must be produced. Moreover, raw data files may be cumbersome and complex in larger surveys. Analysis in such instances is done more efficiently on one or more smaller analysis files. Each analysis file would normally contain only those data items associated with the units of observation required for a proposed set of analyses.

2.8.1   Machine Editing and Coding

What Is Done?

Final checking of the raw data prior to analysis is often done by preparing computer programs which (1) check for items with values that are out of some arbitrary range or inconsistent with other values, and (2) assign codes to certain raw data items that will be later used in analysis. We refer to this phase of the survey as machine editing and coding, which, as can be inferred from the description above, requires a period of preparation in which computer programs to perform the edit checks and to create the codes are developed. Some of the manual editing/coding is repeated here as a final check of the earlier work, while the rest is work that can be done more rapidly and accurately by computer. The outcome of machine editing is a series of "flags" on those questionnaire items which, according to editing criteria, are possibly incorrect. The correctness of each flagged item must be verified, a process that may call for the questionnaire to be reexamined or, in some cases, for the respondent to be recontacted.

What May Cause Problems?

- Specifications for machine editing and coding may be misapplied by the computer programmer, thus creating problems for the analyst. For example, the programmer may incorrectly assign codes for responses to the question on the annual salary that the respondent believes an experienced athletic trainer should make. Instructions to the programmer may call for assigning a salary between $10,000 and $14,999 the code "2", but the programmer may actually assign a salary between $10,000 and $19,999 the code "2".
- Edits may fail to detect all remaining errors on the raw data file. For example, some important checks may be overlooked or, as was the case with manual edit checks, edit checks may fail to "flag" some responses as errors.
- Some coding categories for items used to define subgroups of the population for analysis may represent too small a portion of the population. Assigning a separate code to five-year age subgroups (e.g., 15–19, 20–24, 25–29, etc.) in the telephone survey with a total of 300 responding adults may cause estimates for some individual subgroups to be made from fewer than 10 adults. Precision of such subgroup estimates would be unsatisfactory. A broader age classification might be better.
- The investigator may misinterpret the intentions of the survey's sponsor in developing coding categories, resulting in analyses that are unresponsive to the objectives of the survey and the sponsor's needs.
- Computer malfunction may cause data to be lost or to be made incorrect.

2.8.2   Construction of Analysis Files

What Is Done?

At some point in a study, the survey's objectives are translated into an analysis plan, which may consist of several sets of specific analyses, each using a specific analytic tool (e.g., descriptive statistics, regression analysis, etc.) and requiring a particular set of data items. When the cleaned raw data file is too large to be used efficiently for these analyses, smaller and more manageable analysis files are often created. In addition to survey data analysis, files may contain identification numbers, sampling weights (see Section 2.9.1), and identifiers indicating from what location in the sampling design the respondent was selected (e.g., which stratum, which primary sampling unit, which secondary sampling unit, etc.).

What May Cause Problems?

- An analysis file may not contain all of those variables needed to perform a set of analyses satisfactorily. For example, an important independent variable (included in the survey) may inadvertently be omitted from the analysis file produced for a set of analyses in which linear regression models are being fit. Important explanatory power of the model may be lost by such an omission.
- Linkage of some questionnaire items may be impossible in constructing an analysis file. Suppose that some questionnaire items in the personal interview survey deal with the entire household (e.g., household annual income, length of time at the present address, etc.), while other items apply to individuals only (e.g., opinion on the quality of prior health care, number of days lost from work due to disability, etc.). Suppose, furthermore, that one cleaned file is created for household-level items and another for person-level items. If one wishes to create an analysis file containing a mixture of both types of items, a unique household identifier must exist on both the household and individual files so that items can be merged onto one file. Analysis may be less effective if a mixture like this cannot be constructed.
- Computer malfunction may once again cause data to be lost or made incorrect.

2.9   ANALYSIS

The analysis phase is where prior efforts in the survey culminate, allowing the investigator finally to attempt to answer those research questions raised by the survey's objectives. In analysis we learn such things as whether a majority of high school athletic directors favor the hiring of athletic trainers for their varsity terms, if those who have visited the university hospital know more about it than those who have never visited it, and which things are most strongly related to a person's attitude about the quality of health care received.

A wide variety of tools is available for making statistical inference to the survey population. Each tool involves estimation and estimators that differ in mathematical complexity. Relatively simple descriptive estimators such as totals, means, and ratios, as well as more complex relationship measures such as regression or correlation coefficients, may be used in exploratory analysis, whose primary aim is to make the characteristics of the population being studied more understandable. Some of these tools may also be used in confirmatory analysis when the objective is to test statistical models or assumptions indicated by exploratory work. At the analysis phase of the study, confidence intervals are often developed for some of the important survey estimates.

2.9.1 Computation of Sampling Weights

What Is Done?
One well-known and commonly held view of statistical inference from sample survey data is the fixed-population view, in which inferential statements are made and limited to a measurable characteristic of a finite population. The basis for this inference is a probability sample of members in the population. Under this view, estimators of population characteristics must take into account the manner in which each survey respondent was sampled as well as the response probability (i.e., the likelihood of one's agreeing to participate at the time that one was approached).

Two qualities of each respondent are identified under the fixed-population view. One is *structural identity*, indicating which part of the design structure (i.e., stratum, primary sampling unit, secondary sampling unit, etc.) the person came from. The other is *sampling weight*, indicating the relative likelihood of being selected and responding in the survey. These weights are often adjusted to improve the statistical quality of estimators.

When the survey analyst follows the fixed-population view of statistical inference, computation of sampling weights for each respondent is an important activity prior to analysis. Using records documenting the sampling process, a provisional weight is calculated as the reciprocal of each respondent's original probability of selection. The provisional weight is then often multiplied times a nonresponse adjustment factor, which is assumed to be the reciprocal of the respondent's response probability (see Section 9.1). The response probability is usually estimated for some group of survey eligibles as the proportion who responded in the survey. The product of the provisional weight and the nonresponse adjustment for each respondent is therefore the approximate reciprocal of the probability that the respondent was chosen and participated in the study. To reduce the sampling error of estimates further, a final poststratification or ratio adjustment might be multiplied times the provisional weight and the nonresponse adjustment to yield the final sampling weight.

What May Cause Problems?

- The nonresponse adjustment may poorly reflect the respondent's response probability, thus making it less likely that the adjustment will serve its intended use of reducing nonresponse bias. In the telephone interview survey, for example, estimating response probabilities is made more difficult because it is virtually impossible to determine whether an eligible person is present when no one answers the telephone after repeated attempts.
- The sampling process may have been poorly documented, forcing the original selection probability to be estimated or conjectured. Errors in computing the selection probably contribute to an increase in the bias of survey estimates.
- Computational errors may be made in computing provisional weights and later adjustments.

2.9.2 Preparation

What Is Done?
Three other tasks are often completed before analysis begins. One is to develop an analysis plan, in which the strategy for meeting the study's analytic objectives is outlined. Such a plan may be quite explicit, including such details as what resulting tables will look like if proposed analyses are largely descriptive. The second task is to locate or create any needed computer software. Useful software will often be available, especially when required analysis tools are simple or when the complete design is simple (e.g., designs in which sampling weights are nearly equal). In the event that complex analyses are proposed for complex samples, special software may have to be developed. A third task done in preparation for analysis is a simple unweighted descriptive profile of all data items found on analysis files. This helps to identify extraneous values and provides a check for later analyses.

What May Cause Problems?

- Failure to use sampling weights can cause survey estimates to be biased. Suppose that in the population covered by the telephone interview survey, for example, persons living in households with few adults tend to visit hospitals more frequently than persons living in households with many adults. Suppose further that the analyst ignores the sampling weights (implicitly assumes equal weights) in estimating the average number of hospital visits per person during the past year. In reality, however, weights are roughly proportional to the number of adults in a household since one adult at random is selected in households that have themselves been chosen with roughly equal probabilities. By ignoring the weights in a sample that by design overrepresents persons from households with few adults, the estimate of the average number of hospital visits is likely to be biased on the high side. Using the weights, on the other hand, compensates for the underrepresentation of persons from the many-adults households by giving them relatively larger weights than persons from the few-adults households.

- By ignoring the structural identity of respondents, measures of the statistical precision of estimates may be inaccurate. Consider, for example, the problem in the personal interview survey of estimating the proportion of persons who are generally satisfied with the quality of health care received in the past year. If persons within the same clusters tend to have more similar attitudes on this subject than the population at large, treating the respondents as members of a simple random sample will in most cases produce variances that are too small (overstating the actual level of precision).

- Some surveys take a long time to finish. As a result, the people who planned and conducted the survey may not do the analysis. The analysis staff may make mistakes because those involved in collecting the data did not adequately document their work.

- The analysis plan may answer the wrong questions. This problem could arise when survey objectives are vague and thereby misinterpreted by the analyst, or when the objectives are clear but the analyst constructs an inappropriate plan for meeting them.

- Undetected programming errors may exist in computer software developed in preparation for analysis.

- Errors on the analysis files will have little or no chance of being detected unless the unweighted descriptive profiles are done.

2.9.3   Operations

What Is Done?
The analysis plan is put into practice using available computer software during the operation phase of analysis. Here answers or the frustration of being unable to obtain answers appears. In well-conceived descriptive studies with straightforward objectives, the analyst may follow the plan to the letter. In exploratory studies, on the other hand, drastic departures from the plan may be dictated by early findings, so that the original and final strategies barely resemble each other.

What May Cause Problems?

- The analyst may make mistakes in following the analysis plan. For example, the frequency distribution of opinions about the quality of office-based physician care may be produced when the real intent was to analyze responses to a similar question on hospital-based physician care.
- Errors in professional judgement by the analyst may misdirect analysis in exploratory study, thus causing important findings to be missed.
- Programming and syntax errors by those using the analysis tools may lead to erroneous findings and misguide conclusions.

2.10 FINAL DOCUMENTATION

Although working papers, progress reports, memoranda, and other forms of correspondence may provide a thorough (albeit piecemeal) historical account of the study, a final report of the survey from start to finish is often produced. The final report, containing a detailed accounting of all phases of the study, becomes one of the principal means by which the study can be evaluated by the research community. It should therefore contain an objective reporting of achievements as well as problems.

2.10.1 Report Preparation

What Is Done?
The final report is often written by the investigator with assistance from those coordinating the major phases of work. Its content may vary somewhat, but three general topics are usually covered. First, the report will discuss the background and objectives of the study to indicate why the survey was needed, what questions it tried to answer, and what previous work had been done on the subject. Second, a description of the procedures used to conduct the study usually appears in the report, including the sample design, estimation methods, training strategy, data collection protocol, and data processing protocol. Third, the report also typically includes a section in which findings are presented and discussed. This section may point to further work dictated by the findings and discuss the study's shortcomings.

What May Cause Problems?

- The statistical precision of survey estimates may not be computed or reported, thus making it difficult to know how good the estimates are. Because of the relatively small sample size in the telephone interview survey, for example, it is important to measure the precision of estimates for the entire population and for important population subgroups. Computed precision measures, although subject to imprecision themselves for some of the smaller subgroups, nonetheless, give the analyst an indication of statistical quality for all of the reported findings.

- The study's findings may be incorrectly interpreted. Consider the personal interview survey, for example, where the analyst may use factor analysis (see Mulaik, 1972) to reduce 25 rating scores obtained from questions on the perceived quality of specific health services down to, say, three underlying factors. If the analyst does not understand the meaning of some of the important concepts related to factor analysis (e.g., factor loadings, axis rotation, etc.), computer output from a preprogrammed factor analysis subroutine may be misunderstood, thus causing presented findings to be misleading.

- Some of the study's major procedural problems may be left out of the final report, thereby making a thorough qualitative assessment impossible. In the personal interview survey, for example, dwellings must be listed to select the households in the final stage of sampling. Suppose that the report fails to mention that many of those doing the listing had incorrectly excluded off-base military housing from the lists. Without mention of this problem and some attempt to assess its impact, incorrect inferences may be drawn by those using the study's findings. Similar problems may arise if other shortcomings are not discussed (e.g., bogus interviews, data-entry keystroke errors, nonresponse, interviewer mistakes, etc.).

2.10.2   External Review

What Is Done?
Knowledgeable experts able to make an objective appraisal are sometimes asked to review the survey's final report after a complete draft has been written. The reviewer may be chosen to represent those who will ultimately use the study's findings. The reviewer's major duties would be to judge the quality of the presentation of the study to the reader and to point out weakness in the study itself which at that point can be clarified (e.g., changes in the analysis, computation of production rates for data entry).

What May Cause Problems?

- The investigator may not arrange for a review of the final report, thus failing to recognize things that might improve the study and its presentation.
- The reviewer may do a poor job in fulfilling reviewing duties.

2.11 CONCLUDING REMARKS

In this chapter we have sketched the process of conducting surveys and pointed out some of the things that can go wrong. We have supplemented a purposely brief discussion of each study phase with lists of some of the things that in our experience at the Research Triangle Institute have in some way either contributed to or actually caused survey errors. Although most of the obvious problems have been mentioned, we have undoubtedly left out many of the more subtle problems or those encountered in other areas of survey research.

CHAPTER 3

Frames: Definitions of Frames and Frame Errors

The frame plays a fundamental role in survey sampling. Probability sampling involves selecting a subset of units from a finite collection of units in a manner that lets one determine the probability of obtaining that subset. The *frame* is the finite population of units to which the probability sampling mechanism is applied. The population of frame units is not necessarily equivalent to the population for which information is to be collected. For example, one may wish to do a survey of people living in apartments. However, instead of directly selecting a sample of these people, a sample of apartments may be selected. The population to be surveyed is apartment dwellers; the finite population to which the sampling is applied is the list of apartments.

Several terms have been used to denote the population to be studied in a survey. Following a discussion by Kish (1978), we refer to the population the researcher wishes to measure as the *target population*. Kish distinguishes the target population from inferential populations, those to which statistical inference is to be made. There is a single target population, but there may be several inferential populations. For example, a survey of high schools within a state might be conducted to determine the number of students enrolled in chemistry and physics classes during the current school year. The target population would consist of students within the state's high schools.

To assist in planning for needed science equipment, the researcher might use the results to predict the number of students who will enroll in chemistry and physics during the next five years. Staff of the state universities might use the data to estimate the proportion of matriculating freshman with training in chemistry and physics. Thus there can be several inferential populations. Of course, the target population can also be an inferential population, and one would expect the researcher to estimate the number of students enrolled in chemistry and physics during the study year.

The structure of the frame, the information it contains, and the quality of that information will determine the type of sample designs and estimation procedures that can be used in a survey. Simple frames lacking auxiliary information support simple sample designs; complex frames containing auxiliary information support more complex designs. The simplest type of frame is a list that clearly identifies each element of the target population. If the list contains no information other than the identify of the elements, typically very simple sample designs are used for selecting the sample. A simple random sample may be selected, or if the list is large, a systematic sample or a systematic sample of clusters may be used.

Suppose that one were to conduct a survey of the students in a particular elementary school. A simple list of students by name that could be constructed from information provided by the school's administration would be adequate for selecting the sample. To contract the sample students, some additional information other than the name of the student would be needed. One might need to know home addresses if the survey were to be conducted outside the school. Alternatively, if in-school data collection were planned, information on class schedules would be adequate for locating students. This example of the simplest type of frame contains information that allows one to distinguish the elements of the population—here, names of pupils. In addition, it contains information that allows one to find or make contact with sample members of the target population.

In other cases more complex frames are required. Consider the U.S. National Hospital Discharge Survey (Pearce, 1981), aimed at describing the demographic and medical characteristics of discharges from U.S. hospitals in a particular calendar year. Although it is theoretically possible to construct a list of all discharges, it would take too much time and money to construct such a frame. Thus, access to the population of discharges is achieved in a two-step process. First, a list of hospitals is constructed and a sample selected. Then, within each sample hospital, a list of discharges is constructed and a sample selected within each hospital. Thus we have a first-stage sampling frame, the hospital list, and a second-stage sampling frame, the discharge list within hospitals.

Many sample designs use auxiliary data to produce more efficient samples. The relative efficiency of one sampling procedure compared to an alternative one is given by the inverse of the ratio of the variances under the two designs (Sukhatme and Sukhatme, 1970:106). The design with the smaller variance is said to be more efficient or to have higher precision (Raj, 1968:85).

Complex sample designs that are more efficient than simple random sampling, such as those employing stratification, probability proportional to size sample selection, or special estimation techniques such as ratio regression estimators, require additional information beyond the identity of the target elements. The list of hospitals used for the U.S. National Hospital Discharge Survey contains information that allows an improvement in the efficiency of the sample design. The list contains information on number of beds in the hospital, its location, and type. Information on number of beds is used to stratify the hospital by size. Differential sampling rates are applied in each size stratum so that, overall, large hospitals had a higher chance of being in the sample. During the estimation procedure, information on total number of beds for all hospitals in a particular stratum and total number of beds associated with sample hospitals is used to construct a ratio estimator for the total number of discharges. Correlation between bed size and total number of hospital discharges improves the efficiency of estimation. The increased efficiency is made possible because the sampling frame contained auxiliary information on size of the hospital.

Construction of the sampling frame may be one of the most difficult tasks survey designers face. More than one survey has been abandoned because of an inability to construct a frame. Ideally, the nature of the target population determines the type of frame; however, in practice the type of frames available may also determine the target population for the survey. In the survey of hospital discharges, the large size of this population precluded the use of a simple list as a frame. Similarly, the nature of the frame determined the population actually covered in the survey. In a hospital discharge survey, one would like to study the characteristics of *people* who were discharged from U.S. hospitals. Estimates of the number of people hospitalized, average number of hospitalizations per hospitalized person, number of people hospitalized for heart disease, and so on, would be desirable. However, because of the way records are maintained within hospitals, it is easy to obtain a list of discharges during a particular year but difficult to obtain one of unique people who were discharged. Since individual people can be hospitalized more than once in a particular year, the list of discharges within a hospital does not have a one-to-one correspondence with the list of people discharged from the hospital. The problem is compounded by a person's going to more than one hospital during a calendar year. Thus, even if a list of unique people for a particular hospital could be constructed, it is likely that some of the people on this list would also have been discharged from another hospital. The difficulty of constructing a frame of persons discharged from hospitals has, so far, precluded making estimates of the number of people discharged. Instead, the population of discharges is described in reports from the survey, and the available frame has determined the nature of the population studied.

3.1 DEFINITIONS OF FRAMES

In the preceding section we indicated the general nature of sampling frames and their role in surveys. In this section we present some definitions and descriptions of frames in the survey research literature. These definitions differ as to the number of concepts and/or operations the definition or description attempts to encompass. Names given to various concepts also vary. Different definitions of frames encompass different concepts, including:

1. The target population is a finite population of identifiable elements.
2. Sampling is done on some sets of units, but this set is not necessarily the target population.

3. Some mechanism must exist for linking the target population and the set that is sampled.

4. To be able to collect information from elements, it must be possible to locate and distinguish them from each other.

5. More than one type of linkage can exist between target elements and the sample set. These linkages determine the type of sample design and estimation procedures that can be used for the survey.

6. Some sample designs and estimation procedures require auxiliary information about the population elements. This information must be known for every member of the target population.

We cite some examples of definitions below. For a more complete list of definitions, see the Compendium of Nonsampling Error Terminology following Chapter 12.

Zarkovich (1966:97) gives a simple definition of a frame that incorporates the concept of a finite collection of identifiable elements that compose the population and is used for selecting samples: "By ... 'frame' we understand the list of units. These lists are used in sample surveys for the selection of samples."

Other definitions include this concept as well as the idea that sampling units are not necessarily equivalent to population elements. For example, Scheaffer et al. (1979) define a *frame* as a list of sampling units. *Sampling units* are then defined as nonoverlapping collections of elements from the population.[1]

Hansen et al. (1963) add the third idea, that of linkage mechanisms by defining a *list of sampling units*, a *target population* of reporting units, and *rules of association* that link these two sets. Rules of association allow establishment of a linkage between a selection of listed sampling units with known probabilities to a selection of reporting units with known probabilities. Hansen and co-workers did not use the term *frame*. Note the difference between a target population of reporting units as Hansen et al. (1963) define it and the definition of target elements we have adopted. The population for which measurements are to be made may not be the reporting units. Parents may report for their children, medical personnel for patients, and so on.

A definition from Szameitat and Schäffer (1963:518) includes the need to distinguish and locate elements. Namely, a *frame* is

all material which describes the components of the target population (or an adequate part of that population) in such a way that it is possible to determine in the course of the survey individual components and to delimit them from other components.

---

[1]Complete references and exact wording of the authors' definitions appear in the Compendium of Nonsampling Error Terminology for all terms in this chapter printed in italics.

Incorporating the idea of the usefulness of auxiliary information for specific types of sample designs Szameitat and Schäffer also define a *preferred frame*:

> Of particular use for a sample survey is a frame which not only provides a description of the individual components but also supplies additional information, such as the size of the components or their inclusion into specific parts of the target population. If there are several frames of the same level available for a target population, that one will generally be used which contains the most exact information and can, at the same time, be easily applied (p. 519).

After reviewing these definitions, we constructed the following definition which encompasses the six concepts listed above: The *frame* consists of materials, procedures, and devices that identify, distinguish, and allow access to the elements of the target population. The frame is composed of a finite set of units to which the probability sampling scheme is applied. Rules or mechanisms for linking the frame units to the population elements are an integral part of the frame. The frame also includes auxiliary information (measures of size, demographic information) used for (1) special sampling techniques, such as stratification and probability proportional to size sample selections, or (2) special estimation techniques, such as ratio or regression estimation.

3.2   TYPES OF FRAMES / FRAME STRUCTURES

An adequate frame for a survey clearly distinguishes and identifies the sampling units and contains clear rules of association linking each population element to a sampling unit. Some frames are structured with a one-to-one association between units in the frame and those in the population. A *serial list frame* consists of a numbered listing of elements of the target population. Jessen (1978) points out that such a frame is often considered to be the preferred type because all the target elements are included, the total number of elements is known, and it is easy to select random samples by generating a set of random numbers.

Two other types of frames that also exhibit simple one-to-one correspondence between sampling units and target units are frames consisting of *mixable objects* and *count frames* (Jessen, 1978). The former consists of a set of physical objects each associated with a target element. These objects are then mixed to simulate the process of randomization. An example is a deck of cards, each identifying a member of the target population. A count frame consists of an ordered set of unnumbered physical units for which it is possible to identify a particular unit by counting through the set. For example, the frame for a sample of accounts in a business might consist of several file drawers of account records. A random sample of records could be selected by generating a set of random numbers and counting through the file to identify the sample records. Jessen cautions against the use of count frames because it is not possible to verify from the sample that the sampling was done correctly.

Frames need not be so closely associated with the target population as those just described. Neither is it necessary to have a one-to-one correspondence between sampling units and target elements. The following sample selection method illustrates these points.

In 1981 the Research Triangle Institute (RTI) conducted a survey that employed a two-stage sample of pharmacies and prescriptions within them. The large majority of pharmacies maintained serially numbered and serially ordered prescription files which were used for the second-stage sampling frame.

Samples of prescriptions originally filled during the years 1976–1980 were selected. Approximately 880 prescriptions were to be chosen from 1980 and 280 from each of the four previous years, to yield a total sample of approximately 2000 prescriptions. A data collector visited each pharmacy and determined the beginning and ending number of the sequence of prescription numbers for each year. This information was telephoned to a sampler at RTI. Clusters of prescriptions were defined for the entire file with the cluster size being approximately 88 for 1980 and 28 for 1976–1979. Once the total number of clusters was known, a systematic sample of 10 clusters was selected from each calendar year using the fractional interval technique Kish (1965) describes. Beginning and ending numbers of prescriptions in each sample cluster were then relayed to the data collector. The entire process of second-stage frame construction and sample selection took only about 30 minutes.

In one pharmacy, prescriptions were serially numbered but filed by patient's surname. The pharmacy had started operation in late 1976 and had filled 40,050 prescriptions through March 1981. Clusters of prescriptions could be defined by their serial numbers; however, finding the sample prescriptions would have required a search through the entire file. To deal with this problem, alphabetic clusters were defined and selected.

A telephone directory for the area in which the pharmacy was located was available at RTI. Boundaries of the alphabetic clusters were defined using the distribution of names in the telephone book under the assumption that the distribution of names would be nearly the same in the pharmacy. Because the exact number of prescriptions in each alphabetic cluster could not be determined in advance and a minimum sample of 2000 was required, it was decided to aim for a sample of approximately 2500 by selecting 10 clusters with approximately 250 prescriptions in each. This implied that the entire frame (alphabetic file) should be divided into 160 clusters.

Cluster boundaries were defined using the telephone book. The white pages of the telephone book contained 1006 pages of names. Dividing these into 160 clusters gives 6.2875 pages per cluster. The entire telephone book was divided into 160 clusters, 140 of which had 6.3 pages of names and 20 of which had 6.2 pages of names. A systematic sample of 10 clusters was selected, and cluster numbers were converted into page numbers to the nearest tenth of a page.

Each page of the telephone book had five columns, so that half a column represented a tenth of a page. Alphabetic boundaries of each cluster were determined by identifying the names at the boundaries of the page clusters. When a boundary fell within a string of identical names, a random procedure was used to determine if the name fell in or outside the cluster boundary. Once the alphabetic boundaries were known, they were telephoned to the data collector in the pharmacy.

This example illustrates several features of frame construction. The finite set of units to which the sampling mechanism was applied was not the actual list of units in the population but was the list of names in the telephone book. The rule of association that linked the target elements to the sampling units was clear—all prescriptions for patients whose names fell within specific alphabetic cluster boundaries.

The frame created for the alphabetic file just described is an example of a *cluster frame* in which each frame unit may be associated with more than one target element. In a widely used type, the *area-household frame*, sampling units are defined as land areas identified from maps of (less often) aerial photographs. Each land area represents a cluster of housing units. The housing units may be the target elements, or households or persons within them may be the target elements.

A special type of frame being increasingly used is a frame in which the total number of population elements, N, associated with the frame is not predetermined; however, the frame allows selection of a sample in which the units have equal probabilities of selection. The exact probabilities of selection are unknown. The numbers used in Waksberg's method of clustered random digit dialing are an example of such a frame; see Waksberg (1978). Because the probabilities of selection are not known, additional information is needed to estimate target population totals.

3.2.1 Frame Structures

Frames can each be classified into one of four types of frame structures based on the association between frame units and target elements.

1. *One-to-One.* Each frame unit is associated with a unique population element and each population element is associated with a unique frame unit.
2. *One-to-Many.* Frame units may be associated with more than one population element and each population element is associated with a unique frame unit.

3. *Many-to-One.* Each frame unit is associated with a unique population element and each population element may be associated with many frame units.
4. *Many-to-Many.* Each frame unit may be associated with many population elements and each population element may be associated with many frame units.

Determining the probabilities of selection of sample elements is easy in the case of one-to-one or one-to-many association. In both cases, the probability that a particular target element is in the sample is equal to the probability that its single associated frame unit is in the sample.

For the many-to-one and many-to-many types, it is necessary to know the number of frame units the target elements are associated with to produce unbiased estimates. The number of associations between an element and a frame is called the element's *multiplicity* (Sirken and Levy, 1974). This information is sometimes difficult to obtain, and the researcher considers multiplicity to be a nuisance. In other cases it is advantageous to construct a frame with multiplicity. Use of frames with multiplicity is described in Chapter 5.

Several of the types of frames above may be used in a single survey. Multiple-stage samples require a frame at each stage of sample selection. A survey of those eligible to vote might employ three stages of sample selection: areas, housing units within areas, and persons past voting age within the housing units. The first-stage frame would consist of a list of areas that had a one-to-many association with members of the target population (potential voters). The second-stage frame would list housing units within the selected sample areas; each housing unit would also exhibit a one-to-many association. At the final stage of sampling, data collectors might visit the sample housing units and list all residents 18 years old or older. Items on this list would have a one-to-one association with members of the target population, although aliens or others not eligible to vote might be included in the list erroneously.

3.3   THE EFFECT OF THE FRAME ON TOTAL SURVEY ERROR

The type of frame used for the survey and any deficiencies or inefficiencies in it affect the total error of the survey. As with other components of the survey design, the choice of frame for a survey must be based on assessing the cost of using that frame and the total error of the survey estimates when that particular frame is used. It may cost more to construct a frame that contains the information needed to use probability proportional to size sampling; however, the gain in efficiency of the sample may counterbalance the cost of constructing the frame. The choice to be made is whether the cost and time necessary to construct a certain type of frame are justified. Perhaps, it is most cost-effective to use the money to survey a somewhat larger sample.

The efficiency of one type of frame relative to another is intimately related to the efficiencies of the sample designs used in each case. Although the type of available frame limits the type of sample design that can be used, more than one type of sample design can be used for a particular frame, so that the relative efficiency of frames cannot be considered without considering the types of sample designs they will support. Nowhere in the literature is there a discussion of the relative merits of different types of frames per se. Explicit discussions of optimum sample designs may consider the relative cost of constructing the frames needed for the various sample designs under consideration. Thus the researcher interested in assessing the relative efficiency of various frame types must, at present, consult the literature on the efficiency of various sample designs.

In addition to the frame's effect on the survey in terms of relative efficiency to the sample design it permits, certain deficiencies or errors in the frame will have an impact on the variance of the survey estimates. For example, the frame may contain information used to construct strata and optimally allocate the sample to the strata. If there are errors in the stratification information, variances of the estimates may be larger than those that could be achieved if a frame without stratification errors were used.

Deficiencies in the frame can also introduce bias into the survey estimates. In our example of a survey in which the target population was persons eligible to vote, if the frame constructed at the last stage of sample selection included a significant number of people not eligible to vote, survey results could be biased. We have found the following taxonomy of frame problems to be useful for examining how frame errors can potentially bias survey estimates or reduce their efficiency: (1) missing population elements, (2) inclusion of nonpopulation elements, (3) multiplicity problems, (4) failure to account for clustering, (5) incorrect auxiliary information, and (6) incorrect accessing information.

We discuss each of these problems below and present some of the terminology used to describe them.

3.3.1 Missing Population Elements

Failure to link some members of the target population to the frame is perhaps the most serious type of frame error because it cannot be recognized from either the sample or the frame. Failure to recognize and correct for missing population elements may introduce bias into the survey estimates. Totals will always be underestimated, and other statistics may or may not have bias, depending on whether the population missed differs from that covered.

A variety of terminology has been used to refer to this type of error— *undercoverage, noncoverage,* or *incomplete coverage.* These terms also refer to the quality or extent of enumeration in a census. For example, Eckler and Pritzker (1951:31) refer to a *coverage check* as a reenumeration "designed to measure the extent of undercounting and overcounting of the units of enumeration." This multiple use of terminology arises because procedures used to check enumeration in a census are often identical to those used to check the frame for missing population elements. Frames are often constructed by making census-type listings of units (households, people, etc.). Since a census aims to count the entire population of elements, and in frame construction we aim to account for each population element, the procedures for measuring the quality of censuses and frames are similar. Thus Zarkovich (1966) speaks of the *net underenumeration* of a census as the number of persons missed. Similarly, net undercoverage is the number of population elements missed from the frame, be they people or whatever. This distinction, however, is not always adhered to in the literature, and undercoverage may refer to census underenumeration.

Omission of some population elements from the frame has given rise to terms that distinguish various sets of population elements. The terms *target population* and *target universe* designate the population one wishes to measure in the survey. The population linked to the sampling frame may not be exactly identical to the target population. Terms used to refer to the population actually covered by the frame are *sampled population*, *truncated universe*, and *survey population*. Some authors distinguish between the sample and survey population, with the sample population being the one linked to the frame and the survey population being the one actually located. In this case the survey population excludes the nonrespondent subclass.

3.3.2 Inclusion of Nonpopulation Elements

The frame may also contain elements that are not part of the target population, often called *overcoverage*. If this error is not corrected, totals will be overestimated and other statistics may be biased. Relative to undercoverage, there is usually less potential for bias because elements not in the target population can often be recognized and eliminated during the survey. Unbiased estimation for the target population can be achieved by using the theory of domain estimation because the target population is a subclass of the sampled population. If the nontarget elements can be recognized, the chief impact of this type of frame error is a loss of efficiency in the estimates relative to a frame that did not exhibit overcoverage.

In some cases the frame may contain listings or units obviously not linked to a member of the target population, such as a vacant housing unit. In other cases, frame units are linked to elements that are similar to but not part of the target population. For example, in a household survey aimed at studying U.S. citizens, all households on the sampling frame in which aliens resided would not be linked to a member of the target population. In the former case, frame units not linked to any target element are called *blanks*, *empty listings*, or *duds*. The term *foreign elements* has been used to describe the latter case, in which the frame is linked to elements not part of the target population. Another term is *out-of-scope units*.

Overcoverage often occurs during survey field work because elements to be included in the sample are not exactly identified but are, rather, identified through rules or descriptions of the material to be in the sample. Thus overcoverage can occur when the rules of association as defined by Hansen et al. (1963) are incorrectly applied. One particular source of overcoverage is that of *boundary problems*. Boundary problems occur when population elements that lie near the boundary of an area sample plot are considered to be within the boundaries of the sample plot. In a census, this type of error can result in some persons being counted more than once, particularly those living in housing units near the borders of the enumeration areas.

Overcoverage may also occur when a frame contains nonexistent or foreign elements. Field staff who encounter a vacant housing unit or foreign element may assume that a mistake has been made and substitute an occupied adjacent unit. Thus overcoverage will occur, and it is not likely that subsequent analyses will reveal that units have been substituted.

In some types of surveys, overcoverage and undercoverage may tend to cancel each other, as in a census where persons erroneously omitted from one enumeration area are erroneously included in another. Thus one speaks of *net coverage error*, the difference between the number missed in the population and the number incorrectly included. Some authors term this difference *net underenumeration* or *net noncoverage*. There is an implied assumption in the use of these terms that the undercounting will always exceed the overcounting. A net coverage error of zero does not necessarily protect one from bias because the erroneously included units may differ from the erroneously excluded ones for the characteristic being measured.

3.3.3 Multiplicity Problems

Elements of the target population may be linked to more than one frame unit. To determine correctly the inclusion probabilities of sample elements, it is necessary to know how many times each sample element is associated with the frame, or to remove the multiple associations before drawing the sample. The problem is referred to in the literature as *duplicate listings* and erroneous inclusions. The number of frame units a particular population element is linked to has come to be called *element multiplicity*.

3.3.4 Improper Use of Clustered Frames

Often, frame units represent clusters of target population elements, such as housing units that contain more than one element of a target population of persons. If a single element is taken out of the cluster without regard to the cluster's size, sample estimates will be in error because the probability that a particular sample element is in the sample is not known. Often the error arises because one assumes that the probability that the individual elements are in the sample are equal to those of the selected frame units which are clusters of elements. This type of error occurs when a household sample has been selected, and person-level statistics based on a purposively chosen or haphazardly chosen household respondent are calculated using household probabilities of selection.

3.3.5 Incorrect Auxiliary Information

Many frames contain auxiliary information used for special sampling and estimation techniques, such as information used for stratification, measures of size used for probability proportional to size sampling, and characteristics highly correlated with the survey variables of interest which can be used for ratio and regression estimation. This information may be in error or missing for some frame units. Szameitat and Schäffer (1963) term this type of error a *deviation in content*. The principal effect of deviation in content is to decrease the precision of the survey estimates.

3.3.6 Incorrect Accessing Information

A final type of frame error discussed in the literature is *out-of-date* or *inaccurate frames*. In this situation, information about the element does not permit its location, very similar to the problem of undercoverage or nonlinkage. The chief difference is that in this case the undercoverage can be recognized from the frame itself. This situation could occur when a university wished to survey its alumni. Although it is likely that the university would possess all the names of its alumni, the current addresses of some might not be known. Inability to track down sample members because of outdated addresses is a common source of undercoverage in list frames. A survey in 1986 of Vietnam era veterans based on a frame of military discharge records would clearly suffer from this type of error.

Several points should be noted about the classification of frame errors above. First, some of the errors arise because of poor composition of the frame and others come from inadequate use of the frame. Undercoverage generally arises from poorly constructed frames, those without linkage to all members of the target population. In other cases the frame may provide linkage to all members of the target population, but mistakes in its use may introduce errors. Interviewers who substitute adjacent occupied housing units for sampled vacant units introduce overcoverage into an otherwise adequate sampling frame. Similarly, researchers who fail to account for clustering and select a single person from a household but use the household selection probabilities for estimation have introduced errors into an otherwise adequately constructed frame.

A second point is that some of the problems described above can be recognized and often resolved by collecting specific information during the survey. Overcoverage may be present in the frame; however, it can sometimes be recognized and corrected for by screening the sample prior to data collection. For example, if the target elements are smokers employed by company A, a frame of employees will probably not reveal which are smokers. However, once the sample is selected, nonsmokers can be identified and discarded by an initial screening. Similarly, it may be difficult to determine multiplicity within a frame, but it may be feasible to do so during the data collection. A person who wished to survey a bank's customers might select a sample of accounts and determine during the data collection the number of accounts each customer maintained.

Finally, some of the problems we have classified as frame errors could be characterized as another type of error with a different interpretation of the situation. If units near the boundaries of area segments are erroneously included during construction of a second- or third-stage sampling frame, we would say that a frame error was present, specifically, overcoverage. In the context of a census where boundary problems result in multiple enumeration, the error would be considered a measurement error. Similarly, use of incorrect probabilities for estimation when the researcher fails to account for clustering could be considered a sampling bias caused by use of an incorrect estimator. The context also determines whether failure to locate some of the elements of the target population is considered a frame or a nonresponse problem. One could assume that all organizations on an establishment list without an address will not be contacted and consider the problem as a frame deficiency; alternatively, these elements might be sampled and attempts made to trace them, in which case failure to locate the designated element would be considered a nonresponse error.

CHAPTER 4

Frames: Quantifying Frame Errors

In Chapter 3 we discussed six types of frame errors. In this chapter we discuss methods for quantifying frame errors, including measures of the frequency with which the error is occurring and its impact on survey statistics. Various error measures are defined, along with the means of quantifying the measures.

Frame problems can be illustrated using figures depicting the structural relationship between the frame and units of the target population. Four types of linkages between frame units and target elements were discussed in Chapter 3 and are illustrated in Fig. 4.1. Let j index target elements, i index frame units, N_i be the number of target elements linked to the ith frame unit, and M_j be the number of frame units linked to the jth target element. Figure 4.1a illustrates a one-to-one association or linkage between frame units and target elements. Both N_i and M_j equal 1 for every i and j. In the one-to-many association (Fig. 4.1b) each frame unit is linked to a cluster of target elements. Here N_i represents the cluster sizes; all are greater than 1. M_j is 1 for all target elements. The most complex case is many-to-many linkage (Fig. 4.1d). The frame units are linked to clusters of elements; elements may be linked to more than one cluster; and both N_i and M_j may be greater than 1.

To illustrate the different kinds of frame problems, a more complex diagram is necessary. Figure 4.2 illustrates the structural relationship between a target population and an imperfect frame. The figure and discussion that follow are derived from material originally presented by Szameitat and Schäffer (1963).

Suppose that target elements can be divided into K groups using some vector of auxiliary variables $\mathbf{X}$. For example, hospitals might be divided by region, size, and type of ownership groups. The vector $\mathbf{X}$ in this case consists of three auxiliary variables: (1) hospital location, (2) number of beds it contains, and (3) type of ownership.

Let gj index the jth target element in the gth part of the target population; $g = 1, 2, \ldots, K$. Also, let $\mathbf{X}_{gj}$ denote the auxiliary variable associated with gj. The frame units are also divided into groups. Let hi index the

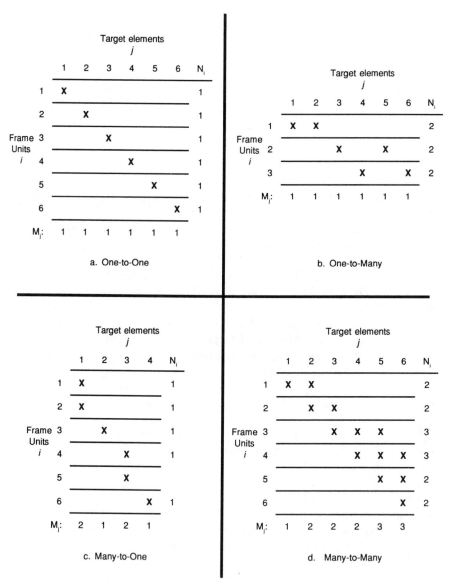

Note: **X**s indicate linkages of target elements to frame units.

Figure 4.1 Alternative structural relationships between a target population and a sampling frame.

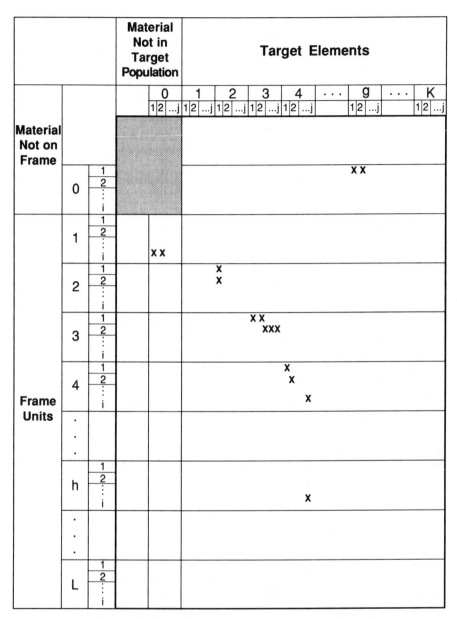

Figure 4.2   Structural relationships between target population and sampling frame.

ith frame unit in the hth part of the frame; $h = 1, 2, \ldots, L$. If the frame is structured one-to-one or many-to-one, auxiliary variables $\mathbf{Z}_{hi}$ can be defined that correspond to those defined for each target element gj.

Some nonpopulation elements may be included in the frame. We will denote these by letting $g = 0$ so that the 0_j denotes the jth nontarget element. In addition, some units containing target elements may be missing from the frame. Letting $h = 0$ for all such units, 0_i denotes the ith unit that should have been included in the frame and was not. This might be housing units erroneously excluded from the frame.

Figure 4.2 shows the structural relationship between the frame and target populations. The shaded area of the upper left corner represents material not included in the target population or frame. Some nontarget elements have been erroneously included on the frame. For example, a frame for a survey of hospitals might erroneously include some extended care nursing facilities. In the diagram the elements associated with frame unit $1i$ are nontarget elements. Some target elements have not been included in the frame. Elements 1 and 2 in group g are an example.

Some frame units represent clusters of elements, as is shown for the units in frame group $h = 3$. Target elements linked to more than one frame unit are shown for the group where $g = 2$. Some target elements are associated with the incorrect stratum. Target element $4j$ is incorrectly associated with stratum h. Stratum 4/group 4 represents target elements associated with the correct stratum with a one-to-one association.

This model may be expressed in algebraic form by defining a series of indicator functions where

$$A_{gj, hi} = \begin{cases} 1 & \text{if population element } gj \text{ is correctly linked} \\ & \text{with the unit } hi; \, g = 1, 2, \ldots, K; \text{ and } h = 1, 2, \ldots, L \\ 0 & \text{otherwise} \end{cases}$$

$$B_{gj, hi} = \begin{cases} 1 & \text{if the element } gj \text{ is incorrectly linked} \\ & \text{with the unit } hi; \, g = 1, 2, \ldots, K; \text{ and } h = 1, 2, \ldots, L \\ 0 & \text{otherwise.} \end{cases}$$

In frames with a one-to-one or many-to-one structure, $A_{gj, hi} = 1$ if $\mathbf{X}_{gj} = \mathbf{Z}_{hi}$ and $\mathbf{B}_{gj, hi} = 1$ when $\mathbf{X}_{gj} \neq \mathbf{Z}_{hi}$. Suppose that in the population there is a woman named Sam Jones at 123 Some St., Ataplace, USA. This woman is linked to the frame in that there is a person on (or linked to) the frame identified as Sam Jones at 123 Some St., Ataplace, USA. However, this linkage is incorrect because the Sam Jones on the frame has been classified as male and placed in the incorrect stratum. In this case, the A indicator equals 0 and B equals 1. Note that $B_{gj, 0}$ indexes target elements that are not linked to the frame; and $B_{0j, hi}$ indexes nontarget elements linked to the frame. Thus if Sam Jones is listed on the frame but no address is available, then she is, as far as the survey goes, unlocatable and is missing from the frame, so that $B_{gj,0} = 1$. Note that both A and B can be 0 for a certain cell in Fig. 4.2; both cannot be 1 however. Using these indicator functions, we have

$\sum_{gj} A_{gj,hi} = N_{hi}$ = Number of elements correctly associated with frame unit hi

$\sum_{gj} B_{gj,hi} = N_{hi}^*$ = Number of elements incorrectly associated with frame unit hi

$\sum_{hi} A_{gj,hi} = M_{gj}$ = Number of frame units target element gj is correctly associated with

$\sum_{hi} B_{gj,hi} = M_{gj}^*$ = Number of frame units target element gj is incorrectly associated with

$\sum_{gji} B_{gi,0} = N_0$ = Number of target elements missing from (not linked to) frame

$\sum_{hij} B_{0j,hi} = M_0$ = Number of times nontarget elements included in frame (not necessarily the number of nontarget elements linked because some nontarget elements may have multiple linkages to the frame).

The total number of target elements in the population is N, and is the sum of the number of missing elements and the number of elements included in the frame. If there is no multiplicity ($M_{gi} + M_{gi}^* = 1$), then

$$N = N_0 + \sum_{hi} (N_{hi} + N_{hi}^*).$$

The six types of error discussed in Chapter 3 can now be described in terms of these sums.

1. Missing Elements. Any frame for which $N_0 \neq 0$.
2. Inclusion of Nonpopulation Elements. Any frame for which $M_0 \neq 0$.
3. Multiplicity Problems. Any frame for which $M_{gj}^+ = M_{gj} + M_{gj}^* \neq 1$ and M_{gj}^+ is not determined and accounted for in the estimation procedures.
4. Improper Use of a Clustered Frame. Any frame in which $N_{hi}^+ = N_{hi} + N_{hi}^* \neq 1$ and it is assumed to be 1.
5. Incorrect Auxiliary Information. Any frame for which $N^*(hi) \neq 0$.
6. Inadequate Accessing Information. This situation is similar to the one for missing or nonlinked elements, because target elements for which accessing information is inadequate and for which it is known in advance that the information is inadequate are effectively not linked to the frame. A major difference is that in this case, the value of N_0 can be determined from the frame (assuming no other types of nonlinkages).

In Chapter 3 we classified frame errors into six types, and in this chapter we have illustrated these errors by means of a diagram depicting the structural relationship between a frame and the target population. We discuss below the effect of each type of error on estimates of totals and means. We will assume that the target population contains N elements indexed by j, and that Y_j is to be measured for each member of the sample. The total and mean for the target population are, respectively,

$$Y = \sum_{j=1}^{N} Y_j$$

and

$$\bar{Y} = \frac{1}{N} \sum_{j=1}^{N} Y_j.$$

Measures of error and experimental procedures for quantifying the error are discussed in each case. We begin with missing elements, the most serious type of frame error because it cannot be recognized from examining the frame.

4.1 MISSING ELEMENTS

If population elements are missing from the frame, the total Y will always be underestimated. In terms of the model above, assume that $M^+(gj) = 1$ for every element gj; that is, we assume each element is linked to a single frame unit and let

$$N_A = \sum_{hi} (N_{hi} + N_{hi}^*)$$

be the number of population elements linked to the frame. Then the target population is composed of N_0 elements missing from the frame and N_A elements linked to the frame, that is, $N = N_0 + N_A$. Similarly, the population total Y is the sum of the total for the linked elements and the total of the nonlinked elements,

$$Y = \sum_{j=1}^{N_0} Y_j + \sum_{j=1}^{N_A} Y_j = Y_0 + Y_A.$$

If no nontarget elements are included in the frame and if the sampling and estimation procedure correctly accounts for clustering, the mean square error (MSE) of an estimate, y_A, from a survey design using the frame with nonassociated elements is

$$MSE(y_A) = Var(y_A) + (Y - Y_A)^2.$$

This assumes that there is no bias in y_A as an estimate of Y_A.

Several measures of the impact of missing elements on the estimates are:

$$Net\ bias(y_A) = Y_A - Y = -Y_0$$

$$Relative\ bias(y_A) = \frac{-Y_0}{Y}$$

$$Ratio\ of\ squared\ bias\ to\ MSE(y_A) = \frac{(Y_0)^2}{MSE(y_A)}.$$

On of the most extensive discussions of frame errors in given in a dissertation by Kiranandana (1976). Kiranandana expresses the relative bias in estimates of the total in terms of the ratio of the population means for the omitted elements and the associated elements. Letting r denote this ratio and W_0 denote the proportion of omitted elements, we have

$$r = \frac{\bar{Y}_0}{\bar{Y}_A}$$

and

$$W_0 = \frac{N_0}{N},$$

which gives

$$Relative\ bias(y_A) = \frac{-W_0 r}{rW_0 + (1 - W_0)}.$$

If $r = 1$, that is, if the mean for the omitted elements is the same as that for the included elements, the relative bias equals the negative of the proportion of the population not associated with the frame. If the characteristic being measured does not take negative values, r is always nonnegative. In this case, the relative bias of y_A is negative and decreasing in r. Thus, when estimating totals, omitting elements with small values of Y_j has less relative effect than omitting elements with large values.

Similar measures of the extent and impact of undercoverage can be defined for the mean. Assume that $\bar{y}_A$ is an unbiased estimate of $\bar{Y}_A$. In the presence of a frame that does not include all the target population elements,

$$MSE(\bar{y}_A) = Var(\bar{y}_A) + (\bar{Y}_A - \bar{Y})^2.$$

Again, net bias, relative bias, and ratio of squared bias to the mean square error can all be used to measure the impact of undercoverage, with

$$\text{Net bias}(\bar{y}_A) = \bar{Y}_A - \bar{Y} = W_0(\bar{Y}_A - \bar{Y}_0)$$

$$\text{Relative bias}(\bar{y}_A) = \frac{W_0(\bar{Y}_A - \bar{Y}_0)}{\bar{Y}}$$

$$\frac{(\text{Bias})^2}{\text{MSE}(\bar{y}_A)} = \frac{\left[W_0(\bar{Y}_A - \bar{Y}_0)\right]^2}{\text{MSE}(\bar{y}_A)}.$$

Using Kiranandana's (1976) notation, relative bias in the estimate of the mean in the presence of noncoverage is

$$\text{Relative bias} = \frac{W_0(1 - r)}{(1 - W_0) + rW_0}.$$

Note that there is no bias in estimates of the mean because of noncoverage if the means of the associated and nonassociated target groups are the same. Again, if $Y_j \geq 0$ for all i, the relative bias is strictly decreasing in r. If r is < 1, the relative bias is positive; it is 0 when $r = 1$; and it is negative if $r > 1$.

In Table 4.1 we show the relative bias in the estimate of a total for various proportions of missing elements and various values for the ratio of the mean of the missing elements to the mean of the included elements. The table clearly illustrates that the impact on estimates of totals is relatively small if the elements missing from the frame tend to have small values of Y_j. In Table 4.2 we show the relative bias in estimates of the mean for the same values of W_0 and r. Missing relatively small values of Y_j do not have the same relatively benign effect on the mean as they did on the total, and missing

Table 4.1   Relative Bias of a Total for Various Ratios of the Mean of Missing to the Mean of Nonmissing Elements and Differing Proportion Missing

Proportion Missing W_0	Ratio of Mean of Missing Elements to Mean of Nonmissing Elements $r = \bar{Y}_0/\bar{Y}_A$				
	0.1	0.5	1.0	2.0	5.0
0.01	−0.0010	−0.0050	−0.01	−0.0198	−0.0481
0.025	−0.0026	−0.0127	−0.025	−0.0488	−0.1136
0.05	−0.0052	−0.0256	−0.05	−0.0952	−0.2083
0.10	−0.0110	−0.0526	−0.10	−0.1818	−0.3571
0.15	−0.0173	−0.0811	−0.15	−0.2609	−0.4688
0.25	−0.0323	−0.1429	−0.25	−0.4000	−0.6250
0.50	−0.0909	−0.3333	−0.50	−0.6667	−0.8333

Table 4.2 Relative Bias of a Mean for Various Ratios of the Mean of Missing Elements to the Mean of Nonmissing Elements and Differing Proportion Missing

Proportion Missing W_0	Ratio of Mean of Missing Elements to Mean of Nonmissing Elements $r = \bar{Y}_0/\bar{Y}_A$				
	0.1	0.5	1.0	2.0	5.0
0.01	0.0091	0.005	0	-0.0099	-0.0385
0.025	0.0230	0.0127	0	-0.0244	-0.909
0.05	0.0471	0.0256	0	-0.0476	-0.1667
0.10	0.0989	0.0526	0	-0.0909	-0.2857
0.15	0.1561	0.0811	0	-0.1304	-0.3750
0.25	0.2903	0.1429	0	-0.2000	-0.5000
0.50	0.8181	0.3333	0	-0.3333	-0.6667

small values are just as serious as missing large values. Biases in estimates of the mean may be positive or negative; however, if the means of the missing and nonmissing elements are identical, there is no bias. All researchers interested in estimating means fervently hope that this is true, although it is unlikely that this hope often is realized.

In the preceding discussion we made a distinction between determining the number of missing elements and measuring their impact on the estimates. Any estimate of N_0 or W_0 is a measure of the extent of undercoverage. The net bias, relative bias, mean square error, and ratio of squared bias to MSE are all measures of the impact of undercoverage.

Actual reports from surveys vary in their discussion of undercoverage. Some ignore the problem altogether. Others provide only a qualitative description of the target elements missing from the frame. For example, one may see statements such as

> The survey covers only those members of the target population whose telephone numbers were listed in the current directory. Persons with unlisted numbers and new listings are excluded;

or

> one limitation in the present design is that it does not include in the target populations those young men and women who drop out of high school before graduation (or before the last few months of the senior year, to be more precise) (Hindelang et al., 1981:597).

Other survey reports may provide an estimate of the extent of the missing group, and a very few will have investigated the impact of the missing elements on the estimates.

The key to measuring the effect of nonassociated target elements is obtaining an estimate of Y_0 and for measuring the extent, an estimate of N_0 is needed. Neither can be estimated from the sample. Some alternative source of information must be used. That this type of frame error cannot be recognized from the sample probably makes it the most serious type of frame error. One can remain ignorant of the magnitude of the problems unless precautions are taken.

Researchers have used several procedures to determine the magnitude of frame errors. These procedures can be divided into two groups: those using external data and those employing special data collection procedures during the course of the survey. External data procedures include comparison with an external figure; in-flow, out-flow techniques; and reverse record checks (Eckler and Pritzker, 1951) or dual record systems. Data collection procedures include quality check approaches and linkage methods.

4.1.1 Comparison with an External Figure

Another survey or census may have measured Y. Comparison of Y with y_A may indicate whether or not noncoverage has occurred. This, of course, is only the beginning step. If the two estimates are not approximately equal, either one or both may be in error, and even if they are equal, one cannot be sure that the comparison source did not have coverage problems similar to the current survey. If they are not nearly equal, however, this is a clear indication that something is amiss with one of the figures. It is necessary to determine which figure is more accurate to decide if the missing elements have had a serious impact on the current estimates.

4.1.2 In-Flow, Out-Flow Technique

For some types of estimates the total should be equal to sums and differences of certain quantities. For example, the population at time t_2 should equal the population at t_1, plus births in the interval $t_2 - t_1$, minus deaths in the interval, plus net migration. Again the conclusion one draws from the comparison depends on the assumptions about the relative accuracy of the two figures.

Both the in-flow, out-flow technique and the comparison with external data can be applied at the aggregate level or for subdomains of the population. Missing population elements for certain age groups can be detected by comparing the age distribution obtained in a survey to that expected from the aging of the population during a previous study. Basically, these types of studies are consistency checks that examine whether or not the data are consistent with the results from other sources or exhibit certain relationships known to exist between subgroups.

Each method can be a measure of either the impact or extent of undercoverage. As an example, suppose that we were interested in knowing the number of people in the county and their total expenditure for health care. Comparing two population totals would measure the impact of undercoverage for the variable number of people; however, it would be an estimate of the extent of undercoverage for total expenditure for health care and would not totally answer the question of the impact of undercoverage on estimates of expenditures because those not included may have very different health care expenditures. Note also that the procedures used for checking for frame noncoverage can also be used to check for undernumeration if the survey is a census. An example is the following technique, often used to check the quality of a census, but also appropriate for checking for frame undercoverage.

4.1.3 Reverse Record Check or Dual Record System

In a reverse record check, target elements that should be included in the frame are identified from a set of records. The frame is then examined to see if the elements were included in it. If this method is used in a *sample* survey, it requires the use of compact clusters (Zarkovich, 1966) in which all elements clustered within a certain area, time period, and so on, are included in the sample. An element identified in the records must be slated to be in the sample, or we would not know whether it was missing from the survey because of noncoverage or because of not being selected in the sample. If one or more of the survey variables can be obtained from the records, an estimate of impact of undercoverage can also be obtained.

A method Sekar and Deming (1949) describe for estimating birth and death rates and the extent of vital events registration can also be used for estimating the extent of undercoverage in a frame. The method is known variously as *dual record systems* (Wells, 1971), *population growth estimation* procedures (Marks et al., 1974), and *capture–recapture methods* (Wolter, 1983). The method is particularly appropriate for the situation in which a list frame has been used for a survey. Suppose that a second independently constructed list is available for the target population or for the sampling units if the frame is not an element list. Let

N_{A1} = number of elements associated with the first frame

N_{A2} = number of elements associated with the second frame

$N_{A1,\,A2}$ = number of elements associated with both frames (also called the number of matched elements or events).

Assuming that the probability that an element is included on one frame is independent of the probability that it is included on the other, then under certain conditions [see Wolter (1983) for a summary],

$$\hat{N} = \frac{N_{A1}N_{A2}}{N_{A1,\,A2}}$$

is the maximum likelihood estimator of the total elements in the population. The estimated proportion of elements missing from the first frame is

$$\hat{W}_{01} = 1 - \frac{N_{A1,\,A2}}{N_{A2}}.$$

For a more extensive discussion of dual record systems, see Chapter 10, Section 10.6.

4.1.4 Quality Check Approach

In the quality check approach, portions of a frame are independently constructed at a time close to the survey. This procedure is usually done for a subset of areas in a multistage survey in which a frame was constructed for several stages of a sampling structure. In this method, the frame is reconstructed using more accurate procedures, such as better maps and materials, more highly qualified personnel, better training, and closer supervision. Presumably such improvements will result in a more accurate frame that can be used to estimate the undercoverage in the original procedures. If, in addition, elements missed in the original survey are measured for the survey variables, estimates of impact can be calculated.

A similar procedure is use of the *predecessor–successor method* (Dalenius, 1974). The method is similar to the relisting or quality check approach except that it is not carried out independently of the original frame construction. A sample of areas is drawn. The elements listed in the survey are provided. The check is done by picking a point (a listed element) and proceeding along the defined path until the next listed unit is reached. The number of missed elements is noted. Denote this m_1. Then one proceeds in the opposite direction along the path from the selected listed element, noting the number, m_2, of missed elements until a listed element is reached. The estimate of the total missed elements is then

$$m = km_1 + (1 - k)m_2.$$

Here k and $1 - k$ are simply weights that sum to 1. If the missed elements identified in this procedure are surveyed and the value of Y_j obtained, the impact of the missing elements on the estimates can be examined.

In recent years the similarity between dual frame procedures and capture–recapture methods used in biology has received study. These methods have been adapted for the study of census coverage. Wolter (1983) presents an extensive discussion of such coverage error models for censuses. Although these models are described for checking the quality of a census, they are applicable for investigating frame coverage. Two frames or lists are available, as described in the paragraphs above on dual frames. A general coverage error model is defined as follows: It is assumed that the population is fixed and contains N members. It is further assumed that neither frame exhibits multiplicity or contains out-of-scope elements. We define a new indicator variable

$$C_{jf} = \begin{cases} 1 & \text{if the } j\text{th element linked to frame } f, f = 1, 2 \\ 0 & \text{otherwise.} \end{cases}$$

For each individual in the target population, it is assumed that presence or absence of linkages to the two frames follows a multinomial distribution $C_{jff'}$, with

$$\Pr(C_{j1} = 1, C_{j2} = 1) = P_{j11}$$

$$\Pr(C_{j1} = 0, C_{j2} = 1) = P_{j01}$$

$$\Pr(C_{j1} = 1, C_{j2} = 0) = P_{j10}$$

$$\Pr(C_{j1} = 0, C_{j2} = 0) = P_{j00}$$

$$\Pr(C_{j1} = 1) = P_{j1+} = P_{j11} + P_{j10}.$$

$P_{j0+}, P_{j+1}, P_{j+0}$ defined similarly;   and $P_{j1+} = 1 - P_{j0+}$.

The two frames are the result of N mutually independent multinomial events $C_{1ff'}, C_{2ff'}, \ldots, C_{Nff'}$. Thus N_{A1}, the number of elements linked to the first frame, is a random variable with

$$E(N_{A1}) = \sum_{j=1}^{N} P_{j1+}$$

$$\text{Var}(N_{A1}) = \sum_{j=1}^{N} P_{j1+}(1 - P_{j1+})$$

$$= \sum_{j=1}^{N} P_{j1+} P_{j0+}.$$

Several additional assumptions are made for this general model, namely that it is possible to determine correctly the frame linkages for all elements. This procedure is called the *matching assumption* because determining if $C_{j1} = 1$ and $C_{j2} = 1$ (that the element is linked to both frames) requires that one be able to match elements linked to the two frames. It is also assumed that the necessary matching information will be available for each element and each source frame and that no nontarget elements are linked to either frame.

Additional assumptions are needed to estimate the true population size, and thus the extent of coverage and Wolter (1983) discusses several of them. One special case is equivalent to the dual frame method above and requires the additional assumptions that the events of being linked to the two frames are independent and that $P_{j1+} = P_{1+}$ and $P_{j+1} = P_{+1}$.

Now suppose that instead of evaluating coverage by matching the entire contents of each frame, the evaluation is done on a sample basis. Frame 2 is divided into K clusters that completely cover the target population. Every member of the target population is linked to a cluster, but every member may not be linked to frame 2. For example, if frame 2 was an area housing unit frame and well-defined areas comprised the clusters, all elements (housing units) would theoretically be covered by a cluster and only one cluster, but some of these may have been omitted from the within-cluster frame. A sample of clusters is selected and elements contained on frame 1 are matched to elements identified in each cluster. This information is used to estimate N_{A2} and $N_{A1, A2}$. The estimator from the sample is then

$$\hat{N}_S = \frac{N_{A1}\hat{N}_{A2}}{\hat{N}_{A1, A2}}.$$

Wolter gives the expected value and variance of this estimator by first taking the expected value of the estimator under the assumed multinomial distribution and then over the sample design. The Taylor series linearization method is used. The estimator is of the form UW/V, and the first-order approximation to the expected value is

$$E\left(\frac{UW}{V}\right) = \frac{EU \cdot EV}{EW} + \frac{\text{Var}(V)}{(EV)^2}\frac{EU \cdot EW}{EV} - \text{Cov}(U,V)\frac{EW}{(EV)^2}$$
$$+ \text{Cov}(U,W)\frac{1}{EV} - \text{Cov}(V,W)\frac{EU}{(EV)^2}.$$

Using the formula above, we have for the multinomial distribution

$$EU = NP_{1+}$$
$$EW = NP_{+1}$$
$$EV = NP_{11} \quad \text{where } P_{11} = P_{1+}P_{+1}$$
$$\text{Var } V = NP_{11}(1 - P_{11})$$
$$\text{Cov}(U,V) = NP_{11}(1 - P_{11}) - NP_{11}P_{10} \quad \text{where } P_{10} = P_{1+}P_{+0}$$
$$\text{Cov}(W,V) = NP_{11}(1 - P_{11}) - NP_{11}P_{01} \quad \text{where } P_{01} = P_{0+}P_{+1}.$$

Substituting these into the expression above gives

$$E(\hat{N}) \doteq N + \frac{P_{0+}P_{+0}}{P_{1+}P_{+1}}$$

for the expected value with respect to the multinomial model.

In the case Wolter discusses in which the second frame is not enumerated entirely, similar methods are used to arrive at expressions for the variance under the multinomial, considering the sample design with

$$E(\hat{N}_S) = N + \frac{P_{0+}P_{+0}}{P_{1+}P_{+1}} + \frac{1-f}{f}\frac{P_{0+}}{P_{1+}P_{+1}},$$

where f is the sampling fraction for a simple random sample of k from K clusters.

$$\begin{aligned}\text{Var}(\hat{N}_s) &= \text{Var}(\hat{N}) + N\frac{1-f}{f}\frac{P_{0+}}{P_{1+}P_{+1}} \\ &= N\frac{P_{0+}P_{+0}}{P_{1+}P_{+1}} + N\frac{1-f}{f}\frac{P_{0+}}{P_{1+}P_{+1}}.\end{aligned}$$

An interesting addition to Wolter's paper is that he shows the relationship of this method to the commonly used Hansen et al. (1961) measurement error model (see Chapter 11).

4.2 NONPOPULATION ELEMENTS ASSOCIATED WITH THE FRAME

When nonpopulation elements are associated with the frame, the total Y will always be overestimated unless it is possible to recognize the nonpopulation elements in the sample. In terms of the previous model there are, for all elements (population and nonpopulation) associated with the frame, the following number of associations between target elements and frame units:

$$M = M_0 + \Sigma(M_{gj} + M_{gj}^*) = M_0 + M_T$$

where M_0 is the total number of associations of nontarget elements with the frame and M_T is the total associations of target elements. Assume that there is no multiplicity in the frame and that all target elements are included in the frame. In such a case,

$$M_T = N.$$

The measurement for the jth element is Y_j as before, but some of the j are nontarget elements. The total for the frame population or surveyed population is

$$Y_F = \sum_{j=1}^{M} Y_j = Y_{\bar{T}} + Y_T,$$

where we have summed over all elements linked to the frame not just the N elements in the target population. Here

$$Y_{\bar{T}} = \sum_{j=1}^{M_0} Y_j$$

is the total for the nontarget elements included in the frame. If we use a sampling and estimation procedure that gives unbiased estimates of Y_F, the following measures can be defined for measuring the impact of including nontarget elements on the frame:

$$\text{Bias}(y_F) = Y_F - Y_T = Y_{\bar{T}}$$

$$\text{Relative bias}(y_F) = \frac{Y_{\bar{T}}}{Y_T}.$$

If we use a formulation similar to that of Section 4.1, which depicts the relative bias in terms of the extent of overcoverage and the ratio of the means of the nontarget elements and target, we get

$$\text{Relative bias(total)} = \frac{Q_0 v}{1 - Q_0},$$

where $Q_0 = M_0/M$, the proportion of elements associated with the frame that are nontarget elements, and $v = \bar{Y}_{\bar{T}}/\bar{Y}_T$, the ratio of the mean of nontarget elements to mean of target elements. Similar statistics for the estimates of means are:

$$\text{Bias}(\bar{y}_F) = Q_0\left(\bar{Y}_{\bar{T}} - \bar{Y}_T\right)$$

$$\text{Relative bias}(\bar{y}_F) = \frac{Q_0\left(\bar{Y}_{\bar{T}} - \bar{Y}_T\right)}{\bar{Y}_T}$$

$$= Q_0(v - 1).$$

Estimates of Q_0 measure the extent of overcoverage, and estimates of both v and Q_0 are needed to measure the impact of overcoverage.

Inclusion of nontarget elements on the frame is likely to be a problem in multistage samples in which higher stage units are geographic areas. The final frame of elements within selected areas is constructed in the field. Only those target elements within the selected sample areas should be included in the later stage frame. Because of difficulties in determining boundaries in the field, target elements situated outside the selected geographic area may be included. This problem can also arise in a complete census, where it causes some elements to be enumerated more than once, resulting in overnumeration.

Boundary problems are not the only way in which this type of overcoverage can occur. Essentially, any failure in the application of rules of association that causes target elements to be incorrectly linked to a sample unit selected in a previous sampling stage will cause overcoverage. In many surveys nontarget elements included in the sample can be recognized from the sample data themselves (discussed below in detail). However, incorrect linkages to a higher stage sample unit may not be revealed by the data.

It is possible to conduct surveys using frames that include nontarget population elements. If one can recognize which members of the sample are not elements of the target population, the theory of domain estimation can be used to make unbiased estimates. The target population is, in this case, a subpopulation or domain of the frame population. As an example, suppose that one wished to gather information about children who walked to school on a regular basis. A sample of children could be selected from the list of enrollees (frame) maintained by the school. Each member of the sample could be contacted and those who walked could be identified and included in the survey.

A loss of efficiency occurs relative to a frame that did not include the nontarget elements [see Szameitat and Schäffer (1963)]. The loss of efficiency can be formulated as follows for simple random samples of m from the M frame elements. Let

$$Y'_j = \begin{cases} Y_j & \text{if } j \text{ is an element of the target population} \\ 0 & \text{otherwise} \end{cases}$$

$$\hat{Y}_T = \frac{M}{m} \sum_{j=1}^{m} Y'_j.$$

Letting

$$S_y^2 = \frac{1}{M_T - 1} \sum_{j=1}^{M_T} \left(Y_j - \bar{Y}_T \right)^2,$$

we have

$$V(\hat{Y}_T) = \frac{M^2}{m}\left[(1 - Q_0)S_y^2 + Q_0(1 - Q_0)\bar{Y}_T^2\right].$$

The efficiency of the frame with nontarget elements relative to a frame with only target elements is

$$\text{Relative efficiency} = \frac{M^2\left[(1 - Q_0)S_y^2 + Q_0(1 - Q_0)\bar{Y}_T^2\right]}{M_T^2 S_y^2}.$$

If the nonpopulation elements can be recognized from the sample data, domain estimation can be used to produce unbiased estimates of target population characteristics and to quantify the various measures of error discussed above. Although unbiased estimates can be constructed in this situation, one may still wish to make estimates of the loss of efficiency to evaluate whether or not it is cost-effective to remove the nontarget elements from the frame before sampling. Similarly, if it is costly to determine exactly which elements are nontarget elements, it might be advantageous to estimate the bias from overcoverage to determine if it is worthwhile in future surveys to collect the information needed to correct for this overcoverage. Hansen et al. (1963) suggest that one may achieve gains in efficiency by stratifying the frame units into groups that contain few nontarget elements and groups that contain large numbers of nontarget elements and varying the sampling rates.

If the usual sample data do not reveal the nontarget elements, both external data procedures and quality check approaches can be used for detecting overcoverage. In terms of the external data procedures discussed above, the comparison to external figures and the inflow–outflow technique can be used to detect overcoverage. The reverse record check is not appropriate for checking for overcoverage unless the records contain the entire target population in which case the record system would most likely have been used as the frame.

Marks et al. (1953) described three methods (direct, indirect, and sampling) for detecting overcoverage that results from erroneous inclusions of units on the boundary of areas in a multistage area sample. Each method is a variation of the quality check approach.

4.2.1 Direct Method

A subsample of the higher-stage sample units is drawn and relisted using more accurate methods. Within each higher-stage unit in the subsample, the number of elements in the original survey and the quality check are compared. This procedure can be used to estimate Q_0. If it is possible to identify exactly which elements were erroneously included in the original survey after the relisting, Y_T can be estimated from the original survey data. Also, $Y_{\bar{T}}$ can be estimated. Thus the impact can also be measured. Because it is not always desirable (or possible) to include all of the original area in the quality check, an indirect method can also be used.

4.2.2 Indirect Method

In this method subareas of subparts of sampled higher stage units are selected and resurveyed using the accurate method: relisting, sampling, and measurement are all repeated. Presumably no overcoverage occurs in the quality check survey. Assuming no other source of error, M_T and Y_T can be estimated. They are then compared to the original survey estimates of numbers of elements and total—that is, M and Y_F, respectively—to estimate the extent and impact of overcoverage.

4.2.3 Sampling Method

In this method a subsample of the elements included in the original survey is selected. One then visits the higher stage unit they were from and determines if they should have been linked with that unit. Data from the subsample and results of the check can then be used to estimate Q_0 and Y_F and $Y_{\bar{T}}$. Note that this sampling method cannot be used to detect undercoverage. In that case, compact clusters are required. Also, when using the quality check approaches to detect overcoverage, one is in a better position than when detecting undercoverage, because once the erroneously included units are identified, the data needed to estimate the bias are already available from the main survey. When erroneously excluded elements are identified, the survey measurements remain to be taken if one is to estimate the bias.

4.3 SIMULTANEOUS INVESTIGATION OF UNDER- AND OVERCOVERAGE

Often one will want to check for erroneous inclusions and omissions simultaneously. Zarkovich (1966) describes the use of compact cluster sampling to check the quality of the frame. He calls this procedure a listing check and defines three measures for the quality of listing. A sample of compact clusters is drawn. Elements in the cluster are relisted. It is necessary to have compact clusters, because to detect omissions we must know that every unit in a particular area (or other cluster) should have been included in the survey.

Let

$$
Z_j = \begin{cases} 1 & \text{if the } j\text{th element were listed (i.e.,} \\ & \text{on the frame for the original survey)} \\ 0 & \text{otherwise} \end{cases}
$$

and

$$
X_j = \begin{cases} 1 & \text{if the } j\text{th element were found to be a} \\ & \text{member of the target population in the relisting} \\ 0 & \text{otherwise} \end{cases}
$$

and

$$
G = \sum_j (Z_j - X_j)^2
$$

$$
= \text{gross error of listing.}
$$

Net error of listing,

$$
D = \sum_j (Z_j - X_j)
$$

$$
\text{Percentage accurate listing} = 100\left(1 - \frac{G}{N}\right).
$$

Zarkovich suggests that one then estimate the total net error of listing and perform an approximate test of significant difference from 0. If a simple random sample of k clusters from K clusters was chosen, an estimator for the total error is

$$
D' = \frac{K}{k} \sum_{i=1}^{k} D_i.
$$

The statistic D'/S'_D is used to test for significance assuming an approximately normal distribution for the statistic. If the actual values of the Y_j are substituted for Z_j and X_j whenever they equal 1, D will be a measure of the net bias from erroneous omissions and inclusions. This is not likely to be possible because when $Z_j = 0$ and $X_j = 1$, then Y_i is not available.

Zarkovich describes this method for what he terms Type I and Type II designs. A Type I design is a single-stage cluster sample; a Type II design is a two-stage cluster design. An example of a Type II design is: Stage 1 consists of compact areas that are clusters of households and Stage 2 is composed of households that are compact clusters of persons. The net error of listing then has two components, one from errors in listing households and one from errors in listing persons within households.

4.4 FRAMES WITH UNRECOGNIZED MULTIPLICITY

Population elements may be associated with frame units more than once in either the many-to-one or the many-to-many patterns discussed in Chapter 3. In the many-to-one case each frame unit is linked to only one element, and a single element is linked to more than one frame unit. In the many-to-many situation each frame unit is a cluster of population elements, and a particular population element is associated with more than one frame unit (or cluster). When frames have a one-to-one or a one-to-many pattern of association between the elements and the units, the probability that a particular element is in the sample is equal to the probability that its associated frame unit is in the sample. This is not the case when elements are associated with more than one frame unit.

Sometimes one deliberately constructs frames with multiplicity to increase the chance of accessing a member of a rare target population. For example, if one wished to identify people with a rare disease by means of a household sample, rules that linked members of this target population to their own household and to households of their parents and siblings may be established. In other cases, frames with multiplicity are used because other frames are not available. In an agricultural survey one may wish to select a sample of farms, and only lists of farm owners and operators may be available. Because many farms have more than one owner, some adjustments need to be made. Procedures for conducting surveys using frames with multiplicity are discussed in Chapter 5.

In some cases it may not be feasible to determine multiplicity. For example, one may wish to survey users of a certain product; however, it may only be feasible to sample people at the time of purchase. This is a sample of uses rather than users. People will vary in terms of the number of times they use the product, and more frequent users will have greater chance of being in the sample.

Kiranandana (1976) examined the effect of using a frame with multiplicity and not adjusting for its presence. These methods will be illustrated by considering a frame with a single stratum and no undercoverage or overcoverage.

We assume that

$$N_0 = 0$$
$$M_0 = 0$$
$$N_i = 1$$
$$N_i^* = 0$$

and

$$M_j \geq 1$$
$$M_j^* = 0.$$

In this case, the number of units on the frame is larger than the number of elements in the population and the total number of frame units is

$$M = \sum_j M_j = \sum_i N_i = \sum_i 1 \qquad i = 1, 2, \ldots, M.$$

Suppose that the maximum value for M_j is Γ, and M_j ranges from 1 to Γ. Let T_γ equal the number of target elements for which $M_j = \gamma, \gamma = 1, 2, \ldots, \Gamma$. Then the total number of target elements, N, can be expressed as

$$N = \sum_{\gamma=1}^{\Gamma} T_\gamma.$$

The total frame units $M = \sum_{\gamma=1}^{\Gamma} \gamma T_\gamma$. Now

$$S_Y^2 = \frac{1}{N-1} \sum_{j=1}^{N} \left(Y_j - \bar{Y} \right)^2.$$

If we let $S_{Y\gamma}^2$ denote the variance of the T_γ unique target elements that appear γ times on the frame, then

$$S_{Y\gamma}^2 = \frac{1}{T_\gamma - 1} \sum_{j=1}^{T_\gamma} \left(Y_j - \bar{Y}_\gamma \right)^2$$

$$\doteq \frac{1}{T_\gamma} \sum_{j=1}^{T_\gamma} \left(Y_j - \bar{Y}_\gamma \right)^2.$$

Letting $a_\gamma = T_\gamma / N$, we have that

$$S_Y^2 \doteq \sum_{\gamma=1}^{\Gamma} a_\gamma \left[S_{Y\gamma}^2 + \left(\bar{Y}_\gamma - \bar{Y} \right)^2 \right].$$

Similarly, Kiranandana defines Y_γ as the total for the elements that have multiplicity γ, so that

$$Y_F = \sum_{\gamma=1}^{\Gamma} \gamma Y_\gamma$$

is the frame total with the multiple occurrences, and the total for the target population is

$$Y = \sum_{\gamma=1}^{\Gamma} Y_\gamma.$$

If a simple random sample of m units is selected from the M frame units, the bias in the estimate of the total unadjusted for multiplicity is

$$\text{Bias}(\hat{y}_F) = Y_F - Y = \sum_{\gamma=1}^{\Gamma} (\gamma - 1)Y_\gamma.$$

The relative bias is

$$\text{Relative bias}(\hat{y}_F) = \frac{\sum_{\gamma=1}^{\Gamma}(\gamma - 1)a_\gamma \bar{Y}_\gamma}{\sum_{\gamma=1}^{\Gamma} a_\gamma \bar{Y}_\gamma}.$$

If the means $\bar{Y}_\gamma$ are all equal,

$$\text{Relative bias}(\hat{y}_F) = \frac{M - N}{N}.$$

The mean square error of $\hat{y}_F$ as an estimate of Y is

$$\text{MSE}(\hat{y}_F) = \frac{M - m}{M} \frac{M^2}{m} \frac{1}{M - 1}\left[\sum_{i=1}^{M} (Y_i - \bar{Y}_F)^2\right] + \left[\sum_{\gamma=1}^{\Gamma} (\gamma - 1)Y_\gamma\right]^2$$

$$\doteq \frac{M - m}{M} \frac{M^2}{m} \frac{1}{\sum_\gamma \gamma T_\gamma}\left\{\sum_{\gamma=1}^{\Gamma} \gamma T_\gamma\left[S_{\bar{Y}\gamma}^2 + \left(\bar{Y}_\gamma - \bar{Y}_F\right)^2\right]\right\}$$

$$+ \left[\sum_{\gamma=1}^{\Gamma} (\gamma - 1)Y_\gamma\right]^2.$$

Even for fairly moderate sample sizes, the bias will tend to dominate the mean square error. Kiranandana examined the case when the maximum multiplicity was $\Gamma = 2$. Assuming that 87 percent of the population had multiplicity 1 and 13 percent were associated with the frame twice and that $S_y^2 = 4$, $(\bar{Y}_1, \bar{Y}_2) = (4, 4)$, $(3, 4)$, and $(4, 2)$, the ratio of the bias component to the variance component of the MSE is about 5.3, 5.1, and 1.1 for samples of size $m = 100$. Thus, if multiplicities of any substantial degrees are suspected, some steps should be taken to correct the estimates of totals.

Kiranandana gives similar expressions for measuring the impact of multiplicity on estimates of the mean. They are (for $\bar{y}_F = (1/m)\sum_{i=1}^{m} Y_i$)

$$\text{Bias}(\bar{y}_F) = \bar{Y}_F - \bar{Y} = \sum_{\gamma=1}^{\Gamma} \left(\frac{\gamma}{M} - \frac{1}{N}\right)Y_\gamma.$$

This bias will be 0 when the $\bar{Y}_\gamma$ are all equal.

$$\text{Relative bias}(\bar{y}_F) = \frac{\sum_{\gamma=1}^{\Gamma} \gamma T_\gamma \bar{Y}_\gamma}{M \sum_{\gamma=1}^{\Gamma} a_\gamma \bar{Y}_\gamma} - 1$$

and for the MSE of $\bar{y}_F$ as an estimate of $\bar{Y}$, we have

$$\text{MSE}(\bar{y}_F) = \frac{M-m}{M} \frac{1}{m} \left(\frac{1}{\sum_{\gamma=1}^{\Gamma} \gamma T_\gamma} \left\{ \sum_{\gamma=1}^{\Gamma} \gamma T_\gamma \left[S_{Y_\gamma}^2 + \left(\bar{Y}_\gamma - \bar{Y}_F \right)^2 \right] \right\} \right)$$

$$+ \left[\sum_{\gamma=1}^{\Gamma} \left(\frac{\gamma}{M} - \frac{1}{T} \right) T_\gamma \bar{Y}_\gamma \right]^2.$$

In the absence of knowledge about the multiplicities, there are no known methods for quantifying the measures of error. In the situation where it is known that the maximum value is $\Gamma = 2$, Kish (1965) presents a method for estimating N and Y. If a simple random sample of m units is selected from the M frame units and if there are t_2 duplicates in the sample, then

$$M - t_2 \frac{M^2}{m^2} = \hat{N}.$$

Similarly, $\hat{Y}$ may be estimated as

$$\hat{Y} = y \frac{M}{m} - y_2 \frac{M^2}{m^2},$$

where y is the sample total and y_2 is the sample total for the elements appearing twice.

4.5   IMPROPER USE OF CLUSTERED FRAMES

Procedures for conducting surveys when the frame units represent clusters of target elements are now well developed in the survey methodology literature. It appears that failure to account for clustering is an error made in the early days of surveys before textbooks on sampling were widely available or by those new to the practice of surveys. For example, a sample of housing units may be selected with equal probability. When the housing unit is visited, only one person resident in it may be interviewed. Thus the people entering the sample have varying chances of selection, depending on the number of people in the household. Those in large households have smaller chances relative to those in small households. In some cases the researcher lists within the household and randomly selects a single person but ignores this second stage of sampling. In other instances, a haphazard selection is made.

Hansen et al. (1953a) suggest that if a single element has been measured for a cluster and the size of the clusters can be determined, measurements for the elements should be weighted by the size of the cluster even if the element was not selected at random. The estimates are not strictly unbiased; however, the weighting procedure would more nearly approximate an unbiased estimate than would the unweighted procedure. A self-weighting sample can be produced by selecting the clusters with probability proportional to size and then selecting randomly one element from the cluster for inclusion in the survey.

4.6 INCORRECT AUXILIARY INFORMATION

Use of incorrect auxiliary information in a frame decreases the precision of the survey estimates. The auxiliary information is used for stratification, probability proportional to size sampling, ratio estimation, or regression estimation.

Suppose that one wished to estimate the expenditures for health insurance for a population of textile-manufacturing companies and report this information for companies in five different size classes. Using information on the size of each company, the entire population of companies is stratified into the five size classes and a sample selected from each class. If information on company size was incorrect, some companies would be placed in an incorrect stratum. During data collection, current information on company size would be collected. This information could be used to form a reporting domain for each of the five size classes; these domains would span several strata. These estimates would be unbiased; however, they are likely to have higher variances than the estimates that could have been produced if all companies were correctly stratified into the reporting domains when the frame was constructed. Szameitat and Schäffer (1963) derive the effect of errors in the stratification variables on the subgroup estimates in this instance. On the basis of the auxiliary variables one stratifies the entire population into strata. Assume simple random sampling without replacement from the L strata of the frame, and that one wishes to estimate Y_g, the total for the gth part of the target population. Let

$N_{gh, +}$ = number of target elements in g associated with the hth group of the frame

$$N_{gh, +} = N_{gh} + N_{gh}^*,$$

where

$$N_{gh} = \text{number of elements in } g \text{ correctly assigned to } h$$

$$N_{gh}^* = \text{number of elements in } g \text{ incorrectly assigned to } h$$

$$N_{h,\,+} = \text{total elements associated with stratum } h.$$

Assume a one-to-one correspondence between elements and units. Also, let

$$_gY_{hi}' = \begin{cases} Y_{hi} & \text{if the unit is an element of } g \\ 0 & \text{otherwise} \end{cases}$$

$$_gY_h' = \sum_{i=1}^{n_h} {}_gY_{hi}'.$$

Then

$$_g\hat{Y} = \sum_h \frac{N_{h,\,+}}{n_h} {}_gY_h'.$$

The variance is

$$\text{Var}\left({}_g\hat{Y}' \right) = \sum_h \left(1 - \frac{n_h}{N_{h,\,+}} \right) \frac{N_{h,\,+}^2}{n_h} \left(\frac{N_{gh}}{N_{h,\,+}} S_{{}_gY_h}^2 + \frac{N_{gh}}{N_{h,\,+}} \frac{1 - N_{gh}}{N_{h,\,+}} \overline{Y}_{gh}^2 \right).$$

Note that this is an extension of the formula for the inefficiency of a frame with overcoverage. What we have in this instance are several within-stratum frames that exhibit overcoverage relative to group g. If the auxiliary information had been of sufficient quality so that one could form strata that contained only units from g, a more precise estimate of Y_g could be obtained. If H is the collection of strata that cover g, then

$$V\left({}_g\hat{Y} \right) = \sum_{h \in H} \left(1 - \frac{n_h}{N_{h,\,+}} \right) \frac{N_{h,\,+}^2}{n_h} S_{{}_gY_h}^2,$$

with $N_{gh} = N_{h,\,+}$ for all $h \in H$.

4.7   INACCESSIBLE ELEMENTS

When target elements cannot be accessed from the frame information but are known to exist, we are in the situation discussed in Section 4.1 on undercoverage except that N_0 is known. All methods described in that section can be used for measuring the impact of nonaccessible elements. The extent of undercoverage is known of course.

The discussion above of the impact of frame errors focused on each error individually and presented no comprehensive model for assessing the impact of frame structure on the total error of survey estimates. Comprehensive models for frame errors are not available in the literature. Ideally, a complete model for the impact of frame errors would allow one to examine the cost effectiveness of various frame structures. In doing so, some of the things that the model would need to consider are:

- The bias and increased variability that would result from allowing the frame to exhibit certain imperfections
- The cost and time required to create a frame of a certain quality, such as the cost of eliminating multiplicity, obtaining auxiliary information, forming strata, and so on
- The number of times the frame could be used
- Types of sample designs and estimation procedures that can be supported by the frame; cost of carrying out such
- Types of information that would need to be collected during the survey to determine the frame–target population linkages; the cost and quality of such information

Obviously, since every survey conducted uses some type of frame for accessing the target population, the items listed above are considered to some extent during the process of choosing and constructing the frame. However, this procedure is often done informally, and a formal investigation into the quality of various frames could be helpful, especially when the frame being constructed will be used for a variety of surveys.

CHAPTER 5

Frames: Conducting Surveys with Imperfect Frames

It is possible, often necessary, and sometimes desirable to use imperfect frames when conducting surveys. Good results can still be achieved. Researchers are required to use imperfect frames because the expense or time required to build a better frame is too great. For example, in a survey of achievement levels the Research Triangle Institute (RTI) selected a sample of fourth-grade public school students. Because of time constraints on the survey, it was not possible to wait and select the sample from lists of schools and fourth-grade classes within schools for the year in which the survey was to be conducted (Refior, 1980). Instead, the previous year's list of classes was used to construct the sampling frame. Information on the number of fourth graders in each school in the previous year was available along with information on the number of teachers teaching fourth-grade classes during the first period of each day. This information was used to construct a list of fourth-grade classes within each school.

Schools were selected with probabilities proportional to the number of fourth-grade classes, and a preliminary sample of classes was selected for each school to yield an equally probable sample of classes. This sample was designated by indicating the sequence number for the sample members.

The schools falling in the sample were then visited. Rules for listing the classes were established. If there were no changes in the number of classes from the number in the preceding year, classrooms with preselected sequence numbers were in the sample. For example, suppose that a school had 10 fourth-grade classrooms in the prior year and the sequence numbers 2 and 9 were randomly selected for the survey. If the school had 10 classrooms in the current year, classes 2 and 9 on the current year's list were in the sample.

If the actual number of classes was larger, a new sample was selected. Thus, in our example, if there were 13 fourth-grade classrooms in the current year, a new sample would be selected. If the actual number of classes in the current year was smaller than the previous year and if none of the sequence numbers of the preliminary sample was larger than the current number of classes, no reselections were made. However, if one or more of the sequence numbers was greater than the actual number of classes, the sample was reselected. For our example, this means that if there were nine classes in the current year, classes 2 and 9 were in the sample; however, if there were eight classrooms in the current year, a new sample would then be selected.

This procedure introduced some complexities in the analysis in that classrooms did not have equal probability of selection because of the change in the number of classrooms from the frame numbers. It did have the advantage, however, of allowing the survey to proceed in a more timely manner, an advantage that outweighed the complexities in the analysis.

In Chapter 4 we described methods for measuring error associated with the use of frames with various types of imperfections and indicated briefly how one might adjust estimation methods to reduce the impact of each type of error. As stated in the first paragraph of this chapter, there are often advantages to using frames with certain types of imperfections even when better frames could be constructed. These advantages have simulated development of appropriate techniques for using imperfect frames. In this chapter we focus in detail on methods developed to accommodate two problems or imperfections: frames with missing elements and frames with multiplicity. The goal of the methods is to permit valid (unbiased) estimates of target population characteristics.

5.1 USE OF INCOMPLETE FRAMES

Recall that an incomplete frame is one for which some members of the target population are not linked to the frame by rules of association. We discuss three methods used for conducting surveys using incomplete frames: redefinition of the target population, linking procedures, and use of multiple frames.

5.1.1 Redefinition of the Target Population

One way of dealing with the problem of an incomplete frame is to simply "define it away." In this case the target population is simply considered to be that population which can be accessed by the frame. This is probably the cheapest method for treating the problem of an incomplete frame and may be widely used because of its simplicity and low cost. This may seem like an abdication of responsibility, but it is not always a bad technique. For example, in the mid-1970s RTI conducted a survey for the National Institute of Neurological and Communicative Disorders and Stroke (NINCDS) to measure the incidence, prevalence, and economic costs of head and spinal cord injury. Several alternatives for accessing the population of head and spinal cord injury victims were considered. An area household frame was considered to be very inefficient because most households would not contain a person who had recently experienced a head or spinal cord injury. An area household frame would also be incomplete because people who had died as a result of their injury or had been institutionalized in nursing homes, for example, might not be linked to a surviving household. It was finally decided to use hospital discharges as a frame for the survey. Thus the target population was redefined to consist of people who had had a head or spinal cord injury serious enough to require hospitalization. Restricting the survey to this subgroup of the target population did not seriously undermine the ability of the survey to meet NINCDS goals because hospitalized cases represent the major component of serious injury and account for most of the economic costs of the injuries.

5.1.2 Linking Procedure

One method that can be used for supplementing an incomplete frame during sampling or data collection is called variously the half-open interval or linking procedure. It is basically the same as the predecessor–successor method discussed previously for measuring the extent and impact of under-coverage. The difference is that instead of carrying out the method to evaluate a completed survey, the linking procedure is carried out at the time of the survey and measurements are made on the identified unassociated elements.

In this situation, target elements not included in the frame are linked by means of some counting rule or rule of association to target elements that are included in the frame. Rules of association must be unambiguously defined before the start of the survey. For example, one might select a sample of school children from a list prepared several months before the survey. Students that enrolled in school since the list was prepared would not have a chance to be in the survey. However, before the school was visited one could define a rule which said that each new student was to be linked to the last person listed in his or her classroom.

If such a rule links all elements missing from the frame to elements on the frame, the frame will actually cover all the elements. When an element included in the frame is selected into the sample, it and all its linked elements are included. The inclusion probabilities for the missing elements are equal to those of the listed element to which they are linked. Essentially, what has been done is to construct a frame in which some of the included elements actually represent a cluster, and the linkage rule identifies elements in the cluster. In our example, if the last student in a particular classroom was selected into the sample, that pupil and all the new students in that classroom would be included in the survey.

When the method is used in an area sample it is usually called the half-open interval. New houses may be missing from a list of housing units, or the listers may have neglected to include some units. These will be physically scattered among the listed houses. If a path from listed housing unit to listed housing unit is defined, any nonlisted housing units that fall between the listed units are linked to the housing unit that immediately precedes them in the path.

5.1.3 Multiple Frames

Another method for conducting surveys in the presence of frames with missing elements is to use multiple frames (Hartley, 1962). This technique is referred to by several different names in the literature. Sometimes the term *supplemental frame* or *complementary frame* is used. There are differences implied by the various names. Multiple frame procedures allow the frames to overlap; no attempt is made to eliminate elements from one frame that are also included in another frame; and samples are drawn from each frame.

In the other cases, elements included in the original frame are eliminated when the complementary frames or supplementary frames are constructed. Each frame can then be considered a stratum of a combined frame. In the combined frame elements are linked to one and only one stratum. No special estimation procedures are needed. Estimates of totals are simply the sum of the stratum estimates. Variances are the sums of the separate stratum variances. Combined ratio estimates can be used for estimating means or proportions. Since it is likely that different sampling plans would be used in the two frames, care must be taken when constructing the data file and data-processing procedures to assure that the variances are appropriately estimated. Kish (1965) discusses several procedures for constructing unique lists from one or more overlapping lists. Because this situation is equivalent to using stratified sampling, it will not be discussed further in this chapter.

The theory of domain estimation is used to make estimates in the presence of multiple frames. Hartley (1962) describes the method for two frames (see Fig. 5.1). To achieve complete coverage and make unbiased estimates, it is necessary that:

1. Every member of the target population be associated with at least one of the frames.
2. One can determine the frame membership for each member of the sample.

Note that if several surveys of the same target population are envisioned, one could use the information from one multiframe survey to estimate the bias that would result from using a single frame for future surveys rather than using multiple-frame procedures. The population is then divided into three domains:

Domain (a): Elements associated with frame A only

Domain (b): Elements associated with frame B only

Domain (ab): Elements associated with both frames A and B

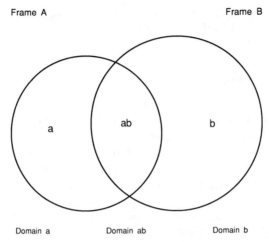

Figure 5.1 Two overlapping frames.

Hartley considered three cases: (1) domain membership can be determined in advance, (2) domain membership cannot be determined but domain sizes are known, and (3) no information on domain sizes or membership is available.

When the domain membership of each target element can be determined in advance of sampling, the three domains can be considered separate strata and separate samples drawn from each stratum. A unique list can be constructed for the overlap stratum, or one could leave the list as is and consider the frame for the overlap stratum to be a frame with multiplicity where each element has a known multiplicity of $\gamma = 2$. Procedures for making estimates for frames with multiplicity are discussed in Section 5.2.

If the domain sizes are known but the elements cannot be allocated to domains prior to sampling, separate samples are drawn from each frame. A poststratified estimate of the total and mean can then be used. In frame A two poststrata are formed, one corresponding to domain a and one to domain ab. Similarly, in frame B we will have a poststratum corresponding to domain b and one corresponding to domain ab. These must be weighted together such that the sum of the weights equals 1 to achieve unbiased estimates.

Let

$$N_A = \text{number of elements associated with frame } A$$
$$N_B = \text{number of elements associated with frame } B$$
$$N_a = \text{number of elements associated with domain } a$$
$$N_b = \text{number of elements associated with domain } b$$
$$N_{ab} = \text{number of elements associated with domain } ab.$$

Assuming no multiplicity within frames, complete coverage by the combination of the two frames, no associated nontarget elements, and a simple random sample of sizes n_A and n_B from the two frames, an unbiased poststratified estimate from frame A for the total of domain a is

$$\hat{Y}_A(a) = \frac{N_a}{n_a} y_A(a),$$

where n_a is the number of units in the sample from domain a and $y_A(a)$ is the sample total for domain a. Similarly, $\hat{Y}_A(ab) = (N_{ab}/n_{ab})y_A(ab)$, and so on. The poststratified estimate is

$$\hat{Y} = \hat{Y}_A(a) + W_A\hat{Y}_A(ab) + W_B\hat{Y}_B(ab) + \hat{Y}_B(b),$$

with $W_A + W_B = 1$.

Now the total for frame A, $Y(A) = Y(a) + Y(ab)$, which can be estimated by $\hat{Y}(A) = \hat{Y}_A(a) + \hat{Y}_A(ab)$. The variance for a poststratified estimate (Cochran, 1977; Sukhatme and Sukhatme, 1970) is approximated by

$$\mathrm{Var}\left[\hat{Y}(A)\right] = \frac{N_A^2}{n_A}\left(1 - \frac{n_A}{N_A}\right)\left(\frac{N_a}{N_A}S_a^2 + \frac{N_{ab}}{N_A}S_{ab}^2\right)$$
$$+ \frac{N_A}{n_A^2}\left[\left(1 - \frac{N_a}{N_A}\right)S_a^2 + \left(1 - \frac{N_{ab}}{N_A}\right)S_{ab}^2\right].$$

Assuming that the n_A and n_B are large enough that the terms of order $1/n_A^2$ and $1/n_B^2$ are negligible and ignoring the finite population corrections, Hartley gives the following for the variance of $\hat{Y}$:

$$\mathrm{Var}(\hat{Y}) \doteq \frac{N_A^2}{n_A}\left[S_a^2(1 - \alpha) + \alpha W_A^2 S_{ab}^2\right] + \frac{N_B^2}{n_B}\left[(1 - \beta)S_b^2 + \beta W_B^2 S_{ab}^2\right].$$

Here $\alpha = N_{ab}/N_A$ and $\beta = N_{ab}/N_B$, that is, the relative sizes of the overlap domains to the frame sizes.

Hartley gives expressions for the optimum value of W_A and the optimum sampling fractions. These are based on the use of a linear cost model that contrasts the costs of sampling and data collection with the two frames. Let

$$c_A = \text{cost of sampling and data collection from frame } A$$

$$c_B = \text{cost for frame } B$$

$$C = n_A c_A + n_B c_B.$$

The expression for the optimum value of W_A given by Hartley is

$$\frac{c_A W_A^2}{c_B W_B^2} = \frac{S_a^2(1 - \alpha) + S_{ab}^2 \alpha W_A^2}{S_b^2(1 - \beta) + S_{ab}^2 \beta W_B^2}.$$

Once W_A is known, the optimum sampling fractions for fixed cost C are given by

$$\frac{n_A}{N_A} = \frac{C[S_a^2(1 - \alpha) + S_{ab}^2 \alpha W_A^2]^{1/2}}{c_A}$$

$$\frac{n_B}{N_B} = \frac{C[S_b^2(1 - \beta) + S_{ab}^2 \beta W_B^2]^{1/2}}{c_B}.$$

This formulation has assumed that both frames are list frames and that a simple random sample is selected from each frame. In this case, Lund (1968) has shown that the optimum value of W_A is given by

$$W_A = \frac{\alpha n_A}{\alpha n_A + \beta n_B}.$$

The expression for n_A, n_B has a complex form and Lund suggests an iterative solution.

In some cases one may wish to use multiple frames even when frames that cover the entire target population are available because sampling and data collection are less costly under the incomplete frame. Examples often occur in agricultural surveys in which incomplete lists of farms are available. These lists are supplemented by an area sample that covers the entire target population.

In this case, frame B is a subset of frame A. The estimator for Y is

$$\hat{Y} = \hat{Y}_A + W_A \hat{Y}_A(ab) + W_B \hat{Y}_B(ab).$$

Using Hartley's formulation, the optimum value of W_A is now

$$W_A = \left[\frac{(1 - \alpha)S_a^2/S_{ab}^2}{(c_A/c_B) - \alpha}\right]^{1/2}.$$

Hartley gives some examples of the relative efficiency of a survey that used a multiple frame approach to a survey using only the expensive frame A.

The estimator above assumes two one-to-one list frames so that $S_B^2 = S_{ab}^2$. Many applications, however, use a combination of an area frame and a list frame. This is a common use in agricultural surveys in which lists of farms are available. These lists, however, are not complete because of changes in ownership, and so on. Thus the area frame is used to provide complete coverage of farm populations. The sample design may be a complex, multistage design so that $S_{ab}^2 \neq S_B^2$. In this case, Hartley's formula does not apply. Others have examined this case. Armstrong (1979) presents the optimum value of W_A but does not consider the optimum allocation of the samples to the two frames. Bosecker and Ford (1976) give results for use of a stratified list frame.

If the domain sizes are not known, one must use the theory of domain estimation to form estimates. The estimate is given by

$$\hat{Y} = \frac{N_A}{n_A}\hat{Y}_A(a) + W_A\frac{N_A}{n_A}\hat{Y}_A(ab) + \frac{N_B}{n_B}\hat{Y}_B(b) + W_B\frac{N_B}{n_B}\hat{Y}_B(ab),$$

where $W_A + W_B = 1$. The variance for the case Hartley discusses is

$$\text{Var}(\hat{Y}) = \frac{N_A^2}{n_A}\left[(1 - \alpha)S_a^2 + \alpha W_A^2 S_{ab}^2 + \alpha(1 - \alpha)\left(\bar{Y}_a - W_A\bar{Y}_{ab}\right)^2\right]$$

$$+ \frac{N_B^2}{n_B}\left[(1 - \beta)S_b^2 + \beta W_B^2 S_{ab}^2 + \beta(1 - \beta)\left(\bar{Y}_b - W_B\bar{Y}_{ab}\right)^2\right].$$

The lack of information on the size of the domains has resulted in a loss of efficiency as shown by the two terms involving $\alpha(1 - \alpha)$ and $\beta(1 - \beta)$.

Hartley (1974) discusses multiple-frame methodology in general without reference to simple random sampling designs in each frame. He develops a method for finding an optimum two-frame design that minimizes variance subject to fixed expected cost. The method is quite instructive.

Multiple frames were used for several years in the National Assessment of Educational Progress (Moore and Jones, 1973). One goal of this survey in year 2 and year 3 was to measure the education attainments of 17-year-olds who were not enrolled in school. Some 600 to 800 out-of-school 17-year-old respondents were needed. Another component of the survey involved a multistage area sample of household residents aged 26 to 35. The number of out-of-school 17-year-olds identified in this area household sample was only 100; thus some means was needed to increase the number. Increasing the area sample was considered too expensive because nearly 100 households had to be screened to find one out-of-school 17-year-old. Thus a multiple-frame approach was tried.

In year 2 four frames that overlapped to form seven domains were formed. The frames were (1) the area household frame; (2) a list of school dropouts for grades 9, 10, 11, and 12; (3) lists of members of neighborhood Youth Corp projects; and (4) enrollees in Job Corps centers. The results indicated that although the last two frames were easy to construct and cheap to use in that they permitted group administration of the questionnaires, the additional coverage obtained was negligible.

In year 3 the multiple-frame approach was also used, but the frames were restricted to the area frame and to lists of out-of-school 17-year-olds obtained from the schools. To increase coverage of the list frames somewhat, it was expanded to include lists of early high school graduates and dropouts from grades 7 and 8.

In recent years the interest of multiple-frame designs has increased because of more use of telephone interviewing. Telephone interviews are generally cheaper than personal interviews, and a more centralized location of the interviewing staff permits greater supervision. In addition, the advent of computer-assisted telephone interviewing has allowed some of the survey data-entry tasks to be merged with data collection tasks. Thus there are many perceived advantages to using telephone interviewing; however, telephone frames usually do not cover the entire population of interest. Consider, for example, the National Health Interview Survey, which is conducted annually by the National Center for Health Statistics. The target population for this survey is the U.S. household population. Conducting this survey entirely by telephone using a frame of telephone numbers for sampling would exclude people in nontelephone households from the survey. However, complete coverage could be achieved by a dual-frame design in which a telephone frame and telephone interviewing are used in conjunction with an area household frame and personal interviewing. Such designs have been called multiframe–multimode designs to distinguish them from multiple-frame designs in which a single mode of data collection is used regardless of the frame from which the units were selected.

The use of two different modes of data collection introduces additional complexity into the assessment of the survey error and into the decision as to whether a single- or multiple-frame design is preferred. Telephone surveys in general have lower response rates that do personal interview surveys. Some types of respondents (such as the elderly and those with language or hearing problems) are believed to have greater difficulty answering questions on the telephone than in person. This raises questions about the quality of the data collected under the two modes. Casady et al. (1981) investigated the use of dual-frame surveys for the National Health Interview Survey in which a complete area household frame is combined with a telephone frame. Sirken and Casady (1982) examined the effect of various nonresponse rates on the results. This analysis was extended by Groves and Lepkowski (1982), who explored the advantages of multiframe–multimode designs considering different administrative structures necessary to use two modes of data collection simultaneously.

5.2 USE OF FRAMES WITH MULTIPLICITY

Frames with multiplicity are those for which some target elements are linked by rules of association to more than one frame unit. For example, suppose that one used a list of accounts to sample the customers of a bank. Those customers who had more than one account with the bank would be associated with the list several times. The advantage of using a frame without multiplicity, either frames with one-to-one associations or cluster frames, is that the probability of any particular element's being in the sample is equal to the probability that its associated frame unit is in the sample. Frames with multiplicity do not have this advantage, and some adjustments must be made to produce unbiased estimates.

In some cases, frames with multiplicity are a nuisance. In other cases they are deliberately created. This occurs when the population under study is very sparsely distributed throughout some larger population, such as households that recently experienced certain rare events such as a birth or death or persons with a rare disease. In these cases researchers deliberately define rules of association that link elements of the rare population to more than one frame unit. In this context, the rule of association is often called a *counting rule* or a *multiplicity rule*. For example, suppose that one wished to estimate the number of people with a rare disease. An area household frame might be the only frame available for use. Because of the rarity of the condition, most households visited would not contain a member of the target population. In other words, the overcoverage of the frame is extreme. One, however, could define a rule of association which says that the members of the target population are linked to their own household and the households of their parents or siblings. Each sample household member would then be queried as to his disease status and whether or not any of his children or siblings who resided elsewhere had the disease.

Nathan (1976) reported on such a survey designed to estimate births and marriages. He examined the use of three counting rules: (1) a conventional rule that linked the marriages and births to a unique household, that of the married persons or the mother; (2) an intermediate counting rule that in addition linked marriages to the households of the married persons' parents and births to the household of the mother's mother; and (3) a wide counting rule in which births were linked to the households of the mother, the mother's mother, and the mother's sisters. Marriages were linked to the households of the marriage couple, their parents, and siblings. A special study of the measurement and sampling errors of the estimates under each rule was undertaken. Results showed that the wide counting rule was to be preferred. Measurement errors were somewhat higher, but they were offset by smaller sampling variances associated with the wide counting rule.

5.2.1 Eliminating Multiplicity

Several procedures for conducting surveys in the presence of frames that exhibit multiplicity are available. One can try to eliminate the multiplicity or one can adjust for it in the estimation procedure. Four methods can be used to eliminate multiplicity: (1) removal by sorting, (2) definition of a unique counting rule, (3) redefinition of the target population, and (4) introduction of an additional stage of sampling. These are illustrated as follows.

5.2.1.1 Sorting

Sorting the frame before sampling can remove the multiplicity (Hansen et al., 1953a, 1963; Kish, 1965). This is a particularly useful technique when the frame units are in one-to-one correspondence, and simple procedures such as alphabetizing or sorting by identification numbers can be used to discover the multiple associations. Once the frame is sorted, each element is retained only once.

5.2.1.2 Definition of a Unique Counting Rule

A number of researchers (Hansen et al., 1953; Kish, 1965; and Sirkin, 1970) have redefined the rules of association so that each element has a unique association with the frame. For example, suppose that one wished to select a sample of clients from a set of client records, and some clients had more than one record; a unique set of records corresponding to the population of clients could be defined by saying that a client is associated with his or her earliest record. This, of course, requires that it is possible to determine if a particular record is the earliest record.

5.2.1.3 Redefinition of the Target Population

Kish (1965) suggests that redefinition of the target populations is a possibility for some frames that exhibit multiplicity. One may have a file of client visits instead of clients. Sometimes the purposes of the survey can be accomplished by making estimates for the population of visits instead of the population of clients.

5.2.1.4 Introducing an Additional Stage of Sampling

Kish (1965) has also discussed another method of dealing with a frame that exhibits multiplicity. Suppose that when an element is drawn from the frame it is possible to determine the element's multiplicity from the frame's auxiliary information. Suppose that the frame exhibits a many-to-one association between target elements and frame units. If frame units are drawn with probability p, the expected frequency of selection of element j is $M(j)p$. If element j is selected $m(j)$ times, a final stage judgement to retain element j with probability $m(j)/M(j)$ yields a sample of target elements in which each unit has the common sampling weight $1/p$. The final sample size of target elements will be smaller than the sample of frame units because of elements eliminated at the final stage of sampling. To control the final sample size one would need to have an advance idea of the degree of multiplicity.

5.2.2 Adjusting for Multiplicity

If the multiplicities cannot be removed or if they have deliberately been introduced, special multiplicity or network estimators must be used. A variety of methods is available. Most of them require knowledge of the multiplicity of the individual sample elements; however, unbiased estimators have been developed for use when there is no knowledge of the element multiplicities.

For cases where the sample size is greater than the largest value of the multiplicity, Goodman (1949) constructed an unbiased estimate for the total size of the target population. Kiranandana (1976) extended it to estimates of the total for some characteristic of the population. Assume a many-to-one correspondence between the frame units and the elements and a simple random sample of m units from the M frame units. Also assume that $m > \Gamma$, the maximum multiplicity. Let

$$\theta(j) = \text{number of times that the } j\text{th element appears in the sample}$$

$$\Phi(\theta) = \text{number of elements that appear } \theta \text{ times in the sample;}$$

$$\theta = 1, 2, \ldots, \Gamma$$

$$m = \sum_{\theta=1}^{\Gamma} \theta \Phi(\theta) = \text{total number of units in the sample}$$

$$\delta[\theta, \theta(j)] = \begin{cases} 1 & \text{if } \theta(j) = \theta \\ 0 & \text{otherwise.} \end{cases}$$

Then

$$\Phi(\theta) = \sum_{j=1}^{N} \delta[\theta, \theta(j)].$$

Now the probability that an element j with multiplicity γ appears in the sample θ times is

$$\Pr[\theta(j) = \theta | M(j) = \gamma] = \frac{\binom{\gamma}{\theta}\binom{M-\gamma}{m-\theta}}{\binom{M}{m}}.$$

Using this, we find that

$$E[\Phi(\theta)] = \sum_{j=1}^{N} E\{\delta[\theta, \theta(j)]\}$$

$$= \sum_{j=1}^{N} \Pr[\theta(j) = \theta]$$

$$= \sum_{\gamma=\theta}^{\Gamma} \Pr[\theta(j) = \theta | M_j = \gamma] T_\gamma.$$

Now we know that

$$N = \sum_{\gamma=0}^{\Gamma} T_{\gamma}.$$

Therefore, an unbiased estimate of N will be one that has the expected value $\sum_{\gamma=0}^{\Gamma} T_{\gamma}$. Let

$$A(\theta) = 1 - \frac{(-1)^{\theta}(M - m + \theta - 1)!(m - \theta)!}{(M - m - 1)!m!}$$

Now

$$\sum_{\theta=0}^{\gamma} A(\theta)\Pr(\theta|\gamma)$$

$$= 1 - \sum_{\theta=0}^{\gamma} \frac{(-1)^{\theta}(M - m + \theta - 1)!(m - \theta)!}{(M - m - 1)!m!}$$

$$\times \frac{\gamma!}{\theta!(\gamma - \theta)!} \frac{(M - \gamma)!m!(M - m)!}{(M - \gamma + \theta - m)!(m - \theta)!M!}$$

$$= 1$$

because

$$\sum_{\theta=0}^{\gamma} \frac{(-1)^{\theta}(M - m + \theta - 1)!\gamma!}{(m - \gamma + \theta - m)!\theta!(\gamma - \theta)!} = 0.$$

Using this, we have for the estimate of N,

$$\hat{N} = \sum_{\theta=1}^{m} A(\theta)\Phi(\theta)$$

and

$$E(\hat{N}) = \sum_{\theta=1}^{m} A(\theta) \sum_{\gamma=\theta}^{\Gamma} \Pr\big[\theta(j) = \theta|m_j = \gamma\big]T_{\gamma}$$

$$= \sum_{\gamma=1}^{m} T_{\gamma} \sum_{\theta=0}^{\gamma} A(\theta)\Pr(\theta|\gamma)$$

$$= \sum_{\gamma=1}^{\Gamma} T_{\gamma}$$

because $m > \Gamma$. The estimate for the total given by Kiranandana (1976) has a similar form:

$$\hat{Y} = \sum_{\theta=1}^{m} A(\theta)Y(\theta),$$

where $Y(\theta)$ is the sample total for the elements that appear in the sample θ times.

Although these estimates are unbiased, they can be very unreasonable—sometimes giving negative estimates for clearly positive quantities. Thus we suggest using alternative estimates by Goodman and Kiranandana: namely,

$$\hat{N}^1 = \begin{cases} M & \text{if } \hat{N} > M \\ \hat{N} & \text{if } \sum_{\theta=1}^{m} \Phi(\theta) \leq \hat{N} \leq M \\ \sum_{\theta=1}^{m} \Phi(\theta) & \text{if } \hat{N} < \sum \Phi(\theta) \end{cases}$$

$$\hat{Y}^1 = \begin{cases} \dfrac{M}{m} \sum_{i=1}^{m} y_i = \hat{Y}_F & \text{if } \hat{Y} > \hat{Y}_F \\ \hat{Y} & \text{if } \sum_{\theta=1}^{m} Y(\theta) < \hat{Y} < \hat{Y}_F \\ \sum_{\theta=1}^{m} Y(\theta) & \text{if } \hat{Y} < \sum_{\theta=1}^{m} Y(\theta). \end{cases}$$

These two estimators place upper and lower bounds on the estimates, the upper bound being the total for the frame ignoring multiplicity and the lower bound being the sample total. The variances of $\hat{N}$ and $\hat{Y}$ are

$$\text{Var}(\hat{N}) = \sum_{\theta=1}^{m} [A(\theta)]^2 \text{Var}[\Phi(\theta)] + \sum_{\theta \neq \theta'}^{m} A(\theta)A(\theta')\text{Cov}[\Phi(\theta), \Phi(\theta')]$$

$$\text{Var}(\hat{Y}) = \sum_{\theta=1}^{m} [A(\theta)]^2 \text{Var}[Y(\theta)]$$

$$+ \sum_{\theta \neq \theta'}^{m} A(\theta)A(\theta')\text{Cov}[Y(\theta)Y(\theta')].$$

See Gurney and Gonzalez (1972) for additional discussion of this problem.

Most surveys that use frames with multiplicity try to gather information about the multiple associations to use in the estimation procedures. Several multiplicity estimators are available that use knowledge of the multiplicities.

5.2.2.1 Weighting by the Inverse of the Multiplicities

Weighting by the inverse of the multiplicities will produce unbiased estimates. During data collection the values of $M(j)$ are determined for each element in the sample. In the case of simple random sampling of size m from the M frame units and many-to-one association, the estimate of the total Y is

$$\hat{Y} = \frac{M}{m} \sum_{i=1}^{m} \frac{Y_i}{M_i},$$

where $M_i = M_j$ for all frame units i linked to population element j.

No special techniques are necessary to determine the variance of this estimator. All the usual variance formulas and their estimators apply by substituting $Z_i = Y_i/M_i$ as the value associated with the ith frame unit.

A ratio estimate of the population mean must be used since T, the total population elements, is not known. The estimate of T is

$$\hat{T} = \frac{M}{m} \sum_{i=1}^{m} \frac{1}{M_i}.$$

In each of these cases, all multiple occurrences of elements in the sample are retained; for example, if selected frame units i and i' both are linked to a single population element j, then $1/M_i$ and $1/M_i$ are both equal to $1/M_j$ and each is retained in the summations above.

Kiranandana (1976) shows that the variance of $\hat{Y}$ is

$$V(\hat{Y}) = \frac{M-m}{M} \frac{MT}{m} \left\{ \sum_{\gamma=1}^{\Gamma} \frac{a_\gamma}{\gamma} \left[S_{Y_\gamma}^2 + \left(\bar{Y}_\gamma - \frac{\gamma \bar{Y}}{\sum_{\gamma=1}^{\Gamma} \gamma a_\gamma} \right)^2 \right] \right\}.$$

The estimator above is appropriate for an unstratified simple random sample. Birnbaum and Sirken (1965) presented three multiplicity estimators for the total number of persons with a characteristic for stratified simple random sampling of clusters. The assumptions are that there are H strata, M_h frame units in the hth stratum, and each frame unit has N_{hi} target elements associated with it; that is, it is a cluster of elements. The goal is to estimate the size $T(r)$ of some subpopulation of rare elements such as those persons with a rare disease.

This procedure represents a change in terminology in that we are no longer referring to the population whose size we wish to estimate as the *target population*. In the strict sense of our definition of the target population as the group on which we wish to make the survey measurements, elements that do not possess the rare characteristic are a member of the target population because an assessment as to whether or not they are a member of the rare population is carried out; that is, a survey measurement is made. However, most of the interest is with the subpopulation $T(r)$.

Estimator 1. Every subpopulation element discovered in the sample is counted and weighted by its number of linkages to the frame units. If there are H strata and M_h frame units in the hth stratum and simple random samples of size m_h are drawn, then

$$\hat{T}(r) = \sum_{h=1}^{H} \frac{M_h}{m_h} \sum_{i=1}^{m_h} n_{hi},$$

where

$$n_{hi} = \sum_{j=1}^{N_{hi}} \frac{X_{jhi}}{M_j},$$

where

$$X_{jhi} = \begin{cases} 1 & \text{if the } j\text{th element is a member of the} \\ & \text{subpopulation; } j = 1, 2, \ldots, N_{hi} \\ 0 & \text{otherwise} \end{cases}$$

and M_j is the multiplicity of the jth element. This is the same as the weighting procedure described above.

Estimator 2. Every subpopulation element discovered in the sample is counted only once and is weighted by its probability of being in the sample. Calculation of the probability of being in the sample requires that one know M_{hj} (the multiplicity of each sample target element within each stratum). Also one must remove multiple occurrences in the sample. The latter requires matching the sample elements. Now

$$\text{Pr} \, (\, j \text{ is not an element of the sample}) = \prod_{h=1}^{H} \frac{\dbinom{M_h - M_{hj}}{m_h}}{\dbinom{M_h}{m_h}} = q_j$$

$$\text{Pr} \, (\, j \text{ is in the sample}) = 1 - q_j = p_j.$$

Thus

$$\hat{T}(r) = \sum_{\substack{j \in s \\ j \in r}} \frac{1}{p_j}.$$

The variance of $\hat{T}(r)$ is given by

$$V(\hat{T}) = \sum_{j=1}^{T(r)} \sum_{j'=1}^{T(r)} \frac{p_{jj'} - p_j p_{j'}}{p_j p_{j'}}.$$

Estimator 3. Every subpopulation element occurring in the sample is counted only once and only in the highest priority in which it occurs. For example, one may wish to link elements to preferred frame units.

$$\hat{T}(r) = \sum_{h=1}^{H} \frac{M_h}{m_h} \sum_{i=1}^{m_h} \delta_{hi},$$

where

$$\delta_{hi} = \sum_{j=1}^{N(hi)} \frac{Z_{jhi}}{M_j \pi_{jhi}},$$

where

$$Z_{jhi} = \begin{cases} 1 & \text{if } j \text{ is an element of the subpopulation and } hi \text{ is the} \\ & \text{highest-priority unit in the sample that contains } j \\ 0 & \text{otherwise} \end{cases}$$

$$\pi_{jhi} = \begin{cases} \text{probability unit } hi \text{ is the highest-priority source } j \text{ given} \\ \text{that } hi \text{ is in the sample}. \end{cases}$$

Birnbaum and Sirken (1965) do not give the variance of this estimator.

Much of the research associated with the various multiplicity estimators and alternate counting rules that have been proposed is directed at determining what is the most cost-effective approach to use in a particular survey situation. There are certain situations in which a multiplicity estimator will always have smaller variance than a conventional estimator using a counting rule that links each element to only a single frame unit. However, survey costs are associated with determining the multiplicities, and measurement errors are associated with this determination. These costs and errors must be balanced against the reductions in sampling errors. Often, one of the most difficult tasks is to design a counting rule and/or an estimation strategy that permits a valid determination of the multiplicity. Many surveys report on the number of events occurring during a time period such as the number of visits to physicians, the number of discharges from hospitals, or the number of arrests rather than reporting on the number of people who visited physicians, who were hospitalized, or who were arrested. The reason for this is that no economic way has been found to determine the multiplicities in these surveys.

Levy (1977) points out that for a stratified sample in which elements can be linked to more than one stratum and more than one unit in a stratum, any set of weights w_{jh} such that $\sum_{h=1}^{H} w_{jkh} \gamma_{jh} = 1$, where γ_{jh} is the multiplicity of the jth element in stratum h, can be used to construct unbiased estimates for the rare population. Lessler (1981) suggests that this idea is the basis of a series of nested multiplicity estimators for multistage surveys. In these methods the multiplicities are determined for each stage or level of a nested sampling structure.

To explain the methods, we depart from the notation we have been using and use the following. Let N be the total number of elements in a population of interest indexed by $\alpha = 1, 2, \ldots, N$. Suppose that we wish to estimate a total for this population:

$$Y = \sum_{\alpha=1}^{N} Y_a,$$

where Y_α is the value for element α. Now suppose that the elements of the target population are linked to a sampling frame with some elements having multiple linkages. Also assume that a stratified three-stage structure has been imposed on the sampling frame, and that

H = Total number of strata indexed by $h = 1, 2, \ldots, H$

F_h = total number of primary sampling units (PSUs) in stratum h, each indexed by $i = 1, 2, \ldots, F_h$

S_{hi} = total number of secondary sampling units (SSUs) in stratum, h, FSU i, each indexed by $j = 1, 2, \ldots, S_{hi}$

T_{hij} = total number of tertiary sampling units (TSUs) in SSU hij, each indexed by $k = 1, 2, \ldots, T_{hij}$.

The method is applicable to any sample design that allows unbiased estimates of totals. For example, suppose that we have measured a characteristic Q_{hijk} associated with unit ijk in stratum h. We have some sample design, be it simple random sampling, unequal probability sampling with or without replacement sampling, or whatever, that allows unbiased estimates of totals. That is, we can construct some function of $\hat{q}$ of the Q_{hijk} that are elements of the sample, so that

$$E(\hat{q}) = \sum_{h=1}^{H} \sum_{i=1}^{F_h} \sum_{j=1}^{S_{hi}} \sum_{k=1}^{T_{hij}} Q_{hijk}. \tag{5.1}$$

Now each target element is linked to one or more of the ultimate sampling units $hijk$ and there may be a cluster of elements linked to the sampling unit. Let $X_{\alpha, hijk}$ be the value for the αth element linked to sampling unit $hijk$. More than one element can be linked to an ultimate sampling unit, and X_{hijk} is the sum over α of the $X_{\alpha, hijk}$. Thus, for $\hat{X}$ to be unbiased for Y, it is only necessary that the sum of $X_{\alpha, hijk}$ over the entire sampling frame be equal to Y_α:

$$Y_\alpha = \sum_{h=1}^{H} \sum_{i=1}^{F_h} \sum_{j=1}^{S_{hi}} \sum_{k=1}^{T_{hij}} X_{\alpha, hijk}.$$

In the survey design, the sample is drawn using the described stratified multistage structure. Each target element associated with the ultimate sampling units is identified and responses to the survey questions are obtained. Three methods that can be used with this structure are the completely nested inverse multiplicity estimator, a superstage estimator, and a counting rule multiplicity.

5.2.2.2 Completely Nested Inverse Multiplicity Estimator

The completely nested inverse multiplicity estimator technique involves weighting the information obtained for each member of the target population by the inverse of the number of associations it has with the various stages of the nested sampling frame structure. The following illustrates that this indeed provides unbiased estimates.

We make use of a series of indicator functions. Let

$$\theta_{\alpha, h} = \begin{cases} 1 & \text{if element } \alpha \text{ is linked to stratum } h \\ 0 & \text{otherwise.} \end{cases}$$

Then

$$\gamma_{1, \alpha} = \sum_{h=1}^{H} \theta_{\alpha, h}$$

$$= \text{total number of strata to which element } \alpha \text{ is linked.}$$

Similarly, let

$$\theta_{\alpha, hi} = \begin{cases} 1 & \text{if element } \alpha \text{ is linked to FSU } i \text{ in stratum } h \\ 0 & \text{otherwise.} \end{cases}$$

Define $\theta_{\alpha, hij}$ and $\theta_{\alpha, hijk}$ similarly. Then the multiplicity at each of these stages of sampling is given by

$$\gamma_{2, h, \alpha} = \sum_{i=1}^{F_h} \theta_{\alpha, hi}$$

and similarly for $\gamma_{3, hi, \alpha}$ and $\gamma_{4, hi, \alpha}$. Thus our sampling frame is a nested structure with four levels of nesting—a stratum level and three levels or stages of sampling within each stratum. (It is not necessary to have the same number of sampling stages within each stratum.) We have defined a multi-plicity measure for each level of the sampling frame as the number of associations or linkages that element α has with the units at a particular level.

Using this, we define

$$X_{\alpha, hijk} = \frac{\theta_{\alpha, h}}{\gamma_{1,\alpha}} \frac{\theta_{\alpha, hi}}{\gamma_{2, h, \alpha}} \frac{\theta_{\alpha, hij}}{\gamma_{3, hi, \alpha}} \frac{\theta_{\alpha, hijk}}{\gamma_{4, hij, \alpha}} Y_\alpha.$$

Thus the expected value of $\hat{X}$ the sample estimate of Y is

$$
\begin{aligned}
E(\hat{X}) &= \sum_{h=1}^{H} \sum_{i=1}^{F_h} \sum_{j=1}^{S_{hi}} \sum_{k=1}^{T_{hij}} \sum_{\alpha=1}^{N} X_{\alpha, hijk} \\
&= \sum_{\alpha=1}^{N} Y_\alpha \sum_{h=1}^{H} \frac{\theta_{\alpha, h}}{\gamma_{1,\alpha}} \sum_{i=1}^{F_h} \frac{\theta_{\alpha, hi}}{\gamma_{2, h\alpha}} \sum_{j=1}^{S_{hi}} \frac{\theta_{\alpha, hi}}{\gamma_{3, hi, \alpha}} \sum_{k=1}^{T_{hij}} \frac{\theta_{\alpha, hij}}{\gamma_{4, hij, \alpha}} \\
&= \sum_{\alpha=1}^{N} Y_\alpha.
\end{aligned}
$$

5.2.2.3 Superstage Inverse Multiplicity Estimator

Any two or more adjacent stages in a sampling structure may be thought of as a single "superstage" for measuring an element's multiplicity with those levels of the sampling frame. For example, the first and second stages of sampling in our example could be considered a superstage with

SP_h = total number of superstage units in stratum h, each indexed by $l = 1, 2, \ldots, SP_h$

$= \sum_{i=1}^{F_h} S_{hi}.$

Then, as before, we let

$$\theta_{SP, \alpha, hl} = \begin{cases} 1 & \text{if element } \alpha \text{ is linked to SPSU } l \text{ in stratum } h \\ 0 & \text{otherwise.} \end{cases}$$

Then multiplicity of element α with the units in the superstage is

$$\gamma_{SP, \alpha, h} = \sum_{l=1}^{SP_h} \theta_{SP, \alpha, hl}.$$

Using this, we define

$$X^1_{\alpha, hijk} = \frac{\theta_{\alpha, h}}{\gamma_{1,\alpha}} \frac{\theta_{\alpha, hi}\theta_{\alpha, hij}}{\gamma_{SP, \alpha, h}} \frac{\theta_{\alpha, hijk}}{\gamma_{4, hij, \alpha}} Y_\alpha.$$

The $E(\hat{X}^1)$ for the sample design used is given by

$$E(\hat{X}^1) = \sum_{h=1}^{H} \sum_{i=1}^{F_h} \sum_{j=1}^{S_{hi}} \sum_{k=1}^{T_{hij}} \sum_{\alpha=1}^{N} X^1_{\alpha, hijk}.$$

This equals Y because summing over the FSUs and SSUs is equivalent to summing over the superstage:

$$\sum_{i=1}^{F_h} \sum_{j=1}^{S_{hi}} \frac{\theta_{\alpha, hi}\theta_{\alpha, hij}}{\gamma_{SP, \alpha, h}} = 1.$$

The traditional estimator that measures an element's total multiplicity is equivalent to considering all parts of the sampling structure a single superstage.

5.2.2.4 Use of Counting Rule Weights
The counting rule is the mechanism linking the target element to the frame unit. It is possible to break some counting rules into parts. For example, Nathan's wide counting rule described above is really composed of three rules: (1) link to household of residence, (2) link to household of parents, and (3) link to household of siblings. Thus we could define a counting rule multiplicity.

Let

$$w_{c, \alpha} = \begin{cases} 1 & \text{if element } \alpha \text{ is linked to the frame by} \\ & \text{means of counting rule } c \\ 0 & \text{otherwise.} \end{cases}$$

Then

$$C_\alpha = \sum_{c=1}^{C} W_{c, \alpha}$$

$\qquad$ = total number of counting rules by which element α is linked to the
$\qquad$ frame.

Now C_α can be defined within any level of the sampling structure or it can be defined overall. Mathematically, the unbiased estimator of Y has the same form as before except that now each of the indicator functions and multiplicity measures refers to a specific subset of the counting rules. The expected value of $\hat{Y}$ has the form

$$E(\hat{Y}) = \sum_{\text{sample}} \sum_{c=1}^{C} \sum_{\alpha=1}^{N} \frac{w_{c,\alpha}}{C_\alpha} \frac{\theta_{h,\alpha,c}}{\gamma_{4,\alpha,c}} \frac{\theta_{hi,\alpha,c}}{\gamma_{2,\alpha,c}} \frac{\theta_{hij,\alpha,C}}{\gamma_{3,\alpha,c}} \frac{\theta_{hijk,\alpha c} Y_\alpha}{\gamma_{4,\alpha,c}}$$

$$= \sum_{\alpha=1}^{N} \sum_{\text{sample}} X_{hijk,\alpha,c} = \sum_{\alpha=1}^{N} Y_\alpha.$$

The estimators are useful because of the flexibility they provide for obtaining the multiplicity information. They were originally designed for a pilot survey that RTI conducted for the National Institute of Neurological Disease and Stroke directed at developing a method for measuring the prevalence of epilepsy in the population (Andersen et al., 1985). Because most epileptics with a serious seizure disorder are taking one or more of a limited set of drugs to control their seizures, a pilot study was conducted that tested a procedure in which epileptics were located through the drugs they were taking. The original design of the pilot called for selecting a sample of pharmacies and a sample of refill logs or prescriptions within pharmacies. Refill logs are lists of prescriptions already on file that are refilled by the pharmacist during a particular day. Using procedures specially designed to maintain confidentiality, people who took one of a group of drugs of interest were traced to the prescribing physician and it was determined whether the patient was epileptic. The information on the proportion of prescriptions that were for control of epileptic seizures could be used to estimate the prevalence of epilepsy. Some methods for adjusting for multiplicity were needed, however, because a particular epileptic will have multiple refills, may use multiple drugs, have multiple prescriptions for these drugs within a pharmacy, and may use several pharmacies, such as one near his or her home and one near the office. The patient was the only source of much of this information.

The pilot study plan called for contacting a sample of patients through their physicians and determining their multiplicities. In the case where refill logs were to be sampled and patients were asked to report their total multiplicity, they would have to recall the total number of refills they had for each of their antiseizure drugs. Since patients were going to be contacted some six to eight months after the reference year, it was felt that they would not be able to recall this information. Thus the search was made for methods for reducing the amount of information needed.

The nested estimators do this. Instead of asking the patient for all his refills, the number of refills within the sample pharmacy of the drug that brought the person into the sample could be counted because they are noted on the back of the original prescription retained by the pharmacist. The number of drugs the patient took could be determined from the physician. This could be used to determine the counting rule multiplicity. Finally, all stages of sampling before pharmacies—namely regions and areas of the country could be combined into a superstage and the patient could simply be asked, for example, to report the number of pharmacies that he had received drug *B* from during the study year.

Consider another example. Suppose that we wished to estimate the number of different customers for a particular firm and describe their characteristics. Also suppose that the firm produced several products that could be purchased at one of several sites. In addition, suppose that records of purchases were kept by each site that sold one of the company's products and that some people are likely to make more than one purchase. The records of purchase are not sorted by person; however, one could look at the billing record of any particular purchaser and determine the number of times he purchased the product during the year. Suppose that customers of the firm were accessed by selecting a sample of sites and a sample of purchases within site. A particular customer's total multiplicity is equal to the number of purchases that individual made of the company's products during the study year. For example, the customer may have used two of the products and made five separate purchases of product *A* and two of product *B*. However, the purchaser is not likely to recall individual purchases when asked.

Now, by using a superstage nested multiplicity estimator with a counting rule weight, one could merely ask the customer for the number of products used (the counting rule multiplicity) and for the number of sites at which he had purchased the product that brought him into the sample (first-stage multiplicity). The within-site multiplicity would be measured from the billing records.

Multiple-frame surveys can be reformulated in terms of a multiplicity survey. The multiple sampling frame that is constructed can be viewed as a single frame with multiplicity. Assuming no duplication within frames, the maximum value of any element's multiplicity is known. Namely, it is the number of sampling frames. One could use the nested estimators suggested by Levy (1977) and Lessler (1981) to make estimates. The various frames are now strata, and the number of frames that an element is associated with is its stratum multiplicity. Casady and Sirken (1980) consider such a case and allow for duplication within frame.

In this three-chapter section on frame errors we have reviewed the methods used for estimating the extent and impact of frame errors and for adjusting for imperfect frames. A 1983 bibliography by Wright and Tsao provides additional references that may be of interest.

CHAPTER 6

Nonresponse: Background and Terminology

One of the most obvious problems in surveys is the inability to obtain useful data on all questionnaire items from all members of the sample. We call this problem *nonresponse*. It indicates a clearly visible "flaw" in the survey operation and has important implications during design and analysis. Because these implications have become widely recognized, hundreds of papers dealing with methods to prevent nonresponse, measure it, or compensate for its effects have appeared in the literature. The breadth of interest in nonresponse is demonstrated by work on the subject appearing in the literature for virtually every profession in which survey research is important. This published research has varied considerably from simple descriptive presentations of effect to complex development of basic theory.

Most of the published work deals separately with unit and item nonresponse. In *unit nonresponse* a unit of importance to the survey fails to participate. A unit as we consider it here may apply to a sampling, data collection, or analysis unit. The second level of nonresponse is *item nonresponse*, in which the unit participates in the survey but, for any of several reasons, data on particular items of the questionnaire are unavailable for analysis.

Chapters 6 through 8 are devoted to nonresponse as a source of survey error. The focus of each chapter is distinct. The major goal of this chapter is to illustrate the great diversity in terminology associated with nonresponse and, in the process, to point out when and why nonresponse occurs. We discuss different ways in which the concept of nonresponse has been viewed in the literature in Chapter 7, with a discussion of how each viewpoint affects the contribution of nonresponse to survey error. Finally, in Chapter 8 we review many of the existing methods for dealing with the problem of survey nonresponse.

We begin this chapter with a discussion of where unit nonresponse occurs during the survey operation. We next consider the terminology used to describe three aspects of nonresponse: the general notion of nonresponse,

103

relative measures of it, and the bias associated with it. In so doing we intend to demonstrate the extent to which these concepts are variously defined by those who have contributed to the literature. We end with a categorization of the many reasons for unit nonresponse. Diversity in terminology is once again indicated.

6.1 SURVEY ACTIVITIES RELATED TO NONRESPONSE

In this section we define and briefly discuss a sequence of three survey activities during which either unit or item nonresponse may occur. We use the terms *location*, *solicitation*, and *data collection* to denote these activities in later sections, where the terminology and affects of nonresponse are considered.

Location refers to that stage during the survey in which the data collector, having been given relevant information, attempts to contact the set of units selected for the sample. Failure to make contact contributes to unit nonresponse. Success in this process depends on several factors. One is the accuracy and completeness of information on the physical location of each member of the sample. For example, the percentage of successfully located dwellings in household surveys will be high when accurate current addresses with accompanying maps are available for sample members. The ability to locate subjects selected from old lists will also depend on the availability of information to help make the process of tracing more efficient (e.g., driver's license number, name and address of a close relative, etc.).

Accessibility of sample members (whether or not planned) is a second factor influencing success in location efforts. Some people tend to be away from home often and therefore are hard to locate in some interview surveys (e.g., college students and traveling salespeople), while others are more likely to be at home (e.g., housewives and senior citizens). People living in areas of controlled access (e.g., apartment buildings or housing subdivisions with security checkpoints) present problems of accessibility in surveys requiring an in-person visit by the interviewer. Those living in a secluded area are less accessible in personal interview surveys because an interviewer may be unable to find them, while some persons, on learning about the survey (e.g., through an earlier letter or telephone call from the investigator) may purposely make themselves less accessible in scheduling an interview.

Also related to accessibility is the timing of attempts to solicit response. For example, call attempts during the weekday in most personal and telephone interview surveys are less likely to find someone at home than attempts made during weekday evenings and weekends. A fourth factor is the number of call attempts since, all else equal, chances of locating a member of the sample are directly related to the number of location attempts made. Finally, there is data collector competence. In telephone surveys, for example, the better interviewers will deal with the various barriers one faces by telephone (e.g., recorded messages, unusual busy signals, etc.), thereby aiding the process of locating members of the sample. More competent interviewers will also deal more successfully with the problem of tracing persons selected from outdated list samples (e.g., by knowing where to look and what to look for in tracking down people who have moved).

The ability to locate sample members is not only affected by the location methods used, but also by the units selected in the sample. A sample from a list of named individuals will, for instance, lead to problems of untraceable movers which can be avoided by sampling dwellings rather than individuals. Similarly with telephone surveys, the problem of tracing can be avoided by selecting a random digit sample of telephone numbers rather than choosing the sample from the listings in the telephone directory where a name is linked to the number. The tracing problem, of course, can be solved by taking whomever happens to be assigned the selected directory listing, but this now creates a frame problem if the individual with the original directory listing now has a new number not appearing elsewhere in the directory.

Unit nonresponse may also occur during *solicitation* in which located sample members are asked to participate in the survey. A member of the sample becomes a participant or respondent in the survey when the data collector succeeds in solicitation attempts. Warwick and Lininger (1975, Chapter 7) describe the solicitation process as one pitting several positive and negative forces against each other. Thought to contribute positively to participation by a member of the sample are (1) the desire for expressing an opinion, (2) the perceived relevance of the study, (3) the desire to share information with a sympathetic listener, (4) the perception that participation would be an intellectual challenge or a source of insight, and (5) the wish to be helpful to those doing the survey. In interview surveys some additional positive factors to successful solicitation include the inteviewer's (1) level of satisfaction in working with people, (2) intellectual curiosity, and (3) professional identification with the survey's objectives. Among those forces known to reduce the likelihood of successful solicitation are the subject's (1) fear of confrontation, (2) perceived feeling of invasion of privacy or disruption of leisure time, (3) hostility toward the sponsor of the survey or the interviewer, (4) sensitivity about the survey's subject and concern about confidentiality of responses, (5) fear of being tested, (6) apathy, and (7) fear of admitting a stranger into the home (in personal interview surveys only). Additional negative forces of relevance to the interviewer's performance include (1) fear, (2) fatigue, (3) difficult survey working conditions and scheduling, (4) poor survey organization, (5) inadequate training and supervision, (6) frustration in locating assigned members of the sample, and (7) a basic lack of personal motivation.

Data collection is defined as the stage in survey in which each participant completes a survey questionnaire and data from it are edited in preparation for analysis. Although one hopes at the end of data collection to have responses to all questionnaire items for each survey participant, this almost never happens. The interviewer in telephone and personal interview surveys may willfully or accidentally fail to ask or record responses to some questions. The respondent may fail to answer certain questions either when interviewed or when completing a self-administered questionnaire, especially questions considered to be sensitive or personal (e.g., on family income, attitudes on abortion, etc.). Finally, unresolvable errors identified while preparing the data for analysis may require that some items on the data file be left blank.

6.2 TERMINOLOGY

Looking through the survey literature, one is particularly struck by the diversity of terminology and associated definitions used in reference to many facets of the nonresponse problem. Different words and phrases describe such aspects as the general notion of nonresponse, relative measures of the extent to which sample members are located and contribute to the survey, and measures of the effect of nonresponse on survey estimates. A review of the diverse terminology used in connection with nonresponse is relevant here because the diversity, while indicating the broad interest in the problem, points up a need for increased efforts to standardize terminology and definitions.

Definitions of many of the concepts related to nonresponse appear in numerous places, such as standard sampling textbooks [e.g., Cochran (1977) and Kish (1965)] and where statistical terminology is discussed [see, e.g., CASRO Task Force on Completion Rates (1982), Kendall and Buckland (1960), Madow et al. (1983a), and the U.S. Bureau of the Census (1975)]. Some definitions result from ad hoc efforts by individuals; others are the outcome of efforts by committees aimed at standardizing terminology and definitions.

The need to standardize terminology related to nonresponse has been addressed with some progress indicated to date. Most attempts at standardization have involved formation of committees to review definitions and recommend new, more broadly encompassing ones. For example, in the mid-1970s, the U.S. Office of Management and Budget (OMB) created a standing task force whose goal was to propose useful definitions for several concepts linked to the study of nonsampling error in surveys. One product of this task force's activities was a *Glossary of Nonsampling Error Terms* (Deighton et al., 1978). Similar committees have been formed by organizations such as the National Research Council, the American Statistical Association, and the Council of American Survey Research Organizations.

6.2.1 The Notion of Nonresponse

Nonresponse as a concept has been defined in a number of ways, some of which we will mention here (see the Compendium of Nonsampling Error Terminology for verbatim illustrations). Most definitions distinguish unit from item nonresponse. In a dictionary of statistical terms, Kendall and Buckland (1960) limit their definition of unit nonresponse to circumstances where individuals are the units being studied. Others, notably the U.S. Bureau of the Census (1975) and Kish (1965), are more general in their reference to units, thus implying that nonresponse may apply to corporate units as well (e.g., businesses, hospitals, and professional organizations). Definitions given for item nonresponse seem to be a bit more consistent with nonresponse, here generally attributed to failure to obtain a response to a particular item when the questionnaire is being completed (U.S. Bureau of the Census, 1976:914; Kalton, 1983:4). Others also consider a collected data item that is later found to be unusable for analysis to be a form of item nonresponse (National Research Council, 1983:3).

Almost invariably some of the units selected for survey samples turn out to be nonmembers of the survey population. A vacant dwelling selected in a sample of dwellings for a personal visit household survey is one common example. The issue, then, is whether the out-of-scope sample member is to be treated as a unit nonrespondent. The Panel on Incomplete Data (National Research Council, 1983:3), which defines nonresponse for sample units which are "eligible for the survey," would clearly not treat ineligibles as nonrespondents. One general definition of nonresponse given by the U.S. Bureau of the Census (1975:50), on the other hand, seems to allow some ineligibles by including vacant houses among the reasons for nonresponse.

Several different words and phrases have been used when referring to the notion of nonresponse. Most researchers, including Kish (1965) and Cochran (1977), use the conventional descriptor, nonresponse; others, including Zarkovich (1966:146), Ford (1976:324), and Kalton (1983:1), also refer to the problem of missing data, possible because of the more general use of the term in classical statistics. The Panel on Incomplete Data (National Research Council, 1983:3) uses the term *incomplete data* to refer to problems resulting from unit nonresponse, item nonresponse, or undercoverage. Sudman (1976:194) talks about biases attributable to "noncooperators"; Suchman (1962:12) refers to the problem of "sampling mortality," and Sukhatme and Sukhatme (1970:147) describe the impact of "incomplete samples." Birnbaum and Sirken (1950:98–99) derive results for the bias due to non-availability of respondents, although their results are applicable to those who do not respond for any reason. Warwick and Lininger (1975) and Deighton et al. (1978) limit their focus to interview surveys by defining the *noninterview* as a type of nonresponse because of not-at-home respondents, refusal, incapacity, or lost schedules (i.e., questionnaires). This definition of nonresponse, while similar to the other, differs in its being applicable to only a certain type of survey. The person, organization, or other unit agreeing to provide survey data, and following through with that commitment has been called a "respondent," "cooperator," "participant," and "completed interview."

Unit nonresponse may occur for any of several reasons: not being at home, not locatable, refusal, communication problems between data collector and sample member (e.g., language barriers or physical impairment), incapacity, scheduling difficulties, and so on. Interpretations of the concept of nonresponse may differ according to which reasons are given for the nonparticipation. For example, Politz and Simmons (1949) deal only with nonresponse of the *not-at-homes*, those selected but not responding because they are unavailable when called on by the data collector. They overlook nonresponse for other reasons. Deming (1953:747) talks of nonresponse arising "from the sources, not at home and refusal" with a distinction being made between "temporary refusals" (e.g., a woman bathing a baby) and "permanent refusals" (e.g., people who would not respond under any circumstances). Kendall and Buckland (1960), along with their definition for nonresponse, suggest that *nonresponse* might best be measured by the number who refuse or are incapacitated, and *nonachievement* by the count of those failing to response for all other reasons.

In some instances, problems of sample coverage and nonresponse have been linked. Specifically, unit nonresponse in those instances is related to problems of undercoverage (where units that should be included are excluded), but not to overcoverage (where units that should be excluded are included). Kish (1965:527) and Kalton (1983:4), for example, view nonresponse and noncoverage as two types of errors of "nonobservation." Similarly, the Panel on Incomplete Data (National Research Council, 1983:3) considers undercoverage and nonresponse to cause problems of "incomplete data." The U.S. Bureau of the Census (1975:3), on the other hand, suggests that one might consider nonresponse error to be a specific type of coverage error, and Cochran (1977:364) prefers to include "noncoverage" (e.g., listing omissions in household samples) as one type of nonresponse.

6.2.2 Relative Measures of Nonresponse

A number of quantifiable measures have been suggested to indicate the extent of nonresponse in surveys. Most apply to unit nonresponse and the vast majority are percentage rates calculated as measures of the number of respondents (or, in some cases, nonrespondents) divided by some count of the number solicited for participation in the study. The relative measures currently being used in practice seem to serve two somewhat different purposes. One is to measure the degree to which nonresponse occurs in the population being studied in the survey. This type of information is useful to the statistician, who knows that the magnitude of nonresponse contributes to the bias of estimates. Relative measures of nonresponse are also useful to the operations supervisor, who will be more interested in knowing how successful the plans for location, solicitation, and data collection were. For example, a high proportion of interviews terminated by the respondent prior to completion would indicate that perhaps the questionnaire was too long or that interviewers did a poor job of conducting the interview. This rate, when computed separately for interviewers during the study, may help to spot those who are having difficulty. Computed at the end, it serves as one indicator of the quality of work done in the survey.

As with the general concept of nonresponse, terminology and associated definitions vary somewhat for relative measures of nonresponse (several definitions are presented in the Compendium of Nonsampling Error Terminology). The most frequently used term in referring to the relative degree of participation is *response rate* (or conversely, for a relative measure of nonparticipation, the *nonresponse rate*). The most widely used and recommended interpretation of these terms [e.g., Bailer and Lanphier (1978), Madow et al. (1983a), and Warwick and Lininger (1975)] defines the response rate as the proportion of participating (or nonparticipating) units among all *eligible* units in the sample, with eligibility tied to membership in the survey population. Other phrases, such as completion rate (CASRO Task Force on Completion Rates, 1982:2) and cooperation rate (Sudman, 1976:194), have been used in conjunction with the same interpretation of relative response.

In addition to using different words to describe the same relative measure, one may observe different interpretations of the same phrase. For example, in defining response rate as the proportion of contacted units that are interviewed, Hauck and Steinkamp (1964:13), are measuring how effective the interviewers were in soliciting responses once they located sample members. By contrast the first response rate defined above measures overall success in locating, soliciting, and collecting the data. For the term *completion rate*, the CASRO Task Force on Completion Rates (1982:2) identifies no fewer than eight different definitions that cover the full spectrum of purposes for which these rates might be used. These interpretations differ according to the type of survey being done (e.g., a census, a telephone survey, an interview survey) and whether success in location, solicitation, data collection, or determining eligibility status is being quantified.

Some relative measures of unit nonresponse apply only to certain categories of nonrespondents. Kendall and Buckland (1960) define a *refusal rate* as the proportion of contacted units that refuse to respond. This measure is designed to reveal how well the data collection team did in soliciting participation once sample members had been located. The National Research Council (1983:31) also defines *refusal rate* as the proportion of *all* eligible sample units that refuse. Kendall and Buckland (1960) use the terms *failure rate* and *nonachievement rate* to refer to the proportion of eligible units that fail to respond for reasons other than refusal or incapacity. Wiseman and McDonald (1978:45) give three different definitions of a refusal rate for surveys in which a specific individual in each member of a household sample is interviewed. These rates are distinguished primarily by whether refusal by the household as a whole or by designated individuals are counted.

Frankel and Dutka (1983:75) suggest a model for surveys with repeated interview attempts in which the concept of participation in a survey is seen as a chance event, with the symbol p representing the probability that a unit would respond if location and solicitation attempts were made. The "proportion reached,"

$$l_{1FD} = \frac{\int_0^1 f(p)\left[1 - (1 - p)^c\right] dp}{\int_0^1 f(p)\, dp}, \tag{6.1}$$

is given as an expected rate of response with $f(p)$ as the continuous density for the beta distribution of response probabilities and c as the allowable number of call attempts in the survey.

In some cases, the nature of the unit response rate is dictated by the setting in which it is used. Filion (1976:484), for example, defines a *cumulative response rate* to be applied to surveys done by mail where response is sought and reported in a series of rates for successive waves of data collection. In each wave, one attempts to obtain a response from all remaining nonrespondents from the previous wave. The cumulative response rate through the Ith wave is expressed as

$$l_{1F} = \sum_{g=1}^{I} \frac{n_{1g}}{n}, \tag{6.2}$$

where n_{1g} is the number of respondents coming in during the gth wave, n is the total number in the sample, and $\sum_{g=1}^{I}$ denotes summation over waves $g = 1, 2, \ldots, I$.

In other rates of response the purpose may be to express the extent of survey response in units other than those sampled. For example, one may wish to know the proportion of hospital beds accounted for by responders in a survey of short-stay hospitals, or the proportion of sales (in dollars) represented by participants in a survey of retail hardware stores. In either case, the object is to measure the degree of response in terms of measures indicating the *size* of units rather than the units themselves. Toward this end, Nisselson (1983:1) discusses a weighted response rate for surveys in which sample members may have unequal selection probabilities. This rate is defined as

$$l_{1N} = \frac{\sum_{i=1}^{n_1} M_i / \pi_i}{\sum_{i=1}^{n} M_i / \pi_i} \tag{6.3}$$

where M_i is the measure of size (e.g., number of hospital beds or amount of retail sales) for the ith sample unit, π_i the selection probability for the ith sample unit, n the number of eligible units in the sample, and n_1 the number of responding sample units. We let $\sum_{i=1}^{n_1}$ and $\sum_{i=1}^{n}$ denote summation over all samples respondents and the full sample, respectively. We see from its definition that Nisselson's rate reduces to the widely used unweighted response rate ($l_n = n_1/n$) when the ith sample number is chosen with probability proportional to size (M_i). When equal-probability sampling (e.g., simple random sampling) is used, l_{1n} is the proportion of the total of size measures in the sample represented by sample respondents. In general, l_{ln} in the hospital survey mentioned above would estimate the proportion of short-stay hospital beds that would have been found in hospitals responding to a complete enumeration of all hospitals.

Rates calculated using Eq. (6.3) will differ from those ignoring M_i and π_i to the extent that these measures are statistically associated with the likelihood to respond. For example, in a simple random sample of 300 hospitals grouped according to small, medium, and large, the response rate may be directly related to hospital size, as illustrated below.

Size	Average Number of Beds	Number of Eligible Hospitals		Response Rate (Percent)
		Responding	Sampled	
Small	25	105	150	70
Medium	100	75	100	75
Large	400	45	50	90
Total		225	300	75

Because of the assumed relationship, the response rate using the number of beds as M_i would be 83 percent if the average number of beds for respondents and nonrespondents in each size category were the same. The corresponding hospital response rate of 75 percent would therefore be lower than the weighted rate.

Other relative measures of unit response and nonresponse are used as performance indicators for the survey data collection. These measures are not only computed for the entire survey but may also be broken down by interviewer, geographic area, and other reporting categories to understand better where problems are occurring if the measures are computed periodically during the survey. When computed after the survey has been completed, these measures may also be used to determine whether weaknesses in location, solicitation, and data collection will require special efforts in later surveys. Reviews by Wiseman and McDonald (1980) and the CASRO Task Force on Completion Rates (1982) identify different so-called *completion rates*. One version—the completion rate Kviz (1977) defines as the number of units with completed interviews divided by the number of eligible units contacted—is a measure of the data collectors' ability to solicit participation once the sample member is located. In mail surveys, the completion rate would be the proportion completing the questionnaire among those who received it.

Computation of some of the rate of unit response and nonresponse discussed above is illustrated using data from the 1978 Virginia Health Survey (VHS) conducted by the Research Triangle Institute. For data summarizing the outcome of field operations for the VHS, see Table 6.1. Rates 1

Table 6.1   Frequency of Final Outcomes for Data Collection in 1978 Virginia Health Survey

	Outcome	Number of Reporting Units[a]	Eligibility Status[b]
F1.	Interview completed	5069	Eligible
F2.	No eligible respondent at home	163	Eligible[c]
F3.	Temporarily absent	45	Eligible[c]
F4.	Refused	302	Eligible[c]
F5.	Vacant	460	Ineligible
F6.	Not a housing unit[d]	90	Ineligible
F7.	Temporary or vacation home; usual residence elsewhere	101	Ineligible
F8.	Other—eligible	26	Eligible
F9.	Other—ineligible	12	Ineligible
	Total	6268	

[a]A single questionnaire was completed for each "reporting unit," which in this survey generally corresponds to a family or household.

[b]Members of the civilian noninstitutionalized population living within the boundaries of the Commonwealth of Virginia at the time of the survey were studied in this survey. A reporting unit was considered eligible if one or more eligible persons currently lived there.

[c]In some interview surveys the eligibility status of subjects in this category may be unknown if direct contact with a member of the household is required to determine whether any eligible people live there (e.g., for a survey of handicapped persons).

[d]Reasons included: merged with another housing unit, demolished, moved away, group quarters, or not used for residential purposes.

to 6 in Table 6.2, as relative measures of the number of interviews completed, vary from 80.9 to 93.1 percent. Variation among these rates occurs because of differences in how the denominators are chosen. The smallest rate includes all members of the original sample in the denominator, while the largest includes only those contacted. Rates with intermediate values represent various attempts to produce a denominator with the number of eligible units in the sample. Rate 1 assumes that the eligibility status has been determined for all members of the original sample. In many surveys, however, there will remain some portion of the sample for which eligibility status is uncertain. We presume in computing rates 2 to 4 that the eligibility of these units in the "other-eligible" and "other-ineligible" categories was unknown. Rate 2 assumes that all those with unknown eligibility are eligible and is thereby the lower bound; while rate 4 assumes all unknowns to be ineligible and is the upper bound. Rate 3, intermediate in value to rates 2 and 4, is produced by assuming that the fraction of eligibles among those whose eligibility is unknown is the same as the fraction of eligibles among those for whom eligibility is known. Variation among rates 2 to 4 is small in the VHS sample because the percentage of units with unknown eligibility is small. The variation would be substantially greater in telephone surveys, where the eligibility status is unknown for a larger portion of the sample (e.g., those not answering after repeated call attempts, those refusing to divulge eligibility information, etc.).

We see in producing the rates for Tables 6.2 that the eligibility status of many sample members may be unknown. Since the denominator of the preferred response rate is the number of eligible members in the sample [e.g., Bailar and Lanphier (1978)], failure to know the eligibility status for some sample members clearly creates a problem. Interview surveys like the VHS, where dwellings are sampled but individuals are interviewed, is one of several circumstances where unknown eligibility occurs. The eligibility status of persons to be interviewed is indeterminant for those dwellings where no contact with anyone in the household is ever made. Unknown eligibility may be an even greater nuisance in telephone surveys with a substantial percentage of those not answering or refusing to answer after repeated call attempts. Finally, survey eligibility in some mail surveys cannot be determined for those failing to return a questionnaire. Interview surveys (e.g., of all adults) in which multiple interviews are possible within selected dwellings cause an additional problem, since for noncontacted dwellings neither their eligibility status nor a count of eligible individuals will be known.

The variation in unit response rates reported for past surveys is as wide as the diversity in terminology surrounding relative measures of nonresponse. Response rates for several surveys from the past 30 years have been compiled by Kish (1965:538–541), Benus and Ackerman (1971), Moser and Kalton (1972:171–173), Love and Turner (1975), Cannell (1977:13–17), Bailer and Lanphier (1978:36–37), the National Research Council (1979:140), and the Panel on Incomplete Data (National Research Council 1983:21–22). The size of reported rates, varying from around 10 to 95 percent, is caused by any of several factors.

First, the size of the rate may be partly explained by how the rate was defined and then how the researchers computed it based on their definition. This explanation would seem plausible in light of a recent study by Wiseman and McDonald (1980) in which 40 survey research organizations were asked to compute a response rate as they would report it. Using the same set of outcome data from a typical telephone survey, 29 different rates ranging from 12 to 90 percent were produced by these organizations. Differences among the reported rates were the result of differences in how both the numerators and denominators were computed.

Second, the amount of unit nonresponse may be determined by the survey protocol. Personal interview surveys tend to have higher response rates than telephone interview surveys, which, in turn, generally have higher rates than self-administered mail surveys done without extensive follow-up. The size of the response rates seems also to be directly related to the following features of the survey protocol: the amount of effort expended during location and solicitation (e.g., the number of follow-up attempts in mail surveys, the use of

Table 6.2 Calculation of Several Rates of Unit Response and Nonresponse Using Data from 1978 Virginia Health Survey

	Rate	Definition	Formula	Value
1.	Response rate[a]	$\dfrac{\text{No. interviewed}}{\text{No. eligibles in sample}}$	$\dfrac{F1}{F1 + F2 + F3 + F4 + F8}$	0.904
2.	Response rate[b]	$\dfrac{\text{No. interviewed}}{\text{No. eligibles in sample}}$	$\dfrac{F1}{F1 + F2 + F3 + F4 + F8 + F9}$	0.902
3.	Response rate[c]	$\dfrac{\text{No. interviewed}}{\text{No. eligibles in sample}}$	$\dfrac{F1}{F1 + F2 + F3 + F4 + g[F8 + F9]}$ $g = \dfrac{F1 + F2 + F3 + F4}{F1 + F2 + \cdots + F7}$	0.903
4.	Response rate[d]	$\dfrac{\text{No. interviewed}}{\text{No. eligibles in sample}}$	$\dfrac{F1}{F1 + F2 + F3 + F4}$	0.909
5.	Completion rate[e]	$\dfrac{\text{No. interviewed}}{\text{No. originally selected}}$	$\dfrac{F1}{F1 + F2 + \cdots + F9}$	0.809
6.	Completion rate[f]	$\dfrac{\text{No. Interviewed}}{\text{No. contacted}}$	$\dfrac{F1}{F1 + F3 + F4 + F8}$	0.931

7.	Completion rate[g]	$\dfrac{\text{No. for whom eligibility was determined}}{\text{No. originally selected}}$	$\dfrac{F1 + F2 + \cdots + F7}{F1 + F2 + \cdots + F9}$	0.994
8.	Refusal rate[h]	$\dfrac{\text{No. refusals}}{\text{No. eligibles in sample}}$	$\dfrac{F4}{F1 + F2 + F3 + F4 + F8}$	0.054
9.	Refusal rate[i]	$\dfrac{\text{No. refusals}}{\text{No. contacted}}$	$\dfrac{F1}{F1 + F3 + F4 + F8}$	0.055

[a]As defined by Warwick and Lininger (1975), Bailer and Lanphier (1978), and the Panel on Incomplete Data (National Research Council, 1983).

[b]A lower bound for rate 1 assuming that the "other" category is not distinguished according to eligibility and that all in this category are *eligible*.

[c]An approximation for rate 1 assuming that the proportion of eligibles in the "other" category is the same as that of eligibles in the remaining categories combined.

[d]An upper bound for rate 1 assuming all "other" categories are *ineligible*.

[e]As defined by Kviz (1977).

[f]As defined by O'Neal et al. (1979).

[g]Among the rates the CASRO Task Force on Completion Rates reviews; assuming that the "other" category has not been subdivided.

[h]As defined by the Panel on Incomplete Data (National Research Council, 1983).

[i]As defined by Kendall and Buckland (1960).

an introductory letter sent to subjects in advance of the first call attempt, the number of call attempts in interview surveys); the level of data collector experience with location and solicitation work; how much incentive (e.g., in the form of endorsements of monetary inducements) potential respondents are offered; and the level of interviewer training and supervision in interview surveys. Finally, the nature of the survey population is often important in determining the rate of unit nonresponse. Some populations are usually receptive to surveys, whereas others are not. For example, one normally expects higher response rates from households in rural areas than from those in urban areas, and surveys of physicians usually have low response rates.

In addition to considering the definitions of relative measures of nonresponse, there are some factors to consider in interpreting reported rates of unit response. For one, it is important to remember that while a rate tells us the extent of nonresponse, it does not explicitly indicate the impact of the nonresponse on survey estimates. Low response rates point only to a *potential* for severely affected estimates (see Section 6.2.3). In fact, the ultimate effect of nonresponse in a survey with a 90 percent response rate but large respondent–nonrespondent difference may be more severe than a survey with an 80 percent response rate but small respondent–nonrespondent differences. Another factor to consider is how good the rate is in light of past experience with similar surveys. For instance, 70 percent would be considered an excellent response rate in a self-administered mail survey, just satisfactory in a telephone interview survey, and fairly low in most personal interview surveys. Some circumstances simply do not lend themselves to high response rates. For example, in a survey of patients visiting hospitals 10 years ago, where the ultimate goal after screening for a particular type of behavior is to interview them in person, a combined 70 percent response rate would be considered excellent. The varying uses and circumstances in which relative measures of unit nonresponse are presented make it difficult to find a formula that will always be appropriate. Instead, perhaps it is more useful to be mindful that since there is no uniformly acceptable definition, it is possible to be misled by reported rates if one assumes that the rate was computed one way when in fact the reported rate portrays the level of nonresponse differently, and perhaps in a more favorable light than justified. Two hypothetical examples might help to illustrate this point. For the first example, suppose that in a telephone survey of adults with 900 completed interviews, 300 noncontacts, and 100 other eligible nonrespondents, the response rate is reported as 90 percent by excluding the noncontacts in the denominator. Since it is likely that most of the noncontacts would be eligible, the reported rate would be deceptively high. The usual rate with all eligibles in the denominators would have been closer to 70 percent.

Another example of deceiving reported response rates happens when surveys are conducted in several stages. The final set of respondents in a survey of women during pregnancy, for instance, might be obtained by first sampling medical providers, then screening the providers' records for pregnant females, and then contacting eligible women for an interview. Properly computed, but separate, rates of 80 percent for providers, 95 percent for screening, and 85 percent for interviews might mistakenly lead one to believe that nonresponse was not a major problem in the study since all three rates are high. This view unfortunately would be naive since when considering these rates together, the important overall response rate for eligible women would be closer to the product of the three separate rates, or about 65 percent.

Nonresponse rates are sometimes also generated for item nonresponse. An item nonresponse rate, computed as the number of processed question-

Table 6.3   Selected Rates of Item Nonresponse for the 1980 National Medical Care Utilization and Expenditure Survey[a]

Item	Item Nonresponse Rate[b] (percent)
Age	0.2
Birth date	0.9[c]
Sex	1.7[d]
Race	22.3[e]
Highest grade completed in school[f]	2.2
Average number of hours worked per week[g]	2.0
Employment income[h]	26.8
Total charge for visit to health care provider[i]	
Hospital emergency room	37.2
Hospital (outpatient clinic)	50.9
Hospital (inpatient care)	36.7
Physicians in hospitals	23.6
Dentist	14.0

Source: Cox and Cohen (1985, Chapt. 9).

[a]An item is a "nonresponse" here if imputation was required.
[b]Calculate in general as

$$\frac{\text{no. of item records with missing or unusable data}}{\text{total no. item records}}$$

[c]With either month, day, or year missing.
[d]Race for 21.7 percent of NMCES participants could be imputed logically from the parents, while the remaining 0.7 percent were assigned a race by hot deck imputation.
[e]Sex for 1.2 percent of NMCES participants could be imputed logically based on the subject's relationship to the head of household. Hot deck imputation was applied to the other 0.5%.
[f]Obtained only for persons 17 years or older.
[g]Missing data for more than one of the five rounds of data collection.
[h]Obtained only for persons 14 years and older.
[i]Out of total number of visits to health care provider.

naires with missing data on a particular item divided by the total number of processed questionnaires, indicates the level of difficulty there was in obtaining usable information for that item during data collection. Item nonresponse rates, while low for most survey questions, are often higher for questions dealing with private or sensitive issues and those requiring the respondent to recall facts or past events that might be confused or unknown. The item may be left blank when the questionnaire is being completed or when an initial response recorded during data collection is found later to be unusable. Item nonresponse rates are also often used to determine questionnaire items for which imputation will be required (see Section 8.2.2).

Response rates for several items from the 1980 National Medical Care Utilization and Expenditure Survey (NMCUES) conducted by the Research Triangle Institute appear in Table 6.3. NMCUES, a nationwide personal interview survey of all persons living in selected households, was conducted in five 12-week rounds of interviewing beginning early in 1980 and ending in midyear 1981. The size of the item nonresponse rates illustrates some of the points just made. The lowest rates were for nonsensitive factual items, such as age, sex, and education, while the larger rates are found for income, which for some is considered private information, and for health care charges, which in the absence of memory aids are less likely to be remembered and reported. The high rate of item nonresponse for race is attributable to the children's race not being recorded directly but rather, inferred from the parents.

6.2.3　Bias Associated with Nonresponse

The third fundamental concept covered in our brief review of basic terminology associated with the problem of nonresponse is the bias or systematic distortion in survey estimates occurring because of an inability to obtain a usable response from some members of the sample. This bias is a key measure of the impact of nonresponse on survey error, and although its terminology is more consistent than that used for the concepts of nonresponse and nonresponse rate, several different terms are used to describe it. The majority of researchers use the term *nonresponse bias*. Goudy (1976:368), on the other hand, calls it *response bias*, which is also the name for a measurement error term (see Section 10.1). Some of the early research studies seemed to deal with the bias arising from certain specific reasons for nonresponse. For example, Birnbaum and Sirken (1950:99) refer to *nonavailability bias*, while Politz and Simmons (1949:12) provide a remedy for what they term the *not-at-home bias*. Cassel et al. (1983:149) use the term *pq bias* when discussing the joint bias because of sampling and nonresponse. Schaible (1983:133) defines *incomplete data bias* as the total bias of an estimator minus the bias that would remain with complete response.

The meaning of the term *nonresponse bias* also seems to differ with the type of researcher. Assuming no errors arising from imperfect sampling frames or measurement, some define it as the difference between the expected value over all possible samples of an estimator (the mathematical formula used to produce the survey estimate) and the population value being estimated. The size of the expected value and thus the presence or absence of nonresponse bias, in this sense, is a statistical property of the estimator and the design followed in producing the sample. Estimators formulated and samples chosen in a certain way can be freed of nonresponse bias. To other researchers nonresponse bias is determined by the "representativeness" of the sample (how its composition of respondents compares with that of the population from which it is chosen). The existence of bias under this view is tested by whether the sample of respondents at face value is a microcosm of the population being studied. The effects of weighting, adjustment, and other methods to compensate for nonresponse are not considered. Those samples no longer thought to represent the population because of higher nonresponse rates in certain important subgroups are seen as exhibiting a measure of nonresponse bias. As Filion (1976:401) says, the sample respondents alone "do not validly depict the population investigated and may result in predictions which are inaccurate."

An expression of nonresponse bias, under a simple set of assumptions, can be derived as follows. Suppose that a population of N elements consists of a stratum of N_0 nonrespondents with mean ($\bar{Y}_0$) and a stratum of N_1 respondents with mean ($\bar{Y}$), in the proportions λ_0 and λ_1, respectively, such that $\lambda_0 + \lambda_1 = 1$. The overall population mean can be written as

$$\bar{Y} = \lambda_0 \bar{Y}_0 + \lambda_1 \bar{Y}_1. \tag{6.4}$$

Assuming no survey errors other than those because of sampling and nonresponse, let $\hat{\bar{Y}}_1$ denote the estimator of $\bar{Y}$ based only on nonrespondents in the sample such that $E_s(\bar{Y}_1) = Y_1$, where $E_s(\cdot)$ denotes expectation over all possible samples. Then the bias of $\bar{Y}_1$ can be shown, using the generalized definition of bias, to be[1]

$$\text{Bias}\left(\hat{\bar{Y}}_1\right) = E_s\left(\hat{\bar{Y}}_1\right) - \bar{Y} = \lambda_0\left(\bar{Y}_1 - \bar{Y}_0\right). \tag{6.5}$$

We thereby see illustrated by Eq. (6.5) that the size of nonresponse bias is jointly a function of the expected rate of nonresponse and the amount of respondent–nonrespondent differences in the population.

[1]There is some disagreement as to the definition of *bias*. Some define it as $E_s(\hat{\bar{Y}}_1) - \bar{Y}_1$ [see, e.g., Sukhatme and Sukhatme (1970:10) or Raj (1968:29)], while others use $Y - E_s(\hat{\bar{Y}}_1)$ [see, e.g., Mood and Graybill (1963:172)]. These alternative definitions differ in sign but not in the size of the contribution of the mean square error.

When estimating the population total, $Y = N\bar{Y}$, the bias attributed to nonresponse of the estimator, $\hat{Y}_1 = N\bar{Y}_1$, will be

$$\text{Bias}(\hat{Y}_1) = N_0(\bar{Y}_1 - \bar{Y}_0). \tag{6.6}$$

The size of the nonresponse bias relative to the population being estimated will for both $\hat{\bar{Y}}_1$ and $\hat{Y}_1$ be

$$\text{Relative bias}(\hat{\bar{Y}}_1) = \text{Relative bias}(\hat{Y}_1) = \frac{\lambda_0(\bar{Y}_1 - \bar{Y}_0)}{\bar{Y}}. \tag{6.7}$$

The size of this bias, relative to the expected value (over all samples) of the estimator will for both estimators be

$$\text{Relative bias*}(\hat{\bar{Y}}_1) = \text{Relative bias*}(\hat{Y}_1) = \lambda_0\left(1 - \frac{\bar{Y}_0}{\bar{Y}_1}\right). \tag{6.8}$$

The view linking nonresponse bias to the representativeness of the sample exhibits an appreciation for the contribution of respondent–nonrespondent differences to survey error through nonresponse. Emphasis here is on uncovering "distributional biases" (Hawkins, 1975) in the survey sample. Analysis often involves comparing respondents and nonrespondents with respect to various social, cultural, economic, and demographic variables (Donald, 1960; Kivlin, 1965; Larson and Catton, 1959; Parten, 1966). Sometimes a profile of respondents is compared directly with recent data for the population sampled. The rationale behind these studies is that respondent–nonrespondent dissimilarity discovered for variables related to those of interest in analysis is likely to indicate presence of respondent–nonrespondent differences on the analysis variables themselves, and thus of nonresponse bias in the more mathematical sense. The appeal of respondent–nonrespondent differences studies may be that they are relatively easy to do, provided that variables correlated to the major study variables are available for respondents and nonrespondents. Moreover, a simple comparison between those who did and did not participate gives one a straightforward indication as to whether nonresponse has thrown the original sample out of balance with respect to the major study variables and therefore would bias estimates. These different studies are especially attractive to those "who are forever indebted to survey research as a valuable tool, but who have neither the desire nor inclination to explore the intricacies of sampling theory" (Hawkins, 1975:482). Unfortunately, one never completely knows how strongly related the auxiliary variables are to the analysis variables.

We conclude this section by discussing briefly how some have viewed the relationship between nonresponse bias and other biases and components of error. Three somewhat different perspectives are illustrated in Fig. 6.1.

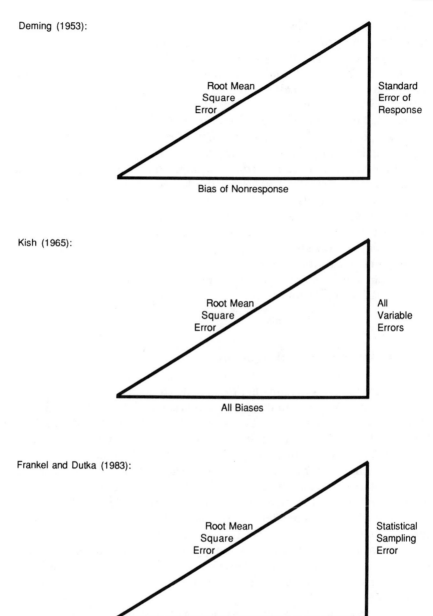

Figure 6.1 Three views relating various components of survey error.

Deming (1953:746) views the "bias of nonresponse," "standard error of response," and "root mean square error" as analogous to the two legs and hypotenuse of a right triangle, respectively. From this perspective, nonresponse bias is isolated while the standard error of response appears to incorporate all other biases and variable nonsampling error components. Kish (1965:521) also uses right triangles but recommends that survey errors be classified into the following two groups: all biases (of which nonresponse is one) that additively determine the length of one leg, and all variable errors (e.g., sampling variance, response variance), which form the other leg. The size of the root mean square error determines the length of the hypotenuse. Frankel and Dutka (1983) suggest a third view where one leg of the right triangle consists exclusively of the square root of the sampling error (the usual variance of an estimator) while the other leg includes all response errors, which are of two types: errors of inadequate response covering a vast array of bias and variable errors arising from measurement and data processing (e.g., all measurement error components, data-entry errors, etc.), and those from nonresponse. The hypotenuse once again represents the root mean square (total) error.

6.3 REASONS FOR NONRESPONSE

Reasons explaining why units fail to respond in a survey are often reported, although the words used to describe them may vary. Terminology here seems to depend on the type of units being studied and the mode of data collection used in the survey. To illustrate this point, consider two surveys, one of households, the other of hospitals. "Not at home" would be a commonly used reason in the former case but not the latter, whereas "proposed study rejected by the research review committee" would be common to the latter but not the former. The method of data collection also affects the reason for nonresponse. In personal interview and random-digit telephone interview surveys "not at home" may be a reason common to both methods, while "unable to locate" is likely to be unique to the first and "persistent busy signal" unique to the second. Despite the diversity in terminology, most reasons for nonresponse seem to fall into one of the five categories presented later in this section.

Ours is not the first attempt to deal with the problem of categorizing nonrespondents. Durbin (1954) and Kish (1965), for example, discuss some of the general reasons for nonresponse in household surveys. More recently, the Panel on Incomplete Data (Madow et al., 1983a) has suggested the concept of an "accountability table" wherein each sample member is assigned to one of a large number of detailed reporting categories. The categories are sufficiently specific so that rates of response and nonresponse can be calculated in any reasonable manner.

We note several things about our categorization of reasons before proceeding. First, most reasons for nonresponse discussed below apply to unit nonresponse, not to item nonresponse, because reasons for item nonresponse, other than refusal, are seldom reported or unknown. One might speculate that the sheer magnitude of the problem of reporting reasons for each individual question asked in the survey explains why reasons for unit nonresponse are usually the only ones reported, At a conceptual level, of course, reasons for unit and item nonresponses are related since unit and item nonresponse are related. All items for units that fail to respond are item nonresponses. Thus reasons for item nonresponse among unit respondents, coupled with the reasons for nonresponse among unit nonrespondents, make up the set of all reasons for nonresponse on individual questionnaire items. Second, reasons for nonresponse given in our categorization may apply separately to more than one type of unit when compiling is done in multiple stages or phases. For example, a national sample of patient medical records in hospitals would have reasons for nonresponse for hospitals and records within hospitals, each selected in separate sampling stages. The final outcome of one of the sample records might be that nonresponse occurred under circumstances in which the hospital agreed to participate but the selected record could not be found. Third, most categories suggested below apply only to cross-sectional (or "one-shot") surveys in which each respondent is interviewed no more than once, or to individual rounds in multiround longitudinal surveys. Finally, it is understood in our categorization that the reason for nonresponse is the one given after the last attempted solicitation. This point is worth mentioning, since other reasons may be indicated during earlier attempts. For instance, the initial reason for unit nonparticipation in a personal interview household survey may be "not at home" if the interviewer stops by but no one is there. The final reasons may be "refusal" after contact is made.

6.3.1 Ineligible

Those units judged to be out of scope for the survey (not a member of the target population and therefore not to be included if selected) are considered jointly to make up one category of nonparticipants even though the number used in this category should not be used in the denominator of rates of response and nonresponse. By way of illustration, a woman 60 years old would be considered ineligible in a survey of women of childbearing age. In household surveys, dwelling units found to be vacant, demolished, or temporary living quarters are also considered out of scope, as are dwellings located outside selected area segments but listed and selected by mistake. A selected dwelling in a civilian residential survey may be considered ineligible because it is a business establishment, a group quarters, an institution, or part of a military facility. An individual selected in the sample but deceased or failing to meet minimum residency requirements would also be considered ineligible.

6.3.2 Not Solicited

The category "not solicited" applies to all units for whom location is unsuccessful so that solicitation is never attempted. In household interview surveys, a good example of this type of dwelling with no one at home when the interviewer calls to make the solicitation attempt. The terminology used in this situation differs. Cochran (1977), Deming (1953), Kish (1965), and Politz and Simmons (1949) all refer to this type of nonrespondent as a *not-at-home*, while Birnbaum and Sirken (1950) call them *nonavailables*. Deighton et al. (1978) speaks of *noncontacts* in similar terms, while the U.S. Bureau of the Census (1975) refers to individuals who fail to respond because of an *absence from home*. Kendall and Buckland (1960) mention nonresponse due to *absence*, while Moser and Kalton (1972) divide this subcategory into those who are *away from home* (*long*) for an extended period and those who *are out at the time of the call* but presumably only for a short time. Durbin (1954) makes a similar distinction by considering *persons known to be away during the field period* and *other not-at-homes*. Kish (1965) suggests an additional category for those *not found*.

Several other specific reasons for nonresponse would be included in this category. For example, Durbin (1954) suggests classification of nonresponse because of *change of address* and *address could not be traced*. Moser and Kalton (1972), with similar intentions, mention the class of nonrespondents who are *movers*. In some instances, *death* could also be included in this general category of reasons for nonresponse. For example, survey eligibility may be based on a past event, at which time a decedent might have been eligible. Finally, persons *lost to follow-up* during an extended period of data collection in a longitudinal study might also be included in this category, although one might reasonably argue that a participant who drops out of a longitudinal study part of the way through the follow-up period could be classified as a partial respondent rather than a nonrespondent. The choice would depend on whether follow-up throughout the study is considered essential.

6.3.3 Solicited but Unwilling

"Solicited but unwilling" and the next category represent two types of unit nonrespondents where the nonrespondents are located and solicited but fail to participate in a survey interview, or in the case of item nonresponse, fail to answer particular questions included in the survey questionnaire. Failure in this first category is specifically linked to the sampled individual's unwillingness to participate, meaning that a conscious decision is made on the part of the unit not to participate. The explanation for the unwillingness may arise from any of several sources, including fear, apathy, distrust, or lack of time.

Nonparticipants in this category are usually called *refusals*. Durbin (1954), Kish (1965), and Moser and Kalton (1972), all use this term, while Deming (1953) divides refusals into *temporary refusals*, who may respond when subsequent solicitation attempts are made, and *permanent refusals*, who will never respond. This latter group might correspond to the so-called *hard-core* refusals mentioned by Cochran (1977), Ericson (1967) and Kish (1965). In mail surveys we might interpret the U.S. Bureau of the Census (1975) category of nonresponse through *failure to return completed questionnarire* as a refusal, although clearly some of these might be considered unintentional refusals by never having received the questionnaire or passive refusals by losing the questionnaire before having a chance to return it. This category would also include those sampled individuals (minors and those mentally retarded for who a signed consent form is not obtained by the person responsible for deciding whether or not to sign (the parent, a guardian, etc.).

6.3.4 Solicited but Unable

Units in the category "solicited but unable" represent the second type of located nonparticipant. The central focus here is the unit's inability to participate, whether unavailable or because of physical, mental, emotional, or language problems. Kish (1965) uses the term *temporarily unavailable* to describe the situation in household surveys where participation is deferred and ultimately denied because those selected are too ill, too tired, unable to find a suitable time, or inaccessible when called on during the period of data collection. Nonresponse by reason of incapacity also falls in this category. Durbin (1954) and Moser and Kalton (1972) refer to those members of the survey sample who are *unsuitable for interview* because of physical incapacity (e.g., deaf or infirm). Deighton et al. (1978) and Kish (1965) refer simply to *incapacity (physical)* in this regard. Finally, the U.S. Department of Health, Education, and Welfare (1977) distinguishes between *incompetent* and *language difficulty*.

6.3.5 Other Reasons

Some among those with a discernible reason for nonresponse do not fit in any of the four previous categories. Thus a fifth catch-all category for eligible nonrespondents is needed. The category of nonresponse through *lost schedules* as suggested by Kish (1965) and Deighton et al. (1978) is one example. For surveys of individuals, another would be the group of what might be called *partial respondents*, *partial participants*, *cutoffs*, or *breakoffs* for whom one observes the "omission of one or more entries" (U.S. Bureau of the Census, 1975).

6.3.6 Illustration

In Table 6.4 we illustrate the frequency distribution of sample members according to the five nonresponse categories discussed above. Four surveys conducted by Research Triangle Institute and one by the West Virginia Department of Health are used for the illustration. In the Head and Spinal Cord Injury Survey, patient medical records were abstracted in a national sample of acute care hospitals. The goal of the study was to determine the incidence and prevalence of trauma to the central nervous system. Experienced field interviewers, armed with professional endorsements of the study, were used to solicit hospital participation. As mentioned earlier, the Virginia Health Survey (VHS), was a personal interview household survey of the civilian, noninstitutionalized population of Virginia designed to measure the current health status of the Virginia population. Reporting units in the survey corresponded roughly to families. Dwellings selected from household listing sheets were contacted for interview by experienced interviewers, and up to four call attempts were allowed during solicitation. Like the VHS, the National Medical Care Expenditure Survey (NMCES) measured health status through a personal interview household survey of the civilian noninstitutionalized population in which dwellings in selected area were chosen from listing sheets. The protocol for soliciting participation in the NMCES was similar to the VHS; however, unlike the VHS, the NMCES was national in scope and operationally much larger. The NMCES was also a panel survey in which the same sample of households was visited in each of five rounds. The Behavioral Risk Factor Survey (BRFS) in West Virginia was done by random-digit telephone interviewing. Its goal was to produce estimates of health risk in the adult residential population. The BRFS is fairly typical of one-time telephone surveys in administering a short uncomplicated questionnaire after up to five variably scheduled call attempts had been made. Attempts to convert initial refusals was made in each of the four interview surveys. The fifth study (Table 6.4) is a mail survey, the National Survey of Licensed Practical Nurses, whose objective was to determine employment status. Response data presented here were observed after an initial mailout and a mail follow-up about six weeks later. Results after two additional follow-ups, one by mail and the other by telephone, are not presented since a single follow-up was considered more typical than were three fixed-mode follow-ups.

A comparison of entries in columns 2 to 6 in Table 6.4 reveals how the importance of various reasons for nonresponse differs among surveys. The proportion of respondents and response rates is highest for surveys done in person and lowest for the mail survey. The proportion of ineligibles is highest in the telephone survey since a sample of random digits is likely to produce large numbers of business, nonworking, and other nonresidential numbers. The relatively large percentage of unsolicited sample members in the telephone survey is because of numbers where the line was busy or no one

Table 6.4 Frequency of Final Outcomes from Five Surveys Using Proposed Categorization of Reasons for Nonresponse

		Outcome Frequency (%)			
	1975	1978	1978	1982	1983
Survey:	National Head and Spinal Cord Injury	Virginia Health Survey	National Medical Care Expenditure Survey (Round 1)[a]	Behavioral Risk Factor Survey in West Virginia	National Survey Of Licensed Practical Nurses (one mail follow-up)[b]
Data Collection Mode:	In person	In person	In person	Telephone	Mail
Unit:	Hospital	Reporting unit	Reporting unit	Household	Nurses
Conducted by:	Research Triangle Institute	Research Triangle Institute	Research Triangle Institute	West Virginia Dept. of Health	Research Triangle Institute

Table 6.4 *(Continued)*

(1)	(2)	(3)	(4)	(5)	(6)
Respondents	81.0	80.9	80.9	53.4	56.3
Nonrespondents					
Ineligible	3.0	10.6	11.5	28.7	0.3
Not solicited	—	2.6	0.7	7.8	4.7
Solicited but unwilling	15.0	4.8	5.0	3.5	0.3
Solicited but unable	1.0	0.7	0.1	5.3	0.0[c]
Other reasons	—	0.4	1.8	1.3	—
Unknown response status	—	—	—	—	38.4
Total sample size	305	6,268	17,521	891	22,004
Response rate[d] (5)	83.5	90.4	91.4	75.0	56.5

[a]The response for those reporting units participating in each of five rounds was 83.5 percent.
[b]The final response rate after a second follow-up reminder by mail and a third contact by telephone was 80.4 percent.
[c]Less than 0.05 percent.
[d]Computed as number of responding units divided by number of originally selected units not found to be ineligible.

answered after repeated attempts. Some of these numbers, had they been contacted, would have been found to be ineligible. The percentage of unwilling sample members was highest in the Head and Spincal Cord Injury Survey, because much more was asked of respondents and the decision to participate often was made jointly by several parties. In many hospitals the hospital administrator, the chief of the medical staff, and a research review committee all had to be favorably disposed to the study before the hospital participated. The percentage of sample members in the "solicited but unwilling" category is relatively low for the mail survey, although a significant proportion of those within an unknown response status (those eligible nonrespondents whose mail-out was not returned undeliverable) were probably eligible refusals. It is likely that many of the remaining members with unknown response status were unavailable to complete the questionnaire. Others may have misplaced the questionnaire and forgotten about it, while others might have completed the questionnaire but failed to return it. The high percentage of those unable to respond in the telephone survey is largely because of the unavailability of selected adults during a relatively short period of data collection (approximately one week versus periods of several weeks in the other studies).

CHAPTER 7

Nonresponse: Statistical Effects of the Problem

Having reviewed the terminology and context of nonresponse in sample surveys, let us now examine how nonresponse contributes to survey error. We consider the types of statistical problems that nonresponse creates for the survey analyst to provide a motivation for Chapter 8, where we discuss methods for dealing with nonresponse. In our review of the statistical implications of nonresponse, we note that the effect of nonresponse partly depends on how one assumes it to take place and on the type of parameter one intends to estimate in the population.

The material presented here and in Chapter 8 draws heavily on recent efforts to compile and summarize the ever-growing literature on survey nonresponse. Most notable among these state-of-the-art reviews is the excellent work done by the Panel on Incomplete Data of the National Research Council. The panel's three-volume report (Madow and Olkin, 1983; Madow et al., 1983a, b) contains a comprehensive review of the theory and recent application of most of the methodology designed to deal with the problem of nonresponse. We also draw on the work of Santos (1981), Kalton (1983), and Cox and Cohen (1985) for our review. Because of the extensive literature on the subject, we can only hope to highlight and summarize, leaving the reader to pursue these and other sources in greater detail.

We begin by defining some terms used in Chapters 7 and 8. Let an existing finite population of N elementary units be denoted by $\mathbf{U} = (U_1, U_2, \ldots, U_N)'$, where the assignment of labels is arbitrary and the subscript i is used to index members of $\mathbf{U}$. For example, in a sample survey on diastolic blood pressure (DBP), where the goal is to profile and explain the levels of DBP in a population of adults, entries in $\mathbf{U}$ would be adults. Associated with each member of $\mathbf{U}$ is a value of importance to the survey. We use $\mathbf{Y} = (Y_1, Y_2, \ldots, Y_N)'$ to denote this set of key survey measurements, with Y_i being the value linked to U_i. Another set of values associated with $\mathbf{U}$ is $\mathbf{D} = (D_1, D_2, \ldots, D_N)'$, where

$$D_i = \begin{cases} 1 & \text{if } U_i \text{ is selected in the sample} \\ 0 & \text{if } U_i \text{ is not selected in the sample.} \end{cases}$$

In surveys where some form of randomization is used to choose the sample, entries in **D** are random variables, with $\Pr(D_i = 1) = \pi_i$, where π_i denotes the selection probability for U_i (or an adult in the example). Another set of variables linked to **U** is $\mathbf{R} = (R_1, R_2, \ldots, R_N)'$, where

$$
R_i = \begin{cases}
1 & \text{if } U_i, \text{ when selected for the sample, would respond} \\
& \text{and provide useful survey data (become a respondent)} \\
\\
0 & \text{if } U_i \text{ would not provide useful data if selected} \\
& \text{(become a nonrespondent).}
\end{cases}
$$

The unit U_i is said to respond if an acceptable measure of the Y variable is obtained as a result of the unit's participation. Associated with each member (adult) in R is a response probability, $p_i = \Pr(R_i = 1)$. Finally, associated with each member of **U** is the $(K + 1)$-dimensional vector of auxiliary variables $X_i = (X_{i0}, X_{i1}, \ldots, X_{iK})'$, assumed not to be subject to nonresponse, from which we define the $N \times (K + 1)$ matrix $\mathbf{X} = (\mathbf{X}_1, \mathbf{X}_2, \ldots, \mathbf{X}_N)'$. In the example, these auxiliary variables might be known predictors of DBP or possibly variables whose relationship to DBP is to be assessed (e.g., age, race, etc.).

A few additional comments are needed before proceeding further. One concerns the convention used to distinguish measures that apply to respondents from those that apply to nonrespondents. We use the subscript 1 with symbols to refer to respondents in the population or sample and the subscript 0 with symbols to refer to nonrespondents in the population or sample. A second point concerns our use of the symbol Σ to indicate summation and the symbol Π to indicate the product of values. The first letter appearing below or as a subscript to a Σ or Π will generally denote the subscript over which summation occurs, while the symbol above or as a superscript to a summation will refer to the size of the set of observations summed together. For example, $\sum_{i=1}^{N} Y_i$ represents summation of the Y variable over all N population members, whereas $\sum_{i=1}^{N_1} Y_i$ denotes summation over just the N_1 population members who would respond if selected. Conventions differing from this one will be defined as they appear. Finally, we assume in our discussion that there is only one variable of interest (Y variable) considered in the survey. Under these circumstances the distinguishable concepts of unit and item nonresponse, as defined in Chapter 6, become equivalent.

In this chapter we follow two lines of discussion. We begin by devoting a section each to the two general views of nonresponse, distinguished by assumptions about response probabilities (p_i). The so-called *deterministic view* of nonresponse assumes for all members of the population that $p_i = 0$ or $p_i = 1$, whereas for all population members $0 \le p_i \le 1$ under what we call the *stochastic view* of nonresponse. Although these two viewpoints are related (the former takes a more narrow view of p_i than does the latter), the current literature does not make clear which is more appropriate. Those following the stochastic view might correctly argue that the deterministic viewpoint is too confined, and perhaps an oversimplification of reality (Cochran, 1977:360), while those advocating the deterministic view could equally assert that failing to confine p_i causes practical problems with analysis, such as the need to estimate p_i (see Section 7.2). The debate over viewpoint seems likely to continue, however, until we better understand what causes people to participate fully in surveys. Unlike the stochastic selection process in probability sampling, which can be attributed explicitly to the randomization mechanism used to choose the sample, the stochastic behavior of **R** must be assumed. When sampling, moreover, the selection probability for each member of the population can in principle be specified before selection and is thereby known to the investigator. With nonresponse, on the other hand, it is impossible to know precisely the chances that a member of the population would (if selected) be located, solicited, and provide usable data. Response probabilities, at the current level of our understanding, can at best only be conjectured or estimated.

Under either view of nonresponse the decision as to whether or not to respond is clearly tied to steps taken by the investigator to locate and solicit participation and then to collect the data. The receptivity to the study among those selected is also crucial. For example, membership in the respondent or nonrespondent subgroup under the deterministic view will depend on such things as the mode of data collection (by mail, telephone, or in person), the number of call attempts, the nature of interviewer training, the subject matter of the survey, and various demographic characteristics (e.g., age) of sample members. On the other hand, the outcome of R_i and the value of p_i in the stochastic view will depend on these same characteristics of the study.

7.1 DETERMINISTIC VIEW OF NONRESPONSE

By presuming that members of the population are either certain to respond ($p_i = 1$) or not respond ($p_i = 0$), the deterministic view of nonresponse removes any uncertainty from the decision as to whether or not each member of the population would provide usable data for the survey, if selected. Each member of the population can be considered as having been labeled "respondent" or "nonrespondent," or assigned to a respondent or nonrespondent subclass, prior to the survey. Thus the response decision is predetermined and each member of the population, if selected in the sample, would either always respond or never respond. One might alternatively consider the deterministic view to be one in which we have conditioned on the outcome of the R variable for each member of the population. The deterministic view then becomes a conditional form of the stochastic view.

Suppose that the population of N units is assumed to consist of two mutually exclusive and exhaustive subgroups, one with $N_1 = \sum_{i=1}^{N} R_i$ units, which with certainty would respond, and $N_0 = N - N_1$ units, which with certainty would not. Since the R_i must always be either zero or 1, N_1 does not vary over repeat applications of the response process. The proportions in the respondent subgroup, $\lambda_1 = N_1/N$, and nonrespondent subgroup, $\lambda_0 = 1 - \lambda_1$, would depend on characteristics of the study (as mentioned earlier). We resist using the term *stratum* to define the two subgroups since neither is a subgroup in which separate independent samples are chosen, as is the case with stratified sampling [see Cochran (1977, Chap. 5)]. Rather, these subgroups are to be viewed more as subclasses or domains of the population.

Historically, much, though not all of the early work in nonresponse adopted a deterministic viewpoint. For example, Hansen and Hurwitz (1946) in their landmark article on the use of double or two-phase sampling to handle nonresponse follow a deterministic view by assuming N_1 and N_0 to be fixed quantities. In the initial sample of size n selected from the population of N units, they treat the number of sample respondents, $n_1 = \sum_{i=1}^{n} R_i$, as variable due to sampling only rather than to both sampling and a stochastic response. Other early writing [e.g., Birnbaum and Sirken (1950)] makes the same assumptions about nonresponse as did Hansen and Hurwitz. The application of Bayesian methods to nonrespondent subsampling by Ericson (1967) and Singh and Sedransk (1978b), use of double sampling in the presence of frame error by Rao (1968), and extension of the idea of nonrespondent subsampling associated with each of several callbacks to respondents by Srinath (1971) are examples of more recent extensions of the work of Hansen and Hurwitz following a deterministic view. Many standard textbooks on sampling and survey methods adopt the deterministic view in their discussion of nonresponse; see, for example, Kish (1965), Zarkovich (1966), Raj (1968), Sukhatme and Sukhatme (1970), and Moser and Kalton (1972). Recent work in nonresponse seems to be more evenly divided between the deterministic and stochastic views, with perhaps more current interest in the latter. Reviews by Ford (1976) and Kalton (1983) present extensive analytic discussions of nonresponse and nonresponse compensation procedures developed from a deterministic perspective.

In assessing the utility of the deterministic view of nonresponse, one might conclude that it is easier to use in analytic work but perhaps somewhat naive in a practical sense. Its simple perspective of R_i with the specification that $p_i = 1$ or 0 allows respondents and nonrespondents to be treated as separate study domains. This situation leads to analytic formulations of survey error from nonresponse, which are more tractable than results from the stochastic view, thus yielding more informative expressions to the methodologist. On the other hand, one might reasonably argue that if the same protocol for location, solicitation, and data collection were repeated, a sample unit might respond on one occasion and then not respond on another occasion. The first time might find the eventual respondent available and in just the right frame of mind to participate fully, while on the second occasion a business trip might make participation impossible. Thus the outcome of R_i, and thus the size of p_i, can be viewed as being dependent on timing as well as other study characteristics mentioned earlier. One might therefore consider the effect of these characteristics on the response decision to be the outcome of a random process, and thereby treat R_i as a random variable. This then becomes the intuitive justification for the stochastic view considered in the next section. The deterministic view can then be treated as a conditional form of the stochastic view, since only one outcome of the assumed decision-making process for response is observed. Population members are assigned to the respondent or nonrespondent subgroup, depending on how they consciously or subconsciously weigh various influences at the moment when the decision to become a respondent or nonrespondent is made.

7.2   STOCHASTIC VIEW OF NONRESPONSE

Under the stochastic view of nonresponse each R_i is a random variable whose outcome is determined by an assumed chance element in the response process. Associated with each R_i is the response probability, p_i, which may differ among units in the population. A key feature of the stochastic view of nonresponse is the relationship between characteristics of the survey and the collective size of response probabilities. In most instances [e.g., Platek et al. (1977)], p_i is treated as a conditional probability since its size depends on circumstances surrounding the survey.

Under the stochastic view of nonresponse statistical inference must accommodate an additional source of randomization because of the stochastic response process. Thus, just as separate components of an estimator's mean square error reflect the stochastic mechanism used in sample selection or the assumed stochastic mechanism involved in generating the measurements Y_i, the mean square error of estimators derived from a stochastic view of nonresponse will contain separate error components because of nonresponse.

Before reviewing developments in the stochastic view of nonresponse, it is instructive to note that there are two alternative views of statistical inference from sample surveys. Both views have been used in dealing with nonresponse. One is the long-established *classical view*, in which the randomization resulting from the method of sample selection is the basis for making statements about the population from the sample. The other is the more recent *model-based view*, in which **Y** is considered a vector of random variables whose outcomes are the result of a stochastic process that can be expressed statistically in terms of an assumed model. In classical inference the population is seen as a fixed set of measurement **Y**, whereas in model-based inference the population is treated as if it were a random sample from some larger, more abstract "superpopulation" in which the Y variable behaves in some assumed stochastic manner (e.g., follows a normal distribution). We include Bayesian methods in discussing model-based inference since prior assumptions about the behavior of population parameters implies a stochastic rather than a fixed quality about a population being studied.

Research in nonresponse following the stochastic viewpoint and a classical approach to inference has been undertaken for some time. Among the first to treat nonresponse as a chance event were Politz and Simmons (1949), who developed [from an earlier idea by Hartley (1946)] an estimator in which crude estimates of individual response probabilities are used to compensate for the nonresponse. In this treatment p_i is seen as directly related to t_i, the number of times during the previous five days that the respondent would have been available to be interviewed, given that the call was made at the same time of day as the interview took place. Thus the estimate for p_i accounts for the likelihood of response attributable to location, but it does not account for the solicitation and data collection problems that may contribute to nonresponse. The Politz–Simmons estimator compensates for nonresponse by weighting each of the n_1 sample observations inversely proportional to its estimated response probability. The premise of the estimate of p_i, formulated specifically as $p_i = t_i + 1$, is that the likelihood of response can be tied to a person's recent availability. Estimating p_i in this way assumes that $p_i > 0$, since with no chance of responding for some, the effect of confirmed nonrespondents could never be addressed.

In another early use of the stochastic view of nonresponse Deming (1953:747) justified the use of multiple callbacks by viewing the population as consisting of "six classes, according to the average proportion of interviews that will be completed successfully out of eight attempts." The estimator of p_i by this view takes on different values from the Politz–Simmons estimator of p_i ($p_i = 0/8, 1/8, 2/8, 4/8, 6/8, 8/8$). Another early mention of a stochastic viewpoint was made by Dalenius (1961:4), who in connection with a discussion of schemes for reducing the effect of nonresponse, suggests that "one may associate with each person a set of probabilities for his being at home at different points in time." Consideration of these response propensities is needed to suggest survey strategies where to maximize response rates, larger sampling rates are applied to population members with large p_i.

Recent adaptations of the stochastic view define p_i more generally while incorporating R_i into various survey error models. Some extend earlier work from a classical view of inference, while others deal with nonresponse from a Bayesian or other model-based perspective.

We examine briefly a few examples of work following a classical view of inference. In a conceptual review of several classical methods for handling nonresponse, Platek et al. (1977) developed a so-called response–nonresponse error model incorporating contributions from measurement and nonresponse errors under the assumption that the R_i are mutually independent. The population here is viewed as a collection of units, each assigned a response probability. Lessler (1983) gives one extension to these results by adding the contributions of sampling to the error model. In another extension, Platek and Gray (1983) allow the members of **R** to be statistically related, as might be expected in practice.

Thomsen and Siring (1983) propose a stochastic nonresponse model within a multiple-call framework. Three possible outcomes are assumed on each call attempt by the interviewer:

Φ_1: Participation

Φ_2: No participation—call again

Φ_3: No participation—classify as nonrespondent

Given these outcomes the following parameters are associated with the ith member of the population:

$$p_i^{(1)} = \text{probability of } \Phi_1 \text{ on the first call}$$

$$f = \text{probability of } \Phi_2 \text{ on any call}$$

$$\Delta_i p_i^{(1)} = \text{probability of } \Phi_3 \text{ on the second or later call.}$$

From this the probability of participation on the cth call for the ith unit will be

$$p_i^{(c)} = \begin{cases} p_i^{(1)} & \text{if } c = 1 \\ \left(1 - p_i^{(1)} - f\right)\left(1 - \Delta_i p_i^{(1)} - f\right)^{c-2} \Delta_i p_i^{(1)} & \text{if } c > 1. \end{cases} \tag{7.1}$$

In the model Frankel and Dutka (1983) proposed, response probabilities in the population are assumed to follow a standard beta distribution with a continuous density function,

$$f(p) = p^{u-1}(1-p)^{v-1}\left[B(u,v)\right]^{-1}, \tag{7.2}$$

where $B(u,v)$ is a beta function with parameters $u > 0$ and $v > 0$ which determine the shape of the density function. Oh and Scheuren (1983) treat response decisions (outcomes for R_i) among members of the population as the outcome of independent Bernoulli processes whose probabilities are constant (i.e., $p_i = p$; $i = 1, 2, \ldots, N$) given these assumptions. The set of sample respondents can be treated as a with-replacement simple random sample of the initial sample. Oh and Scheuren also extend the model to one in which this uniform stochastic response mechanism is applied to various subclasses of the population. The p_i are assumed to be constant within subclasses but may vary between them. Finally, Williams (1978) defines response probabilities in assessing the bias from attrition in panel surveys used to produce measures of employment. Response probabilities are assumed to be equal within various subclasses, defined by a person's employment status in consecutive panels.

Much of the current thinking in work on nonresponse combines the stochastic view of nonresponse with model-based inference. For example, in discussing inference in the presence of nonresponse, Rubin (1983) alludes to Bayesian methods to arrive at a solution. A stochastic view here results from the response vector, $\mathbf{R}$, as a vector of random variables with prior specifications made about the "responding mechanism," $\Pr(\mathbf{R}|\mathbf{Y})$, where $\mathbf{Y}$ is the data vector. Inference may be achieved by determining the distribution of unobserved values of the Y variable given its observed values, the sampling design, and the pattern of missing data.

In another treatment of model-based inference, Cassel et al. (1983) trace the origin of sample-conditioned individual response probabilities through the following general stochastic mechanism. Let s denote a sample from the population of size N. Define $L_i = \{s : i \in s\}$ for each member of the population (U_i), and for each s define the probability of observing the set of sample respondents, s_1, as

$$q(s_1|s) \geq 0 \quad \text{for all } R_s = \{s_1 : s_1 \subset s\} \quad \text{such that} \quad \sum_{s_1 \in R_s} q(s_1|s) = 1.$$

The response probability for U_i, given s, is thus

$$p_{is} = \sum_{s_1 \in R_{is}} q(s_1|s) \quad \text{where } R_{is} = \{s_1 : i \in s_1 \subset s\}.$$

When the U_i respond independently (of each other and of s),

$$q(s_1|s) = \prod_{i \in s_1} p_i \prod_{i \in s_0} (1 - p_i), \tag{7.3}$$

where s_0 is the set of sample nonrespondents and $\prod_{i \in s_j}$ indicates that the product is taken over all members of $s_j (j = 0, 1)$. Through this general response mechanism Cassel et al. propose estimators that rely on a superpopulation model which establishes the stochastic behavior of the data vector, $\mathbf{Y}$, and a response model that describes the relationship between the p_i and certain auxiliary measures, $\mathbf{X}_i$, known also to be related to Y_i. A simplified review of the steps followed in estimation is the following:

1. Parameters of the response model are estimated from available data.
2. Estimates of the p_i are obtained from the fitted response model.
3. Estimates of the p_i are used to estimate parameters of the superpopulation model.
4. The fitted superpopulation model is used to produce the survey estimate.

As is evident from our illustrations, inference based on an assumed stochastic view of nonresponse depends heavily on the concept of the individual response probability. Assumptions about the existence and behavior of the p_i usually lead to estimators that are functions of p_i. Also of concern is that even with reasonable estimates of p_i, use of response probabilities in survey analysis can lead to extremely poor model-based estimates (see Section 8.1.6).

The difficulty involved in estimating p_i is often a major limitation of the stochastic view. For example, Dalenius (1983b) claims that it is unreasonable to expect one to postulate values of p_i when they are often at least partially explained by the circumstances whose effect on the likelihood of response is not quantified, or by factors external to the data (Y_i) and auxiliary measures ($\mathbf{X}_i$). Assuming this assertion to be correct, inference must be made conditional on the circumstances to the survey since different circumstances would contribute to different levels of p_i. Following this argument, estimates of p_i, based on an assumed relationship to $\mathbf{X}_i$, for example, would be expected to differ if things like the interviewer's demographic characteristics and competence or time of day called were changed for individual members of the sample.

Although one might argue that all inference is to some degree conditional, the problem of finding a reasonable approach to estimating p_i remains. Several have tried. Cassel et al., mentioned earlier, proposed fitting simple linear models involving the $\mathbf{X}_i$ (available for both respondents and nonrespondents) as the basis to estimation. Astin and Molm (1972) and Chapman (1976) suggest a similar strategy. Typically, the model involving the dependent variable (R_i) and independent variable ($\mathbf{X}_i$) is assumed to be linear and additive. Concrete justification for such assumptions is usually lacking, although Sarndal and Hui [in Krewski et al., (1981)], in a simulation study on estimating p_i by model fitting, demonstrate that estimates of means or totals will be unbiased (over repeated samplings of the population) if either the response or superpopulation model is correct.

In a somewhat different approach, Anderson (1979) suggests estimating p_i from controlled studies external to the survey, with the simplification that the sample is divided into H cells within which the p_i are assumed equal so that fewer estimates of p_i are needed. For most methods of estimating p_i the effect of using the estimate rather than actual value of p_i on the total mean square error of survey estimates is unknown.

7.3 EFFECT ON SURVEY ERROR WHEN ESTIMATING MEANS AND TOTALS

The object of most survey analyses is to estimate various population parameters. For example, we might wish to estimate for adults in some population the total number expected to buy a new product being marketed, the average number of dental visits during the past year, the proportion favoring a certain political candidate, or the coefficients of regression of several measures of stress on diastolic blood pressure. In our discussion we arbitrarily distinguish between two types of parameters. One type includes simple parameters such as totals and means (with proportions as a special type of mean), used primarily in preliminary descriptive studies of a population. The second type consists of more complex parameters such as correlation and regression coefficients, as well as covariances and variances, which are used in studies of the relationship between variables.

When nonresponse occurs, estimates of any type of parameter are subject to certain effects of nonresponse which may be measured, minimized, or otherwise accommodated by the investigator. Our main purpose in this section is to examine the effects of nonresponse on estimates of totals and means when some nonspecific form of compensation is made. We present general expressions from a stochastic view of nonresponse with the deterministic effects and other special cases used to illustrate how nonresponse effects are influenced by assumptions about the nature of nonresponse and other population characteristics. We present results separately for the variance and bias components of the mean square error of estimators. Our goal in Section 7.4 is similar, except that we consider the effects of nonresponse on estimates of more complex parameters, but exclusively from a deterministic viewpoint.

We hope that several lessons concerning the effects of nonresponse on survey estimates will emerge from the discussion. One is that nonresponse effects will vary depending on two important things: (1) the type of parameter being estimated, and (2) whether one assumes a stochastic or deterministic view of nonresponse. We shall see that the complexity of expressions for nonresponse effects is related directly related to the complexity of the estimator. For example, expressions related to an estimated regression coefficient (a complex estimator) will be found to be more involved than comparable expressions for an estimated mean (a simple estimator). We also demonstrate that nonresponse effects under a stochastic view can be simplified somewhat by considering the deterministic view as a special case. Finally, we demonstrate how criteria for assessing the effects of nonresponse differ depending on whether classical or model-based approaches to inference are followed.

7.3.1 Stochastic and Deterministic Effects on Estimated Totals

Following the approach Platek et al. (1977) suggest, we first present a simple yet illustrative series of findings on survey errors caused by nonresponse when estimating a total. Disregarding for the moment any effects of sampling and measurement error, a general expression for the effect of nonresponse is developed within the following framework. To begin, suppose that we do a complete enumeration of some population to estimate the total, $Y = \sum_{i=1}^{N} Y_i$, where Y_i is the "true" measurement for U_i (not subject to measurement errors). The survey is assumed to include two steps:

Step 1: Location and solicitation to obtain participation in the survey
Step 2: Data collection (measurement) followed by imputation of un-
usuable data

If the ith member of the population fails to provide useful data for Y_i, an imputed value, Z_i, which may be subject to errors, is used instead. Assuming a stochastic view of nonresponse, the observation actually used for the ith member is a random variable that may be expressed as

$$\hat{Y}_i = R_i Y_i + (1 - R_i) Z_i.$$

Also defined are individual random errors from imputation of nonrespondents,

$$\varepsilon_{0i} = Z_i - Y_i,$$

which are assumed to be mutually independent among population members. Finally, we define the simple imputation bias and variance for each population member as

$$B_{0i} = E_z(\varepsilon_{0i})$$

and

$$\sigma_{0i}^2 = \text{Var}_z(\varepsilon_{0i}),$$

respectively, where the operators $E_z(\cdot)$ and $\text{Var}_z(\cdot)$ denote expectation and variance, respectively, over repeat applications of the imputation methods (part of step 2) and observe that

$$E_r(R_i) = p_i$$
$$\text{Var}_r(R_i) = p_i(1 - p_i)$$

and where $E_r(\cdot)$ and $\text{Var}_r(\cdot)$ denote expectation and variance, respectively, over repeat applications of the survey protocol for location and solicitation (step 1).

Assuming the R_i to be mutually independent among population members, the general expression for the variance portion of the mean square error of the Platek et al. (PST) estimator of Y,

$$\hat{Y}_{PST} = \sum_{i=1}^{N} \hat{Y}_i,$$

is

$$\text{Var}(\hat{Y}_{\text{PST}}) = \sum_{i=1}^{N} p_i(1 - p_i)B_{0i}^2 + \sum_{i=1}^{N}(1 - p_i)\sigma_{0i}^2. \qquad (7.4)$$

The first term of Eq. (7.4) is attributed to the stochastic nature of nonresponse, while the second term comes from the stochastic variation in the Z_i, the imputed values. In the special case where $p_i = p(i = 1, 2, \ldots, N)$ we note that given values of B_{0i} and σ_{0i}^2, the first term will be maximum when $p = 0.5$ and minimum when either $p = 0$ (unrealistic) or $p = 1$ (complete participation). On the other hand, the second term is maximum when $p = 0$ and minimum when $p = 1$. The variance expression for $\hat{Y}_{\text{PST}}$ can be simplified somewhat in several realistic special cases. For example, when adopting a deterministic view of nonresponse, where $p_i = 0$ for N_0 population members and $p_i = 1$ for $N_1 = N - N_0$ members, the first term in Eq. (7.4) vanishes and $\text{Var}(\hat{Y}_{\text{PST}}) = \sum_{i=1}^{N_1}\sigma_{0i}^2$. The Var $(\hat{Y}_{\text{PST}})$ is also simplified if all of the Z_i are constant (e.g., all $Z_i = 0$) when the missing data are ignored, in which case

$$\text{Var}(\hat{Y}_{\text{PST}}) = \sum_{i=1}^{N} p_i(1 - p_i)Y_i^2,$$

which in turn becomes zero when the deterministic value is assumed).

The bias portion of the mean square error of $\hat{Y}_{\text{PST}}$ assuming a stochastic view of nonresponse is

$$\text{Bias}(\hat{Y}_{\text{PST}}) = \sum_{i=1}^{N}(1 - p_i)B_{0i}. \qquad (7.5)$$

We note from Eq. (7.5) that the magnitude of the bias of $\hat{Y}_{\text{PST}}$ is jointly dependent on the likelihood of response and the quality of imputation (when necessary) associated with each population member. Considering the event once again where all p_i equal p, the absolute value of the bias associated with $\hat{Y}_{\text{PST}}$ will be minimum (zero) when $p = 1$ and maximum when $p = 0$ for any set of simple imputation biases that are either all positive or negative. Following a deterministic view, we see that Eq. (7.5) reduces to $\text{Bias}(\hat{Y}_{\text{PST}}) = \sum_{i=1}^{N_0}B_{0i}$. In the no-adjustment situation ($Z_i = 0$) in which only unadjusted respondent data are used to estimate Y and the stochastic view is assumed, $\text{Bias}(\hat{Y}_{\text{PST}}) = -\sum_{i=1}^{N}(1 - p_i)Y_i$, which simplifies to $\text{Bias}(\hat{Y}_{\text{PST}}) = -\sum_{i}^{N_0}Y_i$ when assuming the deterministic view.

The bias of $\hat{Y}_{\text{PST}}$ from a deterministic view can be rewritten as

$$\text{Bias}(\hat{Y}_{\text{PST}}) = \sum_{i=1}^{N_0}B_{0i} = N_0(\bar{Z}_0 - \bar{Y}_0) = N(1 - \lambda_1)(\bar{Z}_0 - \bar{Y}_0), \quad (7.6)$$

where $\bar{Z}_0 = \sum_{i=1}^{N_0} Z_i/N_0$. This expression simply illustrates the rationale for a two-pronged attack to reduce nonresponse bias. One objective is to develop a survey protocol where the expected response rate is high (approaching 100 percent), and the other is to identify an imputation scheme in which the imputed value, on average, will equal the mean among nonrespondents. In the event that the mean among respondents is substituted for each nonrespondent ($Z_i = \bar{Y}_i$), Eq. (7.6) reduces to the well-known form

$$\text{Bias}(\hat{Y}_{\text{PST}}) = N(1 - \lambda_1)(\bar{Y}_1 - \bar{Y}_0). \qquad (7.7)$$

Although it is unrealistic to expect that $(\bar{Y}_1 - \bar{Y}_0)$ will be unaffected by λ_1 (Moser and Kalton, 1972, Sec. 7.4), it is usually difficult to predict how increasing the response rate will affect the respondent–nonrespondent difference in means. On the one hand, it may be reasonable to expect that $(\bar{Y}_1 - \bar{Y}_0)$ will vary directly with λ_1 since those remaining as nonrespondents as λ_1 increases would be hard-core respondents, who may be more likely to differ from respondents. On the other hand, an inverse relationship between $(\bar{Y}_1 - \bar{Y}_0)$ and λ_1 might seem reasonable to expect if members of the set of nonrespondents have similar values of the Y variable but differ collectively from the values of the Y variable for respondents. Regardless of its relationship to λ_1, the size of respondent–nonrespondent differences has rarely been the focus of efforts to reduce nonresponse bias. Instead, the more common preventive cure for nonresponse bias has been to develop a survey protocol that will yield the largest possible response rate, and thus to maximize λ_1.

Equation (7.6) also reveals that nonresponse bias becomes zero if those who would respond are a random subset of the population. In this case we expect $\bar{Y}_1 - \bar{Y}_0 = 0$, so that $\text{Bias}(\hat{Y}_{\text{PST}}) = 0$. Unfortunately, the majority of empirical studies believe these assumptions. Respondents tend to be atypical population members who are more highly motivated to respond and thus collectively different from nonrespondents (Kalsbeek and Lessler, 1978).

An upper-limit expression for the size of $\text{Bias}(\hat{Y}_{\text{PST}})$ relative to the size of Y, can be obtained from Eq. (7.7). Assuming that the Z_i are collectively at least as good as substituting Y_1 for all nonrespondents, that is, $|\bar{Z}_0 - \bar{Y}_0| \le |\bar{Y}_1 - \bar{Y}_0|$, we have

$$\text{Relative bias}(\hat{Y}_{\text{PST}}) \le \frac{(1 - \lambda_1)(1 - \theta)}{\lambda_1 + \theta(1 - \lambda_1)},$$

where $\theta = \bar{Y}_0/\bar{Y}_1$. For example, in a survey with an expected response rate of 80 percent and the mean for nonrespondents half that for respondents, $\hat{Y}_{\text{PST}}$ will tend to be an overestimate, but its bias (allowing for the assumption) will be no more than about one-ninth the size of Y.

Of course, the assumptions above are overly simplified. When U_i responds and provides usuable data but the observed value of the Y variable, W_i, is also subject to mutually independent stochastic errors of measurement, $\varepsilon_{1i} = W_i - Y_i$, the complexity of variance and bias components of the mean square error increases. Platek et al. (1977) also define $B_{1i} = E_t(\varepsilon_{1i})$ and the simple response variance, $\sigma_{1i}^2 = \mathrm{Var}_t(\varepsilon_{1i})$, where the operators $E_t(\cdot)$ and $\mathrm{Var}_t(\cdot)$ denote expectation and variance, respectively, over repeat applications of the protocol for survey measurement (part of step 2 in the PST framework) among population members. The bias of the estimated total,

$$\hat{Y}_{\mathrm{PST}}^* = \sum_{i=1}^N \hat{Y}_i = \sum_{i=1}^N \left[R_i W_i + (1 - R_i) Z_i \right],$$

under this set of assumptions has an additional term because of the bias of measurements, while the variance of the estimator has new terms from the variation in measurement errors and the covariance between measurement and imputation errors. When no adjustment for nonresponse is made, $Z_i = 0$ and thus $\varepsilon_{0i} = -Y_i$ for each nonrespondent. Thus

$$\mathrm{Bias}\left(\hat{Y}_{\mathrm{PST}}^*\right) = \sum_{i=1}^N p_i B_{1i} - \sum_{i=1}^N (1 - p_i) Y_i, \qquad (7.8)$$

whose first right-hand term is the additional response bias from measurement error. However, the second term will be larger than the first unless the p_i are generally high and measurement errors are high. With the no-adjustment assumption, the variance component is

$$\mathrm{Var}\left(\hat{Y}_{\mathrm{PST}}^*\right) = \sum_{i=1}^N p_i(1 - p_i)(Y_i + B_{1i})^2 + \sum_{i=1}^N p_i \sigma_{1i}^2, \qquad (7.9)$$

where the first and second terms are the nonresponse and response variances, respectively. There is no specific component because of imputation here since $Z_i = 0$.

A pattern quickly emerges in dealing with these error models. As each source of survey error (e.g., from sampling) is added, that source along with its relationship to the other existing sources contributes more components to the mean square error of estimates. Platek and Gray (1983) consider a model for estimating totals where errors caused by nonresponse, measurement, and sampling are considered jointly with interaction between the three error types assumed to be zero. Both analytic and numerical results are given for several nonresponse compensation strategies in this setting. One is a version of the Horvitz and Thompson (1952) estimator of Y,

$$\hat{Y}_{PG} = \sum_{i=1}^{n} R_i W_i \pi_i^{-1},$$

where π_i is the selection probability for U_i. This estimator is *design unbiased*, meaning that the bias from the method of sampling is zero, or that $E_s(\hat{Y}_{PG})$ = Y, where $E_s(\cdot)$ denotes expectation over all possible samples. Nonzero bias terms from other sources or error may remain, however. Thus $\text{Bias}(\hat{Y}_{PG})$ = $\text{Bias}(\hat{Y}_{PST})$, as given by Eq. (7.8).

The variance of $\hat{Y}_{PG}$ is the sum of the following components:

1. Because of the randomization procedure used in sampling,

$$SV(\hat{Y}_{PG}) = \sum_{i=1}^{N} Y_i^{*2}(\pi_i^{-1} - 1) + \sum_{i \neq j}^{N} Y_i^* Y_j^* \left(\frac{\pi_{ij}}{\pi_i \pi_j} - 1 \right) \quad (7.10)$$

which is the usual variance of the Horvitz–Thompson estimator but with $Y_i^* = p_i(Y_i + B_{1i})$ replacing Y_i, where $\pi_{ij} = \Pr(D_i = 1, D_j = 1)$ is the joint selection probability for U_i and U_j, and where $\sum_{i \neq j}^{N}$ denotes summation over all possible pairs of population members.

2. Because of measurement error,

$$RV(\hat{Y}_{PG}) = \sum_{i=1}^{N} p_i \sigma_{1i}^2 \pi_i^{-1} + \sum_{i \neq j}^{N} p_{ij} \sigma_{1ij} \pi_{ij} \pi_i^{-1} \pi_j^{-1}, \quad (7.11)$$

where p_{ij} is the joint probability that both U_i and U_j respond and $\sigma_{1ij} = \text{Cov}_{r2}(\varepsilon_i, \varepsilon_j)$ is the simple covariance over repeated trials in data collection between measurement errors for two different population members.

3. Because of the assumed stochastic nature of nonresponse,

$$NRV(\hat{Y}_{PG}) = \sum_{i=1}^{N} p_i(1 - p_i)(Y_i + B_{1i})^2 \pi_i^{-1}$$

$$+ \sum_{i \neq j}^{N} (p_{ij} - p_i p_j)(Y_i + B_{1i})(Y_j + B_{1j})\pi_{ij}\pi_i^{-1}\pi_j^{-1}. \quad (7.12)$$

The cross-product terms in Eqs. (7.10) to (7.12) arise from statistical associations in sampling, measurement, and nonresponse errors among different members of the population. If errors from these three sources are assumed to be independent, the second terms disappear, leaving variance components arising only from the sum of individual variances associated with each member of the population.

Assuming uncorrelated terms for individual members of most survey populations is seldom realistic. For example, the response–nonresponse outcomes of R_i and R_j for two population members in close proximity, either geographically or demographically, may have a positive statistical relationship because similar people would tend to respond similarly to attempts at location and solicitation. In like manner, the measurement errors ε_{1i} and ε_{1j} for different population members may be positively related if U_i and U_j are similar in some respect or if the same interviewer is used to obtain measurement from both. Finally, the sampling designs used in practice would rarely justify the assumption that $\pi_{ij} = \pi_i \pi_j (i \neq j)$ for all pairs of population members, although for some designs and some U_i and U_j (e.g., members of different strata in stratified simple random sampling) this relationship would hold.

7.3.2 Stochastic and Deterministic Effects on Estimated Means

The effects of nonresponse on an estimate of the population mean $\bar{Y} = \sum_{i=1}^{N} Y_i / N$ are in close parallel to the effects on an estimated aggregate total, Y. Because of the substantial overlap in the treatment of estimates for Y and $\bar{Y}$, we focus our attention on how these treatments differ.

It should be noted that much of the literature dealing with estimators of $\bar{Y}$ assumes the population size (N) to be known. In the more realistic event that N is unknown, one might use a reasonable estimator of N, such as $\hat{N} = \sum_{i=1}^{n} R_i p_i^{-1} \pi_i^{-1}$, in formulating an estimator for $\bar{Y}$ and then, to an order of approximation, assume that the components of the mean square error given below still hold.

When N is assumed known, the estimators for totals, $\hat{Y}_{\text{PST}}$, $\hat{Y}_{\text{PST}}^*$, and $\hat{Y}_{\text{PG}}$, can simply be divided by N to estimate $\bar{Y}$. Under these circumstances the bias components of the mean square error presented earlier would differ by a factor of N^{-1} and the variance components could be modified by multiplying by N^{-2}. For example, Nargundkar and Joshi (1975), following a stochastic view of nonresponse, examine an adapted Horvitz–Thompson estimator of $\bar{Y}$ for the no-adjustment situation,

$$\hat{\bar{Y}}_{\text{NJ}} = N^{-1} \sum_{i=1}^{n} R_i Y_i \pi_i^{-1}.$$

Since $\bar{Y}_{\text{NJ}}$ is design unbiased, the only remaining bias (from nonresponse) will be

$$\text{Bias}\left(\hat{\bar{Y}}_{\text{NJ}}\right) = -N^{-1} \sum_{i=1}^{N} (1 - p_i) Y_i. \tag{7.13}$$

Ignoring any interaction between sampling and nonresponse, the variance of $\hat{\bar{Y}}_{NJ}$ can be written as

$$\text{Var}\left(\hat{\bar{Y}}_{NJ}\right) = N^{-2}\left[\sum_{i=1}^{N} p_i^2 Y_i^2\left(\pi_i^{-1} - 1\right) + \sum_{i \neq j}^{N} p_i p_j Y_i Y_j\left(\pi_{ij}\pi_i^{-1}\pi_j^{-1} - 1\right)\right.$$

$$\left. + \sum_{i=1}^{N} p_i(1 - p_i)Y_i^2\pi_i^{-1} + \sum_{i \neq j}^{N}(p_{ij} - p_i p_j)Y_i Y_j\pi_{ij}\pi_i^{-1}\pi_j^{-1}\right], \quad (7.14)$$

with the first two terms due to sampling and the last two attributable to nonresponse.

Lessler (1983) considers an expanded version of the Nargundkar–Joshi model, assuming simple random sampling, treating measurement error following the model by Hansen et al. (1961), and allowing for interdependence among errors arising from sampling, measurement, and nonresponse. Although the variance of the estimator,

$$\hat{\bar{Y}}_L = \sum_{i=1}^{n}\left[R_i W_i + (1 - R_i)Z_i\right]n^{-1},$$

is rather formidable, the bias can be expressed simply as

$$\text{Bias}\left(\hat{\bar{Y}}_L\right) = \sum_{i=1}^{N} p_i B_{1i}N^{-1} + \sum_{i=1}^{N}(1 - p_i)B_{0i}N^{-1}, \quad (7.15)$$

where the first term of Eq. (7.15) is the bias component from measurement error and the second term arises from the bias of imputation associated with nonresponse. No bias term is linked to sampling here because $\bar{Y}_L$ is design unbiased when simple random sampling is used. Brooks (1982) considers sampling, measurement, and nonresponse in a model where a no-adjustment estimator is compared with an estimator following the principle of nonrespondent subsampling that Hansen and Hurwitz (1946) suggest and an adapted Politz and Simmons (1949) estimator in which response probabilities (p_i) are taken into account. Analytic expressions for these three estimators are found to differ markedly, with the nonrespondent subsampling estimator exhibiting the most complex mean square error because of the randomized subsampling. The effects associated with the Politz–Simmons estimator, while formulated more simply, are dependent on the p_i.

The size of nonresponse bias is known to depend on the amount of statistical association between response probabilities (p_i) and the measurements of interest (Y_i) in the population. This relationship can be seen most easily from a stochastic view when using $\hat{\bar{Y}}_{NJ} = \sum_{i=1}^{n} R_i Y_i \pi_i N^{-1}$ to estimate $\bar{Y}$. Ignoring the effect of measurement error and defining σ_{pY} to be the covariance between p_i and Y_i in the population, then from Eq. (7.13), we have

$$\text{Bias}\left(\hat{\bar{Y}}_{\text{NJ}}\right) = \sigma_{pY} - (1 - \bar{p})\bar{Y}, \tag{7.16}$$

where $\bar{p} = \sum_{i=1}^{N} p_i N^{-1}$ and $\sigma_p Y = \sum_{i=1}^{N} p_i Y_i N^{-1} - \bar{p}\bar{Y}$. We see from this result that nonresponse bias is positively bias is positively related to the amount of statistical association between p_i and Y_i (σ_{pY}) although not completely determined by it. In the complete-response setting (when $p_i = 1$; $i = 1, 2, \ldots, N$) the bias, as expected, becomes zero since $\sigma_{pY} = 0$ and $\bar{p} = 1$. Moreover, even when p_i and Y_i are uncorrelated in the population, nonresponse bias will be nonzero as long as nonresponse is possible ($\bar{p} < 1$).

The extent of nonresponse and the amount of respondent–nonrespondent difference can be shown to be interrelated in determining the size of nonresponse bias. The interrelationship is most readily established when using $\hat{\bar{Y}}_1$ to estimate $\bar{Y}$ in the deterministic setting. Assuming N_1 members of the population to have $p_i = 1$ and N_0 members with $p_i = 0$,

$$\sigma_{pY} = \sum_{i=1}^{N} Y_i N^{-1} - \lambda_1 \bar{Y} = \lambda_0(1 - \lambda_0)\left(\bar{Y}_1 - \bar{Y}_0\right). \tag{7.17}$$

Solving for $(\bar{Y}_1 - \bar{Y}_0)$ above, we see that the respondent–nonrespondent difference in this setting is functionally related to the extent of nonresponse (λ_0), provided that p_i and Y_i are not uncorrelated.

Discussions of the effect of nonresponse on survey estimates in most standard sampling texts [e.g., Cochran (1977), Kish (1965), Raj (1968); and Sukhatme and Sukhatme (1970)] consider the problem of estimating a mean, where a deterministic view of nonresponse is assumed. Under these circumstances we can more closely examine the nonresponse bias of estimates. First, recall from Section 6.2.3 that the bias of any estimator of $\bar{Y}$ using only respondent data ($\hat{\bar{Y}}_1$), provided that $E(\hat{\bar{Y}}_1) = \bar{Y}_1$, will be

$$\text{Bias}\left(\hat{\bar{Y}}_1\right) = E\left(\hat{\bar{Y}}_1\right) - \bar{Y} = \lambda_0\left(\bar{Y}_1 - \bar{Y}_0\right) \tag{7.18}$$

since $\bar{Y} = \lambda_1 \bar{Y}_1 + \lambda_0 \bar{Y}_0$, where λ with a subscript generally denotes the proportion of the population who would respond (λ_1) or not respond ($\lambda_0 = 1 - \lambda_1$), and $\bar{Y}$ with a subscript denotes the mean in the population among respondents ($\bar{Y}_1$) and nonrespondents ($\bar{Y}_0$). Moreover, if we assume that N is both fixed and known, the nonresponse bias for the estimator, $\hat{Y}_1 = N\hat{\bar{Y}}_1$, of Y will be $\text{Bias}(\hat{Y}_1) = N \text{Bias}(\bar{Y}_1)$.

Direct estimates of $\text{Bias}(\hat{\bar{Y}}_1)$ in Eq. (7.18) are obtainable in a sample survey if, through extraordinary means, a subsample of nonrespondents can be convinced to participate (see Section 8.1.4). Estimates of $\bar{Y}_1$ and $\bar{Y}_0$ can then be obtained using data from the initial respondents and the nonrespondent subsample, respectively. Since λ_0 is not normally known in practice, it too must be estimated from the sample. For equal-probability sampling designs (where $\pi_i = \pi$, for $i = 1, 2, \ldots, N$)$l_0 = 1 - l_1$ can be used to estimate λ_0, where l_1 is the proportion of eligible sample members who responded before nonrespondent subsampling (see Section 6.2.2). When n in these designs is also fixed (e.g., in simple random samples or in certain multistage designs employing probability proportional to size sampling), l_0 will be a design-unbiased estimator of λ_0.

7.3.3   Model-Based Effects

Effects of nonresponse on estimates of totals and means have also been studied when model-based methods are used. As we have already seen, the classical methods of statistical inference differ from model-based methods; with the former, the vector of measurements **Y** is treated as a constant, whereas with the latter, **Y** is a vector of random variables whose outcome depends on a stochastic generation process in the form of a "superpopulation model."

Cassel et al. (1983) make several important points regarding the distinction between classical and model-based approaches for dealing with nonresponse. Other discussions of the classical versus model-based approaches to statistical inference are given by Smith (1976), Cassel et al. (1977), and Hansen et al. (1983).

Under both approaches the statistician is forced to rely on, and is thereby vulnerable to, assumptions made necessary because of nonresponse. The two approaches are distinguished by the type of assumptions they make in dealing with the nonresponse. The classical analyst relies on the randomization mechanism used to select the sample while making assumptions concerning the relationship between respondents and nonrespondents. One common assumption is that the means for respondents and nonrespondents are equal within certain subclasses or weighting classes. To the extent that subclasses can be formed so that this assumption reflects reality, the analyst succeeds in reducing the effects of nonresponse. The model-based analyst, by contrast, deals with the nonresponse problem by relying on assumptions concerning the stochastic nature of **Y**. For example, a simple model might be assumed in which $Y_i = \beta X_i + \varepsilon_i$, with ε_i denoting random modeling error such that $E_m(\varepsilon_i) = 0$ and $\text{Var}_m(Y_i) = \sigma^2 X_i (i = 1, 2, \ldots, N)$, where $E_m(\cdot)$ and $\text{Var}_m(\cdot)$ denote expectation and variance over the model, respectively, $X_i > 0$ is known, and β and σ^2 are unknown. Inference in estimating $\bar{Y}$ might be based exclusively on this model and data from the n_1 respondents, with limited concern as to how the original sample of size n was selected or how nonresponse reduced to n_1 the number of sample observations. Analysis is in effect conditioned on the sample chosen and those among the sample who respond, implying that nonresponse is not a factor in inference, except that lowering the sample size from n to n_1 would decrease the precision of estimates. To the extent that the assumed model holds and its parameters are estimable, the analyst succeeds in dealing with nonresponse. To some, model-based inference also resembles classical inference with regard to the desirability of "balanced" samples, which, by using classical techniques such as stratification and controlled selection, are broadly representative of all important population subgroups.

Little (1983) gives one version of the model-based approach to inference with nonresponse present. His version uses the principle of maximum likelihood through prior assumptions concerning the stochastic behavior of the vector of binary sample indicators (**D**), the vector of binary response indicators (**R**), and the vector of measurements of interest (**Y**). The steps of inference for dealing with totals and means would be as follows:

1. Specify a joint distribution for **D**, **R**, and **Y** with an associated vector of parameters $\boldsymbol{\theta}$.
2. The distribution in step 1 is used to establish the distribution for the set of sample members (s) chosen from the population of units (**U**).
3. The outcome observations for the distribution of step 2 is applied to s.
4. The maximum likelihood principle is applied to establish which values of $\boldsymbol{\theta}$ are most likely to have produced the observations in step 3.
5. The maximum likelihood estimate of $\boldsymbol{\theta}$ is used to predict values for members of **Y** for those who were sampled but did not respond and for those who were not sampled. The new vector **Y** based on sample respondents plus predicted values might be denoted by $\hat{\mathbf{Y}}$.
6. The sum and average among members of $\hat{\mathbf{Y}}$ would be used to estimate Y and $\overline{Y}$, respectively.

The role of the sampling design (the distribution of **D**) in the model above depends on the design's "ignorability." An ignorable model is one in which the sampling design is assumed not to play a direct role in inference. For example, in an ignorable model such as the one given above, an assumed joint distribution for **R** and **Y** (but not **D**) would be all that is needed in step 1. A *nonignorable model*, on the other hand, is one in which some kind of assumption is made concerning the stochastic behavior of **D**. We also note in the version above of model-based inference that the estimate of Y or $\overline{Y}$ is a by-product of prediction arising from out estimate of $\boldsymbol{\theta}$. By contrast, classical estimation of Y or $\overline{Y}$ is the direct result of inference based on the observed sample data and the known distribution of **D**.

Unlike the classical approach, where components of the mean square error can be formulated to indicate the effect of nonresponse on estimates of Y and $\overline{Y}$, our discussion of the effect of nonresponse when the maximum likelihood approach to model-based inference is used will be limited to mentioning two practical limitations of this approach. One is a procedural problem tied to there being certain realistic assumptions about the distributions of **Y** and **R** in the ignorable case that yield a nonunique solution to the system of likelihood equations used to identify the maximum likelihood estimate of θ. When this happens, we do not know if our estimate of θ is the one most likely to have produced the observed data. The second limitation is the vulnerability of inference to misspecified assumptions about the stochastic behavior of **Y**. False assumptions about stochastic process leading to **Y** may in turn lead to unfounded maximum likelihood estimates of Y and $\bar{Y}$. The quality of estimates obtained by model-based methods must therefore be tied to our ability to withstand incorrect assumptions about the stochastic behavior of **Y**. Rubin (1983) suggests that one way of addressing the model-dependency issue is to specify several different models and to check for consistency among estimates—a costly solution in large, complicated surveys.

Cassel et al. (1983) present a version of model-based inference more closely tied to the classical approach. Their version considers errors from sampling and nonresponse but excludes measurement errors. In addition to the assumptions of the classical approach, they add a random source of error arising from the linear model associated with **Y**. In particular, specify Y_i and Y_j independent ($i \neq j$) with $Y_i = \beta \mathbf{X}_i + \varepsilon_i$ such that $E_m(\varepsilon_i) = 0$ and $\mathrm{Var}_m(Y_i) = \sigma^2 v_i$, where for all i the auxiliary measures $\mathbf{X}_i = (X_{i0}, X_{i1}, \ldots, X_{iK})'$ are assumed known; v_i is dependent on $\mathbf{X}_i$; σ^2 and the vector of regression coefficients $\beta' = (\beta_0, \beta_1, \ldots, \beta_K)'$ are unknown; and in models with an intercept, $X_{i0} = 1$ for all i. In no-intercept models, $X_{i0} = 0$. Since because of the assumed models, errors will contribute to the total mean square error of survey estimates, the concept of modeling bias must be added to biases from sampling and nonresponse. An estimator $(\hat{\bar{Y}})$ of $\bar{Y}$ is said to be *model unbiased* if $E_m(\hat{\bar{Y}}) = \bar{Y}$ for any initial sample s and any subset of respondents s_i that may be generated assuming the model specified. Model unbiasedness is seen to be less useful than design or nonresponse unbiasedness since the former is tied to the assumed model, and an estimator that is model unbiased for one model may not be unbiased for another model.

The central feature of the model-based approach is to predict a value for Y_i if the ith population member is not sampled and, when nonresponse is present, to predict a value of Y_i for those sampled but not responding as well. In the Cassel et al. (CSW) framework the prediction estimator of $\bar{Y}$ in the absence of nonresponse and with N known is

$$\hat{\bar{Y}}_{\mathrm{CSW}} = N^{-1} \sum_{i=1}^{N} \hat{\beta}' \mathbf{X}_i. \tag{7.19}$$

It is shown (Cassel et al., 1983:150) that $\hat{\beta}$ is a weighted estimator of β constructed from values of π_i, v_i, and $\mathbf{X}_i$ for members of the sample. When nonresponse is present a similarly weighted estimator based on sample respondents only ($\hat{\beta}_1$) replaces $\hat{\beta}$ in Eq. (7.19). In the event that v_i can be expressed as a linear function of the members of X_i for $i = 1, 2, \ldots, N$, $\bar{Y}_{\mathrm{CSW}}$

becomes equivalent to the classical regression estimator and has several notable properties:

1. Approximate design unbiasedness in the absence of nonresponse
2. Model unbiasedness for any possible sample and subset of respondents, provided that the model is true
3. Zero expectation over outcomes of the model for the bias attributed jointly to the design and nonresponse

The latter property is illustrated by considering a simple no-intercept model with $K = 1$ and simple random sampling (with sampling fraction $f = n/N$) as the design. The prediction estimator in the presence of nonresponse reduces to the usual ratio estimator,

$$\hat{\bar{Y}}_{\text{CSW}} = \bar{X}\left(\frac{\sum_{i=1}^{n_1} Y_i}{\sum_{i=1}^{n_1} X_i} \right) \quad \text{where } \bar{X} = \sum_{i=1}^{N} X_i/N.$$

For large samples the combined bias attributed jointly to design (sampling) and nonresponse will be

$$\text{Bias}_{sr}\left(\hat{\bar{Y}}_{\text{CSW}} \right) \doteq \bar{X}\left(\frac{\sum_{i=1}^{N} p_i Y_i}{\sum_{i=1}^{N} p_i X_i} \right) - \bar{Y}, \qquad (7.20)$$

whose expectation over the model will be zero and variance to order N^{-1} will be zero. Although this is a reassuring result, this model-unbiasedness property considered here unfortunately holds only when the assumed model is correct or the p_i and X_i are uncorrelated in the population. Since the second condition seldom holds in practice, the appropriateness of the model-unbiasedness property relies on the correctness of the model.

As an interesting sidelight, the prediction (ratio) estimator in this simple case can be expressed as $\bar{Y}_{\text{CSW}} = \bar{y}_1 A_s A_r$, where

$$\bar{y}_1 = \sum_{i \in s_1} \frac{Y_i}{n_1},$$

$$A_s = 1 + (1 - f)\left(\frac{\bar{x}_u}{\bar{x}} - 1 \right) \qquad (7.21)$$

is an adjustment for the effect of sampling,

$$A_r = 1 + (1 - l_1)\left(\frac{\bar{x}_0}{\bar{x}_1} \right) - 1 \qquad (7.22)$$

is an adjustment for the effect of nonresponse, $l_1 = n_1/n$ is the sample response rate, and $\bar{x}_u$, $\bar{x}$, $\bar{x}_1$, and $\bar{x}_0$ are, respectively, the means of the x variable for those unsampled, sampled, responding, and not responding.

From another model-based perspective, Schaible (1983) views model bias as conditioned on s and s_1 when nonresponse is present. *Incomplete data bias*, the term he used to define the difference between the model bias of the estimate in the presence of nonresponse ($\hat{\theta}_1$) and the model bias of the estimator that would occur if there was complete response ($\hat{\theta}$), can be expressed as $\text{Bias}_m(\hat{\theta}_1) = E_m(\hat{\theta}_1 - \hat{\theta})$. Following the original CSW framework with $v_i = 1$, $i = 1, 2, \ldots, N$, the incomplete data bias for the estimator, $\hat{\bar{Y}}_S = \sum_{i=1}^{n_1} Y_i/n_1$,

$$\text{Bias}_m\left(\hat{\bar{Y}}_S\right) = E_m\left(\hat{\bar{Y}}_S - \sum_{i=1}^{n} \frac{Y_i}{n}\right)$$

$$= \sum_{j=1}^{K} \beta_j\left(\bar{x}_{1j} - \bar{x}_j\right), \tag{7.23}$$

where $\bar{X}_j$ and $\bar{X}_{1j}$ are for the jth X variable the means for those sampled and responding, respectively. The model variance, $\text{Var}_m(\hat{\theta}_1 - \hat{\theta})$, is similarly defined as the variance, over outcomes of the model, of $\hat{\theta}_1$ minus the estimator, as if there were no sampling and complete response. For $\hat{\bar{Y}}_S$, the model variance would be

$$\text{Var}_m\left(\hat{\bar{Y}}_S\right) = \frac{(1-f)\sigma^2}{n} + \frac{(1-l_1)\sigma^2}{n_1}. \tag{7.24}$$

The first component of Eq. (7.25) is the usual sampling variance, and the second is variance due to nonresponse.

7.3.4 Numerical Studies

In addition to the analytic work discussed up until now in Section 7.3, several empirical studies have quantified the effects of nonresponse on survey estimates. Most of those following the deterministic view have been done in connection with existing surveys, while those following the stochastic view have had a more general application.

Among empirical studies following the deterministic view is one that Kalsbeek and Lessler (1978) report for the National Assessment of Educational Progress (NAEP), a large survey of elementary and secondary school students in which a standardized series of knowledge tests was administered. The goal of this nonresponse study was to determine how much estimates for 17-year-old students (the group with the lowest response rate) were affected by nonresponse. A subsample of nonresponding 17-year-old students was selected and the bias estimated by using data from the original group of sample respondents and the nonrespondent subsample. Findings indicated that the size of the nonresponse bias relative to the variance component of most survey estimates in this survey was high. As is so often the case in large studies, the NAEP nonresponse study demonstrated that bias was a significant component of the total survey of estimates since bias does not depend on sample size, whereas variance diminishes as the sample size increases. This study also demonstrated the degree to which the confidence level for interval estimates dropped when nonresponse was present. A direct relationship was found between the extent of nonresponse bias and the degree to which actual confidence levels were lower than the nominal level of 95 percent.

Several other numerical studies, mentioned here briefly, are discussed more fully in Chapter 8. Following a deterministic viewpoint, Williams and Folsom (1977) have studied the effects of nonresponse on estimates from the National Longitudinal Study, a national prospective survey of a cohort of high school graduates. The purpose of this empirical study was to determine the size of nonresponse bias remaining after substitute schools were used to deal with school nonresponse in the original sample. Some numerical effects of nonresponse have been obtained from studies following a stochastic view of nonresponse. For example, in the studies by Platek and Gray (1983) and Brooks and Kalsbeek (1982), estimates of the various components of bias and nonresponse are obtained from analytical results formulated under an assumed error model. Rubin (1977), on the other hand, assumes a stochastic view and uses Bayesian methods to produce confidence bounds on sample estimators produced as if complete response had occurred.

7.4   EFFECT ON SURVEY ERROR WHEN ESTIMATING MORE COMPLEX PARAMETERS

Having examined the effect of nonresponse on estimated means and totals, we now turn our attention to the impact that nonresponse may have on more complex statistics encountered in practice. The term *complex* is used loosely here to refer to any estimator that would be used to obtain point estimates of things other than simple means and totals. In particular, we examine the difference between two means, the simple element variance, the simple element covariance, the correlation coefficient, and the regression coefficient.

We review analytical and empirical studies of the effect of nonresponse on complex estimators briefly here. The analytic studies, based on the work of Santos (1981) and Kalton (1983), follow the classical approach to survey inference with a deterministic view of nonresponse and therefore focus on the nonresponse bias of each estimator. Simplifying assumptions such as simple random sampling and large sample approximations are also employed, to make formulas more tractable for discussion. Bias expressions presented in this section are limited to the case where only respondent data are used and no specific effort is made to compensate for the nonresponse, although Santos and Kalton determine the bias when various compensation procedures have been used. The aim of empirical studies, on the other hand, is not so much to estimate bias in measures for individual variables as it is to assess the degree to which relationships among several variables are affected by nonresponse. In some empirical studies this assessment is done by examining the degree to which a relationship among variables differs as respondents begin to resemble nonrespondents. For example, one might examine relationships among groups of respondents differing according to the number of call attempts required before a response was obtained [see Goudy (1976)]. As an alternative approach, the effect of nonresponse on complex estimators can be assessed by simulating patterns of missing data in existing surveys [see Santos (1981)].

The classical framework for the analytic results is as described previously for the deterministic view. A finite population of N elementary units, N_1, that would respond with certainty if selected, and $N_0 = N - N_1$, that would fail to respond with certainty. In a without-replacement simple random sample of n units, n_1 would respond and $n_0 = n - n_1$ would not.

Within this setting we consider first the matter of estimating the difference, $\bar{Y}_a - \bar{Y}_b$, between means in two population groups. Comparisons might typically involve the following in a stable population:

1. The same measurement in the same subgroup at two points in time
2. The same measurement in two different subgroups at one point in time
3. Two different measurements in the same subgroup at one point in time

The bias of the no-compensation estimator based on respondent data only, $\bar{y}_{1a} - \bar{y}_{1b}$, is

$$\text{Bias}(\bar{y}_{1a} - \bar{y}_{1b}) = \text{Bias}(\bar{y}_{1a}) - \text{Bias}(\bar{y}_{1b})$$
$$= \lambda_{0a}(\bar{Y}_{1a} - \bar{Y}_{0a}) - \lambda_{0b}(\bar{Y}_{1b} - \bar{Y}_{0b}), \qquad (7.25)$$

where the subscripts a and b extend previous notation to the two groups.

The bias of a type 1 estimated difference will often be small, for example, in a panel study in which the same survey design is applied to a population at two times (e.g., two successive installments of the Current Population Survey done monthly by the U.S. Bureau of the Census, 1978). The biasing effect of nonresponse will be similar at both points since the response rate, as the difference between respondent and nonrespondent means, is not likely to change much from one time to the next. This continuity is possible if the survey methods remain the same and the relationship between the Y variable and membership in the respondent or nonrespondent strata do not change. The nonresponse bias of an estimated type 2 difference may also be small if the two subgroups have similar response rates and respondent–nonrespondent differences. One example would be a study with the object of estimating the difference in the mean number of health care visits during the last year between persons living in two large urban communities. Because of the similarities in the two groups, one might expect the biasing effect of nonresponse to be similar as well. In type 3 comparisons $\lambda_{0a} = \lambda_{0b}$ (provided that there are not large differences in item nonresponse), the bias of estimated differences depends entirely on the relative sizes of the respondent–nonrespondent differences for the two measurements.

Next we consider the effect of nonresponse on the no-compensation estimator of the element variance of the Y variable obtained from respondent data in a simple random sample. The bias of the estimator, $s_{1y}^2 = \sum_{i=1}^{n_1}(y_{1i} - \bar{y}_1)^2/(n_1 - 1)$, can be obtained by first noting that the element variance

$$S_y^2 \doteq \lambda_1 S_{1y}^2 + \lambda_0 S_{0y}^2 + \lambda_0(1 - \lambda_0)(\bar{Y}_1 - \bar{Y}_0)^2, \qquad (7.26)$$

where S_{1y}^2, S_{0y}^2, and S_y^2 are, respectively, the population element variances for the respondents, nonrespondents, and both groups combined. Thus since $E(s_{1y}^2) = S_{1y}^2$, the nonresponse bias of s_{1y}^2 becomes

$$\text{Bias}(s_{1y}^2) \doteq \lambda_0(S_{1y}^2 - S_{0y}^2) - \lambda_0(1 - \lambda_0)(\bar{Y}_1 - \bar{Y}_0)^2. \qquad (7.27)$$

This formula shows that for the element variance for sample respondents to be unbiased, the respondent and nonrespondent subgroups in the population must have identical means and element variances. The nonresponse bias of the mean is zero if only respondent and nonrespondent means were the same. In the event that element variances for the respondent and nonrespondent strata are equal, we see from Eq. (7.27) that s_{1y}^2 will tend to underestimate S_y^2 to the extent that $(\bar{Y}_1 - \bar{Y}_0)$ differs from zero.

When the Y variable is a 0–1 indicator variable of some attribute so that $s_{1y}^2 = n_1 p_1(1 - p_1)/(n_1 - 1)$, where p_1 is the proportion of sample respondents with the attribute as an estimator of the proportion in the population with the attribute, Eq. (7.27) becomes

$$\text{Bias}(s_{1y}^2) \doteq \lambda_0(P_1 - P_0)[(1 - 2P_1) + \lambda_0(P_1 - P_0)]$$

$$= \text{Bias}(p_1)[(1 - 2P_1) + \text{Bias}(p_1)]. \qquad (7.28)$$

where P_1 is the proportion with the attribute in the respondent stratum and P_0 is a comparable proportion in the nonrespondent stratum. In this case the nonresponse bias of s_{1y}^2 is zero if any of three conditions hold:

1. $\lambda_0 = 0$,
2. $P_1 = P_0$, or
3. $P_1 = (1 - \lambda_0 P_0)/(2 - \lambda_0)$.

Condition 3 implies that the attribute proportions among respondents and nonrespondents need not be equal for the nonresponse bias to vanish.

An approximate expression for the bias of an estimated element covariance arising from survey nonresponse is obtained by first noting that we can partition the covariance between the Y and X variables as

$$S_{xy} \doteq (1 - \lambda_0)S_{1xy} + \lambda_0 S_{0xy} + \lambda_0(1 - \lambda_0)(\bar{X}_1 - \bar{X}_0)(\bar{Y}_1 - \bar{Y}_0), \quad (7.29)$$

where S_{1xy} and S_{0xy} are, respectively, the element covariances for the respondent and nonrespondent strata. The respondent stratum here consists of all population members for whom useful data on both variables could be obtained if the member were chosen. All others are considered part of the nonrespondent stratum. Under these circumstances the nonresponse bias of the covariance estimator based solely on sample data from the n_1 sample members from the respondent stratum,

$$s_{1xy} = \frac{\sum_{i=1}^{n_1}(x_{1i} - \bar{x}_1)(y_{1i} - \bar{y}_1)}{n_1 - 1},$$

is to the same order of approximation as the estimated variance,

$$\text{Bias}(s_{1xy}) \doteq \lambda_0(S_{1xy} - S_{0xy}) - \lambda_0(1 - \lambda_0)(\bar{X}_1 - \bar{X}_0)(\bar{Y}_1 - \bar{Y}_0). \quad (7.30)$$

However, unlike the estimated variance, s_{1xy} does not necessarily underestimate S_{xy} if $S_{1xy} = S_{0xy}$ since the second right-hand term in Eq. (7.30) may either be positive or negative.

When both variables are 0–1 attributes, the element covariance among members of the respondent stratum is $S_{1xy} = P_{1xy} - P_{1x}P_{0y}$. The bias of s_{1xy} in this case becomes

$$\text{Bias}(s_{1xy}) = \lambda_0^2(P_{1x} - P_{0x})(P_{1y} - P_{0y})$$
$$\times \left[1 + \frac{P_{1xy} - P_{0xy} + P_{1x}P_{0y} - 2P_{1x}P_{1y} + P_{0x}P_{1y}}{\lambda_0(P_{1x} - P_{0x})(P_{1y} - P_{0y})}\right]. \quad (7.31)$$

Results presented above for estimated element variances and covariances have important implications for studies exploring the relationships among variables. As in the case with means and totals, nonresponse may have a significant effect on the variance–covariance structure of variables, thus affecting analyses such as regression modeling, which rely on estimates of this structure. As with uncompensated estimators of means and totals, the size of nonresponse bias for variance and covariance estimators is linked to respondent–nonrespondent differences. Moreover, the nonresponse bias associated with the no-compensation estimators for all four parameter types is approximately zero when (perhaps naively) we assume that the nonrespondents are a random subset of the sample. We see next that the nonresponse biases of estimated regression and correlation coefficients can be expressed approximately as functions of the biases of estimated element variances and covariances (Santos, 1981).

Suppose that we wish to determine the nonresponse bias for the no-compensation estimator of the simple linear model regressing the Y variable on a single X variable. In classical inference the single regression slope parameter is defined as $\beta_1 = S_{xy}/S_x^2$, which in the absence of nonresponse would be estimated as $\hat{\beta}_1 = s_{xy}/s_x^2$. When the Y variable is missing for some members of the sample (observations on the X variable are assumed available for all sample members), the no-adjustment estimator of β_1 would be $\hat{\beta}_{11} = s_{1xy}/s_x^2$, where S_{1xy} is the element covariance among those sample members with useful data on the Y variable. The relative bias from nonresponse may be expressed using the linear terms of a Taylor series expansion of $\hat{\beta}_{11}$ about S_{1xy}/S_x^2 as

$$
\begin{aligned}
\text{Relative bias}\left(\hat{\beta}_{11}\right) &= \frac{\text{Bias}\left(\hat{\beta}_{11}\right)}{\beta_1} \\
&\doteq \frac{S_{1xy} - S_{xy}}{S_x^2 \beta_1} \\
&= \text{Relative bias}(s_{1xy}),
\end{aligned} \tag{7.32}
$$

thus making $\text{Bias}(\hat{\beta}_{11}) \doteq \text{Bias}(s_{1xy})/S_x^2$. In the case where missing data exist for the X and Y variables in the same regression framework, the analyst may choose to use only those respondents with data on both variables. We may then note the effect of nonresponse for the estimator, $\hat{\beta}_{11}^* = s_{1xy}/s_{1x}^2$, of the simple regression coefficient of the Y variable on the X variable, β_1. Noting that $E(s_{1x}^2) = S_{1x}^2$ and $E(s_{1xy}) = S_{1xy}$ when no compensation for nonresponse is made, the relative bias of β_{11}^* can, to an order of approximation, be expressed as

$$
\begin{aligned}
\text{Relative bias}\left(\hat{\beta}_{11}^*\right) &\doteq \text{Relative bias}(s_{1xy}) \\
&\quad - \frac{\text{Relative bias}(s_{1x}^2)}{1 + \text{Relative bias}(s_{1x}^2)}.
\end{aligned} \tag{7.33}
$$

Before considering the extension to multivariate regression, we consider the nonresponse bias of a no-adjustment estimator for a zero-order coefficient of correlation between the X and Y variables. Assuming once again that only the Y variable has missing data, this correlation, $R_{xy} = S_{xy}/(S_x^2 S_y^2)^{1/2}$, would be estimated using only data from the respondent stratum as $r_{1xy} = s_{1xy}/(s_x^2 s_{1y}^2)^{1/2}$. A Taylor series linear approximation yields

$$\text{Relative bias}(r_{1xy}) \doteq \frac{S_{1xy}/S_x S_{1xy} - S_{xy}/S_x S_y}{R_{xy}}$$

$$= \frac{\text{Relative bias}(s_{1xy}) - \text{Relative bias}(s_{1y})}{1 + \text{Relative bias}(s_{1y})}, \quad (7.34)$$

where S_x, S_y, and S_{1y} denote the square roots of S_x^2, S_y^2, and S_{1y}^2, respectively.

Santos (1981) gave the trivariate extension to the result of Eq. (7.32) for the case where there are two independent variables, X_1 and X_2, for which the corresponding regression coefficients in a classical framework are

$$\beta_1 = \frac{S_{x_1 y} - S_{x_1 x_2} S_{x_2 y}/S_{x_2}^2}{S_{x_1}^2 (1 - R_{x_1 x_2}^2)} \quad (7.35)$$

and

$$\beta_2 = \frac{S_{x_2 y} - S_{x_1 x_2} S_{x_1 y}/S_{x_1}^2}{S_{x_2}^2 (1 - R_{x_1 x_2}^2)}, \quad (7.36)$$

where in general S_{cd} and R_{cd} denote, respectively, the covariance and correlation between c and d in the population. With missing data again limited to the Y variable, the corresponding no-compensation estimator of β_1 would be

$$\hat{\beta}_{11} = \frac{s_{1x_1 y} - s_{x_1 x_2} s_{1x_2 y}/s_{x_2}^2}{s_{x_1}^2 (1 - r_{x_1 x_2}^2)} \quad (7.37)$$

and of B_2 would be

$$\hat{\beta}_{21} = \frac{s_{1x_2 y} - s_{x_1 x_2} s_{1x_1 y}/s_{x_1}^2}{s_{x_2}^2 (1 - r_{x_1 x_2}^2)}, \quad (7.38)$$

where s_{cd} and r_{cd} are the sample counterparts of S_{cd} and R_{cd}, respectively, and s_{1cd} is an estimated covariance based solely on data from the respondent stratum. Using the Taylor series approximation once again, approximations for the relative bias of $\hat{\beta}_{11}$ and $\hat{\beta}_{21}$ are, respectively,

$$\text{Relative bias}(\hat{\beta}_{11}) \doteq \text{Relative bias}(s_{1x_1y})$$
$$- \frac{\text{Relative bias}(s_{1x_2y})R_{x_1x_2}R_{x_2y}/R_{x_1y}}{Q_1} \quad (7.39)$$

and

$$\text{Relative bias}(\hat{\beta}_{21}) \doteq \text{Relative bias}(s_{1x_2y})$$
$$- \frac{\text{Relative bias}(s_{1x_1y})R_{x_1x_2}R_{x_1y}/R_{x_2y}}{Q_2}, \quad (7.40)$$

where

$$Q_1 = \frac{R_{x_1y} - R_{x_1x_2}R_{x_2y}}{R_{x_1y}} \quad \text{and} \quad Q_2 = \frac{R_{x_2y} - R_{x_1x_2}R_{x_1y}}{R_{x_2y}}.$$

As evident for s_{1y}^2 and s_{1xy}, the no-compensation bias expressions given for estimators of regression and correlation coefficients reduce approximately to zero when those with missing data on the Y variable are seen as a random subset of the population. This follows by noting that when Relative bias(s_{1y}^2) are both zero, the relative bias expressions for $\hat{\beta}_{11}$, $\hat{\beta}_{11}^*$, r_{1xy}, and $(\hat{\beta}_{11}, \hat{\beta}_{21})$ reduce to zero.

In reviewing various complex estimators, we have seen that expressions for nonresponse bias are relatively complex and therefore more difficult to estimate than comparable bias expressions for means and totals. Because of these estimation problems, most empirical studies do not estimate the bias directly. Santos (1981), in addition to his analytic work, uses simulation to assess the effect of no-compensation estimates and the effectiveness of several compensating procedures applied to various univariate and multivariate statistics. His approach was to delete real data randomly from the data set of an income survey, while reflecting the original patterns of missing data, and then to apply various procedures to deal with the nonresponse. Mean deviations and relative biases were tabulated and compared among procedures. In another empirical study Goudy (1976) found differences in estimated multiple regression coefficients obtained using cumulative respondent data from successive waves of a mail survey. The premise here was that those responding in later waves of the study are perhaps reluctant respondents and thus more like nonrespondents than those who respond in earlier waves. It was thought that differences in estimates obtained from successive waves would indicate the presence of an effect nonresponse. In a similar study (Suchman, 1962), response reluctance for a sample member was measured by the number of rounds in a panel study that a person failed to respond. The potential for biased conclusions about the interrelationship among variables was investigated here by examining differences among domain estimates broken down by reluctance and other predictors (education, income, and reading frequency) of some measurement of interest (frequent television viewing). Suchman concludes that valid assessments of statistical relationships can be made even in the presence of low response rates. Similar studies have arrived at the same conclusion (Kivlin, 1965; Schwirian and Blaine, 1966), while others have found that nonresponse affects conclusions drawn from analyzed relationships (Lehman, 1963; Mulford et al., 1974; Thomsen and Siring, 1983). The reader is referred to Chapter 8, where many of these studies are discussed more fully.

Nonresponse: Dealing with the Problem

Understanding survey nonresponse and its statistical implications leaves unresolved the question of how to deal with the problems it creates for analysis. In Chapter 7 we saw expressions for nonresponse bias when estimates are computed from nonrespondent data alone without any effort to compensate for the nonresponse. The size of these no-compensation biases under a deterministic view of nonresponse depends on two factors: the expected response rate and the degree to which respondents and nonrespondents differ according to values of the Y variable, the measure of substantive interest in the survey. Therefore, it is not surprising to find that most methods for dealing with nonresponse address either or both of these determinants.

In this chapter we explore some options available to the researcher for dealing with nonresponse. In many surveys several of these options are chosen as part of a general strategy for handling the missing data problem. The options might be described briefly as follows:

1. *Do Nothing.* The researcher elects to produce estimates based only on respondent data. Choosing this option indicates that either the researcher is unaware of the implications of nonresponse or believes them to be negligible. In the latter case the observed response rate may be sufficiently high or the likelihood of respondent–nonrespondent differences in the Y variable sufficiently low to merit the assumption that nonresponse bias will be near zero and thereby ignorable. In some surveys with high response rates this may be a workable option, although an absence of respondent–nonrespondent differences requires the usually naive assumption that respondents are a random subset of the survey sample.

2. *Reduce the Number of Nonrespondents.* Under this option, the researcher, mostly through improvements in the data collection strategy, increases the likelihood of obtaining useful data from members of the sample.

3. *Substitution.* New sample members in this approach are substituted for unit nonrespondents as a means to maintain the intended sample size, although the bias from nonresponse will not usually be reduced.

4. *Assess the Potential for Bias.* This method calls for an analysis of survey response or an investigation of the potential for respondent–nonrespondent differences, possibly involving the calculation of response rates for various domains or a comparison or respondents and nonrespondents on auxiliary data.

5. *Compensate for the Problem.* This option encompasses a broad grouping of methods where the researcher, in consideration of the nonresponse issue, makes some kind of alternation in the sampling design or in how the survey data are treated for analysis. The alternation may take any of the following forms:

a. *Extrapolation.* Estimates applicable to nonrespondents are projected by observing domain estimates from respondents ordered by degree of resemblance to nonrespondents.

b. *Nonrespondent Subsampling.* Final estimates are obtained using data from respondents and a subsample of nonrespondents.

c. *Weighting Adjustment.* Respondent data are adjusted to account for the nonresponse that has occurred, and the weighting adjustments are then used in the analysis. Although theoretically applicable to both unit and item nonresponse, adjustments are most often used to handle the first type of missing data. For simple estimators of means and totals, one type of weighting adjustments is equivalent to imputing the mean of similar respondents, thus making some applications of weighting adjustment methods an implicit form of imputation.

d. *Explicit Imputation.* This class of compensation procedures is applied when some data items obtained from unit respondents are either missing or found to be unusable. Following a planned process, each missing data item is replaced by a numerical value that is intended as a prediction of the missing value.

e. *Model-Based Inference.* A statistical model involving assumptions about nonresponse and the formation of the sampled population is at the heart of this class of compensation strategies. Parameters of the model are estimated using respondent data, and the fitted model is then used to predict data items for nonrespondents and ultimately, to obtain population estimates.

Some of the compensation methods can be distinguished by the assumptions made about differences between respondents and nonrespondents. Following the deterministic view of nonresponse, some presume that the means of all respondents ($\bar{Y}_1$) and nonrespondents ($\bar{Y}_0$) are the same. Others assume survey respondents to be a random subset of the selected sample, in which case the expected value of survey measures, over outcomes of the sampling process, would be equal for respondents and nonrespondents. Finally, for some model-based methods, expected values of $\overline{Y}_1$ and $\overline{Y}_0$, over outcomes of the model, are presumed equal.

Our discussion of methods for dealing with the nonresponse problem generally follows the categories of options given above. We also distinguish between methods designed specifically for dealing with unit nonresponse and item nonresponse because the literature, despite the conceptual similarities between the two types of nonresponse, is largely segregated, for practical reasons. Methods for handling item nonresponse are those in which explicit efforts are made to improve completion rates for individual questionnaire items for respondents or where some well-defined process is followed to replace missing data items. All other methods presented in this chapter deal with the problem of unit nonresponse, although some (e.g., weighting class adjustments) could, although not very practically, be applied at the item level.

As in Chapter 7, our aim is to highlight and briefly explain some of the methods available in the literature. We give the reader a cursory view of what can be done to deal with nonresponse by reviewing material that we believe typifies available options. Expanded discussion on most topics is available through the many excellent articles and books cited in the text. In particular, we recommended the three-volume work by the Panel on Incomplete Data (Madow et al. 1983a, b; Madow and Olkin, 1983), Kalton (1983), and Cox and Cohen (1985).

8.1   UNIT NONRESPONSE

We first consider those approaches that have been designed to deal with nonresponse at the unit level, which in many surveys accounts for a larger portion of the missing data problem than does item nonresponse. The many and varying methods we review in this section generally fall into two classes; those where the goal is to avoid the problem by maximizing the unit response rate among sample members, and those where some form of compensation is used to reduce the effect on survey error of the unit nonresponse that ultimately occurs. Useful reviews on compensation methods have been produced by Chapman (1976), Kalton and Kasprzyk (1982), and Little (1982).

8.1.1   Prevention

Perhaps the best strategy for dealing with any problem is to keep it from becoming too large in the first place. Limiting bias associated with nonresponse requires steps to control either of its two determinants: respondent–nonrespondent differences and the response rate. Although somewhat affected by the response rate, the extent to which sample respondents and nonrespondents differ is difficult to control since the amount of difference depends on which population members would be respondents or nonrespondents. Moreover, membership in these two groups depends on the relationship between a member's likelihood of responding and the measurement of interest.

Although it is usually unreasonable to expect that respondent–nonrespondent differences can be controlled directly, the response rate as the second determinant of bias can be regulated by introducing data collection strategies which reduce the likelihood that a population member, if selected in the survey, would be a nonrespondent. A number of these prevention methods are summarized from more comprehensive discussions given by Kish (1965), Warwick and Lininger (1975), and Mosteller (1978).

More formally, the goal of preventive methods is collectively, and in some instances selectively, to increase response probabilities (p_i) in the population. From a stochastic perspective this has the effect of increasing the expected number of respondents in the population, $N_1 = \sum_{i=1}^{N} p_i$, and the expected response rate, $\lambda_1 = N_1/N$, over the outcomes of the random response process at work in the population. In the deterministic view, increasing response probabilities means convincing some nonrespondents ($p_i = 0$) to become respondents ($p_i = 1$). This activity has the effect of increasing N_1, the size of the respondent subclass, and $\lambda_1 = N_1/N$, the proportion of the population falling into it. Under both views the manifestation of λ_1 in the survey sample is the response rate, $l_1 = n_1/n$, where n is the sample size and n_1 is the number of respondents, which may be interpreted as either the number of sample members who respond as the result of the random response process operating on each sample member (stochastic view) or the number of respondent subclass members chosen in the sample (deterministic view).

While in principle one hopes for complete response in surveys, the realistic practitioner knows that this goal is usually unattainable. There then remains the matter of deciding what level of λ_1 is acceptable or, in more concrete terms, what is an acceptably high response rate. Perhaps the answer to this question cannot be given in absolute terms (e.g., 75 percent or higher as an acceptable response rate) but rather, is best couched in terms of several criteria. First the acceptability of a response rate may be tied to one's prior expectations in view of the survey topic, the population being studied, and the mode of data collection. A 75 percent response rate, for instance, would be considered quite acceptable in a self-administered mail survey of physicians on treatment practices but somewhat low for a personal interview survey of adults in the general population on some relatively innocuous health care issues. The topic, population, and mode of data collection in the latter survey all lend themselves to a greater likelihood of response than do the same three characteristics of the former survey.

Once the mode of data collection is established, the cost of steps to improve response rates is a second criterion that may have some bearing on deciding whether a response rate is acceptable. For example, an 80 percent response rate for a personal interview survey might be considered acceptable if increasing the response rate to 85 percent would cause the per-unit cost of data collection to double because of increased travel and follow-up efforts. The complexity of steps that would have been required to improve the response rate in a given survey must also be considered in resolving the issue of whether a reported rate is good enough.

The acceptability of a reported response rate also depends on whether and what types of methods have been used to compensate for nonresponse. One good illustration may be taken from mail surveys, where a 50 percent response rate would be deemed unacceptable if nothing else were done to deal with the nonresponse. On the other hand, this rate might be viewed quite differently if nonrespondent subsampling (Section 8.1.4) or a weighting class adjustment procedure (Section 8.1.5) were used. These procedures would serve to eliminate at least partially the negative effects of the initial response level, thus placing the 50 percent rate in a somewhat more favorable light.

Finally, the substantive importance of findings from the study must be considered in judging reported rates of response. Response rates of 80 to 90 percent may be unnecessary in low-budget surveys designed, for example, to confirm that a strong front-runner in a political race will be elected or that a popular local referendum will pass. On the other hand, a rate exceeding 90 percent would be essential for any survey that had important implications for a multibillion dollar health care program.

This and our earlier discussion on defining relative measures of nonresponse (Section 6.2.2) point to the need for caution when using the response rate as a criterion for evaluating the quality of a survey. Clearly, one must look beyond this often misreported and potentially misleading number to determine the net effect of nonresponse on data obtained in the study. To put a reported response rate into a proper perspective, it is important to consider how the rate was defined, how amenable the population studied would have been to the survey topic, what mode of data collection was used, whether cost-effective measures were taken to maximize the rate, whether other nonresponse compensation procedures were used, and the impact of the study's findings.

The remainder of this section is devoted to a brief presentation of the preventive strategies often followed in practice. Although we tie these strategies to location, solicitation, and data collection efforts in different types of surveys, the material presented here should be viewed only as a starting point for the survey researcher who is devising a plan to increase the response rate. The reader is referred to texts by Moser and Kalton (1972), Warwick and Lininger (1975), and Dillman (1978) to supplement our discussion.

The basis for discussion in the remainder of this section is Table 8.1. To minimize redundancy among the lists we have omitted the following three obvious strategies, which apply to most situations covered in the table:

Table 8.1 Preventive Methods to Use (When Feasible) in Dealing with Unit Nonresponse

Point of Application	Mode of Data Collection	
	Mail Self-Reporting	Telephone/Personal Interview
Location	More current mailing address[a] High-priority mailing[a]	More call attempts[a] Clearer interview assignment materials[a] Improved call scheduling
Solicitation	Special tracing efforts[a] Follow-up reminders[a] Privacy and confidentiality assurances Incentives[a] Endorsements[a] Proxy respondents[a] Refusal conversion strategies[a] Miscellaneous enhancements[a]	Special tracing efforts[a] More call attempts[a] Privacy and confidentiality assurances Incentives[a] Endorsements[a] Use of female interviewers[a] Lead letter Reduced interviewer work load Refusal conversion strategies[a]
Data collection	Brief and concise interview Simplified questionnaire format Avoidance of sensitive subjects	Brief and concise interview Avoidance of sensitive subjects

[a]Additional discussion in text.

1. *Selective Recruiting.* Hire staff with experience in dealing with survey nonresponse. For example, one usually prefers interviewers with satisfactory past experience in doing surveys similar to the one for which they are being recruited. The reasons for such preference are obvious. These individuals are more likely to understand the problems faced in obtaining a response from sample members and to have found ways to handle these problems.

2. *Improved Training.* Manuals and sessions devoted to interviewer training include some discussion on how to deal with problems related to nonresponse. Overlooking this aspect of training forces the interviewer to improvise, which may be less effective.

3. *Improved Staff Supervision.* Those in a position to solicit participation in the survey are supervised more closely to assure that methods established to prevent nonresponse are applied correctly.

Several methods listed in Table 8.1 are discussed briefly.

8.1.2 Preventive Methods

8.1.2.1 High-Priority Mailing
A mailed questionnaire would have a greater chance of reaching a sample member if the package is sent by registered or certified mail, where someone must sign for it. Mailgrams or telegrams might also be effective in notifying someone of having been selected into a survey. Kernan (1971) and Dillman (1978) discuss the relative merits of various types of postage.

8.1.2.2 More Call Attempts
In interview surveys a *call attempt* refers to an event in which an interviewer tries to contact a selected member of the sample with the aim of ultimately completing an interview. Allowing a larger number of call attempts before considering a sample member to be a nonrespondent usually increases the chances of response, provided that calls are conveniently scheduled for sample members. With each additional attempt we are more likely to find available those who are often unavailable, and to find more willing those who in previous attempts may have been reluctant to participate in the study. Considering that increasing the allowable number of call attempts reduces nonresponse bias (and the corresponding mean square error of an estimated mean) while increasing survey costs, Deming (1953), Durbin (1954), and Thomsen and Siring (1983) use somewhat different approaches to determine the optimum number of call attempts.

8.1.2.3 Clearer Interviewer Assignment Materials
The primary linkage between the sampling and data collection operations is through the information the former provides the latter to determine who in the survey population is to be studied. When this information is unclear or vague, the chances increase that a sample member will eventually become an unlocatable nonrespondent. To avoid this problem, assignment lists given to field staff should include sufficient information to enable the interviewer to distinguish uniquely each sample member from among the rest of the population. Some items that help to clarify interviewer assignments are correctly spelled full name, complete street address with house number or description, home or business telephone number, and name and address of current employer. Maps sent with interviewers will further improve the process of location in personal interview surveys.

8.1.2.4 Special Tracing Efforts

In longitudinal surveys where some study subjects move between rounds or in cross-sectional surveys where the information available to locate sample members is incomplete or outdated, intensive efforts may be needed to find an address or telephone number. *Tracing*, the process of locating sample members in these situations, involves pursuing leads based on information available from one or more of the following: the sampling frame, a previously completed questionnaire, directory assistance, informants, and publicly available records. The chances of successfully locating subjects is clearly related to the amount of available tracing information and the skill with which it is used.

8.1.2.5 Follow-up Reminders

Some time after a questionnaire is first sent to each sample member in mail surveys, a follow-up in the form of a postcard, letter, telegram, or telephone call might be made to encourage those not having yet responded to do so. Several follow-up procedures and their effectiveness in reducing the extent of the nonresponse problem have been reviewed by various authors [see, e.g., Scott (1961), Hochstim and Athanasopoulos (1970), Dillman (1972), Linsky (1975), Kanuk and Berenson (1975), Duncan (1979), and Rao (1983a, b)].

8.1.2.6 Incentives

Because participation in some surveys requires a significant commitment of a respondent's time and energy, some sort of incentive payment to those agreeing to participate may be offered to respondents (e.g., money, a gift, a redeemable coupon, or an offer to send a report of the survey's findings). A number of studies [e.g., Dohrenwend (1970), Ferber and Sudman (1974), Chromy and Horvitz (1978), Gunn and Rhodes (1981)] generally point to improved response rates in most surveys if an incentive of some kind is offered to those sampled.

8.1.2.7 Endorsements

In some surveys, especially those involving professionals and institutions, endorsements of relevant societies and professional organizations are sought to enhance the credibility of the study, thereby reducing suspicion and increasing the likelihood of response. Although endorsements have been widely used, Kalsbeek and Hartwell (1977) have demonstrated a point of diminishing returns, regarding the number of endorsements, after a few key ones have been received.

8.1.2.8 Use of Female Interviewers

For some interview surveys female interviewers will tend to achieve higher response rates than male interviewers because women tend to arouse less suspicion and pose a less of a threat to certain segments of the population (e.g., the elderly). Although many accept this rationalization, there is little strong evidence to suggest a relationship between response rates and the sex of the interviewer (Platek, 1977; Thompsen and Siring, 1983).

8.1.2.9 Lead Letter

A letter sent in advance of the initial contact in some interview surveys may enhance the credibility of the interviewer once the sample member is first contacted and thus increase the likelihood of response. This letter is best signed by someone whose name, title, or affiliation is held in high regard by members of the survey population. It should contain information designed to inform, assure, and motivate.

8.1.2.10 Proxy Respondents

A proxy respondent is a person who, although not selected in a survey sample, is considered a suitable substitute when the person actually selected cannot or should not participate. For example, in an interview survey any adult in the household who speaks English may be an allowable proxy for interviewing a non-English-speaking sample member. By using proxies nonresponse from inability on the part of the respondent will be less likely. On the other hand, use of proxies may contribute to increased measurement error since facts or opinions may be less accurately portrayed [see Roshwalb (1982)].

8.1.2.11 Refusal Conversion Strategies

Sample members refusing to participate based on an initial solicitation effort may in some instances be recontacted if it appears that there is some hope of converting them to respondents. The potential for conversion is not the same for everyone. Someone refusing because the interviewer called at a bad time might agree to participate if a convenient time could be found. A reluctant respondent might also be converted if the investigator can find a way to emphasize the study's credibility. On the other hand, someone emphatically expressing a distaste for surveys would be an unlikely participant under any circumstances.

Successfully converting a reluctant individual is tied to the reasons for refusal and to the investigator's ability to find arguments to refute these reasons. With this in mind, let us briefly examine how a conversion program might be set up in a survey. First, the investigator must attempt to learn *why* a potential respondent has initially refused for interview surveys. This can be done by training interviewers to probe gently for a reason by suggesting one for the respondent to confirm or revise (e.g., "Perhaps I've caught you at a bad time"). We have found that nonbelligerent refusers will often respond to this kind of probe by telling you why they are reluctant. In mail surveys it is difficult to learn who among nonrespondents are refusals, much less why they have refused. For that reason the conversion call must be by telephone or in person. Second, the investigator must develop an argument to deal with each possible reason for refusal and then find skilled interviewers to make the conversions. Those assigned to convert refusals might, for example, be given a script containing hopefully effective counterarguments (e.g., reasons why the survey is important for those who think participating would be a waste of their time). In most instances conversion attempts are most effectively made a few days after the initial refusal, since employing a hard-sell strategy with someone at the time of an initial refusal is often counterproductive. The change in time may find the refuser in a more receptive frame of mind or, in household surveys, another person who will help pave the way for participation.

8.1.2.12 *Miscellaneous Enhancements*
Several other prevention methods may be used to increase the chances of soliciting participation in mail surveys. The utility of some of these methods has been tested. For example, Kahle and Sales (1978) have investigated the appearance of the envelope and Henley (1976) and Roberts et al. (1978) the use of deadlines for returning completed questionnaires, in each case concluding that such features may often improve response. These and similar studies point to the importance of sending attractive personalized material through the mail, of making a firm but reasonable request of the respondent's time, and of making it easy for the respondent to return the questionnaire. Scott (1961) and Dillman (1978) contain numerous suggestions along these lines.

8.1.3 Identification Studies

Another way of dealing with unit nonresponse is to identify the *potential* for nonresponse bias by attempting to determine whether the likelihood of response is somehow related to the major study variables, or equivalently, whether and how respondents differ from nonrespondents. On the basis of these so-called identification studies, the researcher can qualitatively assess the impact of nonresponse on survey estimates by investigating patterns of response rates among subgroups of the sample. Equivalently, these studies may also be done by comparing respondents with nonrespondents or respondents with the total population. Sociodemographic characteristics (e.g., age, ethnicity, gender, and education) available for responding and nonresponding sample members are commonly used.

In studying response-rate differentials among subgroups of the sample, the object is to determine the relationship between the likelihood of response and auxiliary variables thought to be highly correlated with the Y variable. Observing that subgroup response rates are statistically associated with values of the auxiliary variables may indicate that estimates obtained using the Y variable are biased because of nonresponse. This bias occurs because an association between response rates and the auxiliary variables points toward a nonzero correlation between individual response probabilities and values of the Y variable, and ultimately toward bias in survey estimates (see

Section 7.3). More intuitively, a relationship between response rates and the auxiliary variable implies that distributions of the Y variable and auxiliary variables among sample respondents are not the same as their corresponding distributions among members of the full sample. If the distribution of the Y variable among respondents is skewed because of nonresponse, estimates produced from the respondent sample will be skewed as well.

Identification studies are usually more telling for samples chosen from lists of individual population members than for samples chosen in clusters, especially geographic ones, because there is often more useful auxiliary information on lists. This kind of identification study could, for example, be done in a survey on access to health care conducted for a large company. Assuming that demographic data were available for each employee, response rates might be compared for various subgroups defined by age, race, and insurance coverage since all three would presumably be related to accessibility. Response-rate comparisons in area samples, on the other hand, are often limited to categories of qualitative characteristics for the area units within which the sample was chosen (e.g., population density, regional location).

Survey respondents and nonrespondents are compared using auxiliary data in a second type of identification study. Simple descriptive measures of the auxiliary variables (e.g., means) are typically computed in making these comparisons. The criteria for choosing the auxiliary variables are the same as in studies of subgroup response rates. In area samples the distributions by geographic area, population density, or observed racial or selected area units might be compared for respondents and nonrespondents. In other samples one might contrast some descriptive measure, or the distributions of various sociodemographic variables. To illustrate the latter case, consider the results of a study by Kalsbeek et al. (1974) designed to determine the impact of nonresponse on 17-year-olds selected in the National Assessment of Educational Progress (NAEP), a longitudinal survey of the educational competency of primary and secondary school children as determined by performance on standardized packages of test exercises. In this study respondents (students who showed up at scheduled testing sessions) were compared with nonrespondents (no-show students). A comparison of the mean and median for three auxiliary measures is presented in the top half of Table 8.2. Differences found in the table reveal that the respondents were generally better students, absent less often from classes, and taking a heavier course load. Assuming these measures to be reasonable correlates to test performance, we conclude that nontrivial respondent–nonrespondent differences on test performance are to be expected based on these data. Typical of the numerous other identification studies of the second type are those done by Mayer and Pratt (1966), Pucel et al. (1971), Gannon et al. (1971), Pavalko and Lukkerman (1973), and Settergen et al. (1983).

As in all studies of this kind the auxiliary information is used in lieu of data on the major study variables that are obviously unavailable for all nonrespondents. When the survey includes a subsampling of nonrespondents

Table 8.2   A Comparison of Responding and Nonresponding 17-Year-Olds in the National Assessment of Education Progress[a]

	Respondents	Nonrespondents
Auxiliary Variables		
Grade point average[b]		
Mean	2.6	2.2
Median	2.7	2.2
Number of Courses taken during		
most recent semester		
Mean	5.4	4.7
Median	5.5	5.2
Percentage of school days		
absent in current year		
Mean	5.9	11.3
Median	4.2	7.3
Number of Correctly Answered Exercises[c]		
Mathematics Package (total number of exercises)		
01 (16)	9.4	8.1
03 (19)	10.6	9.0
09 (19)	9.5	8.4
13 (22)	12.4	10.3
Science Packages		
01 (23)	12.0	10.7
03 (13)	6.2	5.6
09 (28)	18.9	17.9
13 (7)	3.2	2.9

[a]The response rate was 75 percent.
[b]Based on a four-point scale (A = 4; B = 3; C = 2; D = 1; F = 0).
[c]Entries in the table are weighted estimates of the average number of exercises answered correctly.

(Section 8.1.4), estimates applicable to nonrespondents based on the subsample can be compared with the corresponding estimates for respondents. Results of this kind of comparison for the NAEP study are presented in the lower half of Table 8.2. As suggested by the comparison using the auxiliary measures in the upper half of the table, we consistently find the estimated average number of correctly answered exercises to be higher for respondents than for nonrespondents.

The strategy in a third type of identification study is to look for indications of respondent–nonrespondent differences on the major study variables by comparing subgroups of respondents who differ according to the degree to which they are likely to resemble nonrespondents. This way of dealing with the nonresponse issue is sometimes used in mail surveys where data collection is completed in waves. The first wave is the period between the first mailing and first follow-up, the second wave is the interval between the first and second follow-ups, and so on. The rationale for using waves is that sample members who respond in the cth wave indicate greater reluctance to participate (and thus stronger resemblance to nonrespondents) as c increases. For interview surveys a comparable measure to wave would be the number of call attempts needed to secure participation.

Respondent data and the wave or call attempt in which the response occurred have been used in different ways to identify and deal with the biasing effect of unit nonresponse. Most of the published accounts of this approach have been applied to mail surveys. In some instances trends observed from survey estimates obtained separately from each wave are noted [e.g., Dunkelberg and Day (1973) and Chapman (1976)], thus examining whether later respondents tend to differ from earlier respondents. Estimates using pooled data through each wave have also been compared for the same reason [e.g., Hawkins (1975) and Jones (1983)]. Assuming that the wave in which a person responds is a reasonable indication of likelihood to respond, monotonically increasing or decreasing trends among estimates by wave may indicate a correlation between individual response probabilities and the Y variable, thereby pointing toward biased estimates due to nonresponse.

Sequences of estimates, obtained by sorting respondents according to when during the period of data collection the response was obtained, may also be used to project estimates as if complete response had occurred. The term *extrapolation method* denotes a compensation procedure in which a sequence of estimates as described above is fitted to an assumed model, and the fitted model is used as the basis for extrapolation. The independent variable (x) in these models is some measure of the point during data collection when the questionnaire was completed (e.g., number of calls or time to completion, cumulative response rate), and the dependent variable (y) is a survey estimate (e.g., of a mean) obtained by pooling data from the start of data collection through the point determined by x.

The basic idea of most extrapolation methods is the same. Each respondent is assigned what we will call a *response outcome variable* (considered as an auxiliary X variable), indicating the degree to which the respondent is assumed to resemble a nonrespondent. As assumed model relating the response outcome variable to one of the major study variables (the Y variable) is then fitted using respondent data. Finally, the fitted model is used to extrapolate the final result. Approaches to extrapolation differ according to which measure is used for the X variable and which statistical model is assumed in relating the X variable to the Y variable.

In one of the first documented uses of the extrapolation idea applied to mail surveys, Hendricks (1949) fits a log-linear model,

$$y = \alpha x^{\beta}, \tag{8.1}$$

where x is the number of calls required to obtain the response, and where α

and β are constant regression parameters to be estimated from the respondent data. The definition of x Hendricks used is typical of the earlier extrapolation methods Scott (1961) summarized. Filion (1976) and Jones (1983) assume the linear model,

$$y = \alpha + \beta x, \tag{8.2}$$

where x is the cumulative response rate up to a given wave. Von Riesen and Novotny (1979) use time to completion for x and fit a piecewise linear model to reflect assumed changes in the linearity of the model during data collection. In mail surveys with one slope change this takes the form

$$y = \alpha + \beta_1 x + \beta_2 (x - X_t)\delta, \tag{8.3}$$

where X_t is the point on the x scale where the slope changes (when a follow-up reminder is mailed) from β_1 to $\beta_1 + \beta_2$ and

$$\delta = \begin{cases} 1 & \text{if } (x - X_t) \geq 0 \\ 0 & \text{if otherwise.} \end{cases}$$

Finally, Ognibene (1971), in a study that compares the feasibility of various models relating x (the cumulative response rate) to y, concludes that the hyperbolic model,

$$y = \alpha + \frac{\beta}{x}, \tag{8.4}$$

gives the most useful results.

The definition of x determines the manner in which the fitted model is used for extrapolation. When x is the cumulative response rate, the extrapolated estimate is for the Y variable in the case of complete response (where the cumulative response rate $x = 1$). However, when x is some measure of how much effort was involved in getting a response (wave, call attempt, or length of time to completion), the object of extrapolation is less clear. One must make some arbitrary choice of the response outcome variable that will adequately predict the value of the Y variable for nonrespondents.

The utility of extrapolation in dealing with survey nonresponse is not firmly established. Advocates of extrapolation would argue that the approach is a relatively simple and inexpensive way to deal with survey nonresponse. The results of extrapolation can be obtained without the expense of nonrespondent subsampling (Section 8.1.4) or complicated reweighting scheme (Section 8.1.5). On the other hand, detractors of extrapolation would question the strength of the relationship between measures of the time to completion and important study variables in light of existing evidence to the contrary. Chapman (1976), for example, finds no clear trends evident in a comparison of respondents obtained on successive call attempts in a large health care examination survey conducted by in-person interviewing. Scott (1961) also finds little evidence of a strong relationship when the Y variable is any one of several demographic characteristics, although he mentions some empirical findings which suggest that estimates obtained by extrapolation will sometimes be relatively good.

Despite some practical advantages, it would seem that extrapolation methods will not be widely used until measures of x that better predict values of the Y variable can be found. The problem with many measures is that they fail to represent the respondent's likelihood of participation. For example, the amount of effort required to obtain a response in an interview survey may reflect only the respondent's availability, not willingness or ability to participate. Presumably, an ideal response measure would reflect all three characteristics. In mail surveys, the wave in which response occurs and the time to completion may only indicate the respondent's tendency to procrastinate or forget.

8.1.4 Nonrespondent Substitution

In the event that unit nonresponse must not cause the final count of respondents to fall below the sample size originally intended for the study, one often-used method of dealing with the problem is to replace each nonrespondent in the initial sample with another member of the population, most frequently a member not included in the original sample. This technique, called substitution, is conceptually similar to explicit imputation used to deal with item nonresponse (see Section 8.2.2). We also note that when the substitute for a nonrespondent is chosen from among the respondents in the sample, substitution resembles a weighting adjustment procedure (see Section 8.1.5).

The rules followed in identifying the substitute respondent are generally random or nonrandom (Chapman, 1983). Random substitution rules require that the substitute be chosen by some type of probability sampling method, a feature that facilitates a statistical assessment of resulting identified estimates. In many applications the random substitute is selected from among members of the same subclass as the nonrespondent, where the subclass is defined in terms of variables thought to be correlated with the major study variables or within the same cluster in a cluster sample [see Waksberg (1978)]. The intent is for each nonrespondent and its substitute to be similar with respect to the variables used to define the subclass, and thus similar as to the major study variables. When substitution is done within subclasses, estimators from samples with substitutes have bias properties similar to those obtained from samples where a randomized hot-deck imputation (as discussed in Section 8.2.2) is used (Schaible, 1983).

Substitutes under a nonrandom designation scheme, on the other hand, are identified by applying (often during data collection) a predetermined set of criteria, but not by probability sampling. The substitute might, for example, be the dwelling unit immediately on the right as the interviewer faces the nonresponding dwelling in an area household sample, the next residential household found after adding one to the last digit of the nonrespondent's telephone number in a telephone survey, or more generally the next encountered member of the population with certain characteristics similar to those of the nonrespondent.

Two methods of dealing with survey nonresponse that are similar but not identical to substitution are worth noting. One is a technique suggested for repeated interview surveys by Kish and Hess (1959), in which a similar substitute is identified by random or nonrandom means from nonrespondents (after c call attempts) in a recent survey. Call attempts are then made to solicit participation from the substitute households. To increase the effectiveness of this replacement procedure, the nonrespondent in the current survey should have demographic characteristics similar to its substitute. This procedure may enable one to increase the response rate, since the investigator can afford to make more call attempts on the substitutes than on the sample nonrespondents, who will have had no more than c attempts made when replaced by a substitute with c attempts already made.

A second similar method to substitution is *supplementary sampling*, in which an initial sample and a set of smaller but randomly chosen supplementary samples are available to use in the event that nonresponse in the initial sample is higher than expected. This method is often used in one-time surveys in which there is relatively little prior knowledge of response rates in the population being studied. The procedure works as follows. Given that the expected response rate in the survey is thought likely to be no lower than λ_L but no higher than λ_H, an initial sample of size $n\lambda_H^{-1}$ is chosen. Following the same sampling design as the initial sample, a set of m supplementary samples of size $n(\lambda_L^{-1} - \lambda_H^{-1})/m$ is chosen independently. As the survey progresses, individual supplementary samples are added until the supplementary sample results in a sample size as close to n as desired. A full commitment must be made to obtaining participation from each member as each supplementary sample is added. Clearly, the larger m is allowed to be, the closer one can get to the desired sample size and the more supplementary sampling resembles individual substitution. Supplementary samples are most useful in surveys where making individual substitutions is too difficult or costly.

Substitution reduces nonresponse bias to the extent that the substitutes are collectively identical in every respect to the sample members for whom the substitutions are made. When substitution is done completely at random and the substitutes are subject to the same data collection protocol as the initial sample, nonresponse bias is unlikely to be substantially affected, since substitutes would replace nonrespondents with respondents resembling those already in the sample (Kish, 1965, Section 13.6B). If, on the other hand, each respondent and its substitute can be matched on known correlates of the Y variable, the biasing effect of nonresponse will be diminished for the same reasons that weighting class adjustments (Section 8.1.5) and hot-deck imputation (Section 8.2.2) reduce bias. Unfortunately, as with most remedies for nonresponse, substitution does not completely eliminate its biasing effects. Empirical studies by Cohen (1955) and Williams and Folsom (1977) on the use of matched substitutes have confirmed this finding.

Some of the advantages and disadvantages of substitution may be summarized from the discussion by Chapman (1983). Among its advantages is that targeted sample sizes in matching categories can be achieved, thus controlling overall sample size and associated variance components from sampling error. Stratum sample sizes can also be controlled by matching within strata —thus, for example, avoiding variance estimation problems caused by strata with fewer than two sample members. A second advantage discussed earlier is that nonresponse bias may be reduced, though not eliminated, if nonrespondents and their substitutes are similar with respect to the major study variables. Among the drawbacks to substitution is that the availability of a substitute may diminish the intensity of an interviewer's efforts to obtain a response, knowing that a backup is "waiting in the wings." The result would be a lower response rate for the initial sample than would be expected if no substitutes were available. Thus every effort should be made to obtain a response from members of the initial sample before substitution is allowed. A second drawback of substitution is the potential for overstating the survey response rate by including substitutes in the numerator but excluding the substituted nonrespondents from the denominator of the calculated rate. A more reasonable rate would be calculated from the initial sample alone or from the initial sample plus those designated as substitutes (and either responding or not).

8.1.5 Nonrespondent Subsampling

At the root of the problem of nonresponse is the availability of survey data for respondents and a corresponding lack of data for nonrespondents. One highly visible class of nonresponse compensation procedures was developed with the idea that estimates which are free of nonresponse bias can be obtained by using data from the sample of respondents and a random subsample of initial nonrespondents, to whom more intensive, costly efforts are applied to obtain a response. Nonrespondent subsampling procedures are most often used in mail surveys, where response rates are low and where more intensive efforts (e.g., by telephone or in person) will obtain useful responses from a high percentage (although usually not 100 percent) of a nonrespondent subsample.

The original idea of nonrespondent subsampling, suggested by Hansen and Hurwitz (1946), follows the classical approach to statistical inference.

Their idea has given rise to numerous extensions and revisions, some of which use model-based approaches to inference. Most of the existing theory presumes that responses are obtained from all subsampled nonrespondents. This provision seldom holds completely, since despite extraordinary efforts, some hard-core nonrespondents will remain, such as those adamantly opposed to surveys in general or those who are unavailable during the entire survey.

The approach followed in all classical nonrespondent subsampling methods is as follows. Consider the survey population to consist of $H \geq 2$ mutually exclusive and exhaustive strata. One of these strata contains all the nonrespondents, while the other $H - 1$ strata are distinguished by the intensity of efforts (e.g., number of call attempts, follow-up letters, etc.) required to obtain a response.

In their original article, Hansen and Hurwitz assume that $H = 2$ and use double sampling or two-phase sampling to select those in the nonrespondent stratum. The population of size N consists of N_1 respondents and $N_0 = N - N_1$ nonrespondents. With simple random sampling (without replacement) used in selecting the initial sample (of size n with n_1 initially responding) and the nonresponding subsample, the estimator of the mean $\bar{Y}$ using the Hansen–Hurwitz method is

$$\bar{y}_{HH} = l_1 \bar{y}_1 + (1 - l_1) \bar{y}_{01}, \tag{8.5}$$

where $l_1 = n_1/n$ is the response rate in the initial sample, $\bar{y}_1 = \sum_{i=1}^{n_1} y_i/n_1$ is the mean among initial respondents, and $\bar{y}_{01} = \sum_{i=1}^{n_{01}} y_i/n_{01}$ is the mean among the n_{01} subsampled nonrespondents. The variance of the Hansen–Hurwitz estimator, given the assumed design, will be

$$\text{Var}(\bar{y}_{HH}) = \frac{(1 - f)[S^2 + (g - 1)\lambda_0 S_0^2]}{n}, \tag{8.6}$$

where $f = n/N$, $\lambda_0 = N_0/N$, $g = n_0/n_{01}$, and S^2 and S_0^2 are variances of the Y variable among all population members and all members of the nonrespondent stratum, respectively. Assuming that $S^2 = S_0^2$, we note that Eq. (8.6) reduces to $\text{Var}(\bar{y}_{HH}) = (1 - f)S^2/n'$, where $n' = n/[1 + (g - 1)\lambda_0]$. Since $n' \leq n$ we see that $\text{Var}(\bar{y}_{HH})$ is greater than or equal to the variance of the estimated mean from a simple random sample of size n, assuming complete response. In other words, $\bar{y}_{HH}$ may remove the bias from nonresponse, but its variance will always be greater than the complete-response variance unless all nonrespondents are subsampled ($g = 1$).

Subject to a simple cost model, the optimum rate of nonrespondent subsampling (g^{-1}) can be determined by choosing that value of g which minimizes $\text{Var}(\bar{y}_{HH})$ for fixed expected cost. Allowing for a few simplifying assumptions, this optimum value is

$$g_{\text{opt}} = \left[\frac{C_{01}(S^2 - \lambda_0 S_0^2)}{(C^* + \lambda_1 C_1) S_0^2} \right]^{1/2}, \tag{8.7}$$

where $\lambda_1 = 1 - \lambda_0$, and C^*, C_1, and C_{01} are average per-unit costs for choosing and setting up the initial sample, gathering data for the initial respondents, and gathering data for the nonrespondent subsample, respectively. In the case where one can reasonably assume that $S^2 = S_0^2$ and that C^* is small relative to $\lambda_1 C_1$, we see that $g_{\text{opt}}^{-2} \doteq C_1/C_{01}$, indicating that the optimum subsampling rate is inversely related to the relative increase in the unit cost of obtaining and processing data from the nonrespondent subsample.

As an alternative to g_{opt}, Srinath (1971) suggests an optimum value for g that does not rely as heavily on knowing λ_1 accurately. The optimum subsample size under the alternative rule is computed as

$$n_{01}^* = \frac{n_0^2}{ng_{\text{opt}}^* + n_0}, \tag{8.8}$$

where

$$g_{\text{opt}}^* = \frac{C_{01}\lambda_0^2(S^2 - \lambda_0 S_0^2)}{(C^* + \lambda_1 C_1) S_0^2} - \lambda_0 = (g_{\text{opt}} - 1)\lambda_0. \tag{8.9}$$

The advantage of the Srinath solution is that regardless of what we know about λ_0 before the survey, g_{opt}^* will yield an estimator with the desired precision by allowing the subsampling fraction to vary depending on the response rate in the sample. Using g_{opt} in Eq. (8.7), the precision of estimates may depart from targeted levels when the actual and assumed values of λ_0 differ.

Several extensions that broaden the applicability of nonrespondent subsampling are mentioned, based on more extensive discussions by Zarkovich (1966), Cochran (1977), and Rao (1983a). El-Badry (1956) considers the case where multiple attempts are made through follow-up letters or calls before a nonrespondent subsample is chosen, and the data obtained through each attempt and the subsample are combined to produce the final estimate. Under this framework $H > 2$ strata are defined, one for nonrespondents (which is subsampled) and $H - 1$ strata for the respondents to each of the $H - 1$ attempts made to obtain a response. Rao (1968) considers nonrespondent subsampling with $H = 2$ but where the list from which the sample is drawn contains an unknown number of duplicated entries. Rao (1973) has also applied nonrespondent subsampling to the setting where stratification is used in choosing the initial sample. Finally, Rao and Hughes (1983) present an optimized solution for nonrespondent subsampling applied to mail surveys when the object is to estimate the difference of means between two domains that may not be defined by strata used for sampling. Two alternative sampling schemes are considered in which nonrespondents are subsampled twice.

Bartholomew (1961) has suggested an idea somewhat related to nonrespondent subsampling for studies in which a maximum of two call attempts are allowed. The selected sample thus consists of three distinct groups: respondents after the first call (group R_1), respondents after the second call (group R_2), and nonrespondents (group NR). Assuming that group R_2 is a fair representation of groups R_2 and NR combined, group R_2 is treated as if it were a nonrespondent subsample and the data from it reweighted to produce an estimator comparable to $\bar{y}_{HH}$ in Eq. (8.5), where l_1 is the proportion of the sample in the group, R_1, $\bar{y}_1$ is obtained from R_1, and $\bar{y}_{01}$ is obtained from group R_2. Clearly, the reduction in nonresponse bias from this approach is directly related to how collectively similar members of groups R_2 and NR are.

We turn next to a series of articles in which Bayesian methods are used to obtain estimates when nonrespondent subsampling has been done. Bayesian statistical inference is based on sample data (as in the classical approach) but also on expressions of prior knowledge concerning parameters to be estimated. Applications of these methods are often characterized by the need for two samples: a pilot sample to establish or substantiate prior assumptions about certain important study parameters and to aid in planning subsequent survey sampling, and the main sample where design plans are implemented. We mention these Bayesian methods here, although a later section is devoted entirely to these model-based methods.

The pioneering work by Erickson (1967) is useful to illustrate the Bayesian approach to nonrespondent subsampling, since his work is a direct adaptation of the Hansen–Hurwitz approach. Essential features related to sampling and data collection are the same in both approaches. The steps followed in obtaining estimates from the survey differ markedly, however. Beginning with joint prior distributions for the proportion of respondents in the universe (λ_1) and for the true mean for respondents (μ_1) and nonrespondents (μ_0), Erickson obtains a joint posterior distribution of (μ_1, μ_0, λ_1) through the first phase of sampling from likelihood functions for μ_1 and λ_1, given observed data from the first-phase pilot sample of n_1 respondents. An overall joint posterior distribution for (μ_1, μ_0, λ_1) is then obtained, given data from the combined two-phase sample. The final step consists of establishing the posterior distribution for the overall mean, $\mu = \lambda_1\mu_1 + (1 - \lambda_1)\mu_0$, from the preceding step. From this posterior distribution of μ an optimized Bayes estimator ($\hat{\mu}_E$) can be derived by minimizing the expectation of the loss function,

$$L = K(\mu - \hat{\mu}_E) + C^*n + C_1n_1 + C_{01}n_{01}, \qquad (8.10)$$

over the posterior distribution of μ, given the initial sample and nonrespondent subsample, where K is a constant. In the special case where prior information is vague, it can be shown that $\hat{\mu}_E$ is equivalent to the Hansen–Hurwitz estimator. Optimum values for n_{01} and n are also derived.

Several other uses of Bayesian methods in surveys with nonrespondent subsampling have been suggested. For example, in their Bayesian predictive approach to double sampling, Rao and Ghangurde (1972) suggest the selection of two two-phase samples, a pilot sample and a main sample. Both sampling designs call for less expensive efforts to be used in obtaining response from an initial simple random sample, followed by more expensive efforts to obtain data from a nonrespondent subsample. Assuming a similar design framework, Singh and Sedransk (1978b) consider a Bayesian approach to regression analysis in the presence of unit nonresponse. These same authors [in Namboodiri (1978)] also use Bayesian methods to estimate the population mean. This work is extended to the case where a stratified two-phase sample is used [in Singh and Sedransk (1983)].

To conclude, consider the following summary of the differences between classical and Bayesian approaches to the use of nonrespondent subsamples:

Criterion	Classical	Bayesian
1. Basis for inference	Sampling distribution of the $n_1 + n_{01}$ respondents determined by methods used to select initial sample and subsequent ones	Prior assumptions concerning behavior of key parameters expressed as a stochastic model
2. Optimality criteria for n and subsampling rates (g^{-1})	Minimizing variance of estimator (based on its sampling distribution), gives prespecified cost	Minimizing an expected loss function determined by cost and some measure of difference between parameter and its estimator
3. Basis of choice of n and g	Assumed variance and cost models as well as reasonable measure of parameters associated with these models	Assumed prior distribution and data from pilot sample

8.1.6 Adjustment Procedures

Strategies for dealing with unit nonresponse described to this point have
been those which the investigator invokes at points in the study other than
when the major study analyses are being done. Preventive methods, nonrespondent substitution, and nonrespondent subsampling are applied earlier in
the study, during the period of data collection. Identification studies are
often done as a preliminary step in analysis or as part of an evaluation of the
study protocol after analysis has been done. Methods described in this
section and the next are those the investigator uses in the process of
formulating and producing survey estimates.

We first look at strategies dealing with how the rate of nonresponse varies
somewhat within most survey samples and thus affects the sample's ability to
present a balanced cross section of the survey population. Consider the
following example to illustrate this point. In the absence of nonresponse, an
equal-probability sample for a survey of adults in the United States would
proportionately represent urban and rural residents. If nonresponse occurred
and, as is typically the case, urban response rates were lower than rural
response rates, the final set of respondents would include a disproportion-
ately large share of rural adults. The classical methods discussed next deal
with the imbalance from nonresponse by adjusting the respondent data, in
effect, by computing a numerical weighting factor which when applied to
each respondent's data at least partially returns a balance to the respondent
sample.

Since the concept of adjustment is important for us here, some discussion
of its meaning in the present framework is in order. First we must establish
what is to be adjusted. To do so, consider the simple problem of estimating a
total (Y), where we know that the estimator Horvitz–Thompson $\hat{Y}_{HT} =
\sum_{i=1}^{n} W_i Y_i$, when applied to probability sample of size n, will be unbiased. In
the absence of nonsampling error the sample weight, $W_i = \pi_i^{-1}$, where π_i is
the sampling probability, adequately reflects the likelihood that data from the
ith member of the sample would be used for estimation. When nonresponse
is present and there are data for $n_1 < n$ sample members, the estimator,
$\hat{Y} = \sum_{i=1}^{n_1} W_i Y_i$, will no longer be unbiased (Section 7.3). Unbiasedness is
possible if we use a revised weight that accounts for the selection probability
(π_i) and the conditional probability that the ith sample member would, if
selected, respond and provide useful data for estimation, that is, its response
probability (p_i), which must exceed zero for all members of the population.
Thus, to achieve unbiasedness, the Horvitz–Thompson estimator must be
modified to $\hat{Y}_{HT}^* = \sum_{i=1}^{n_1} W_i^* Y_i$, where $W_i^* = (\pi_i p_i)^{-1}$ is a nonresponse-
adjusted sample weight. Nargundkar and Joshi (1975) present some of the
theory of this estimator in the absence of other nonsampling errors, as do
Platek and Gray (1983) with measurement errors present (see Section 7.3).
The estimator $\hat{Y}_{HT}^*$ has been adjusted in the sense that response probabilities
for the respondents are used to account for the nonresponse in the survey.

This last statement motivates the underlying principle of nonresponse
adjustment procedures in classical survey inference. To minimize the biasing

effect of nonresponse, a measure of each sample member's response probability is used to adjust sample weights. Since the p_i are unknown, they must be estimated in some reasonable manner. The way in which p_i is estimated distinguishes the various adjustment methods.

In one of the first known adjustment methods, attributed to Politz and Simmons (1949) but based on an idea by Hartley (1946), it is necessary to obtain t_i, the number of days during the previous five that the ith respondent would have been available to be interviewed at the same time of day. The estimated response probability for the Politz–Simmons method is then $\hat{p}_i^{(1)} = (t_i + 1)/6$ (see Section 7.2).

A second adjustment method, called the *weighting class adjustment*, estimates response probabilities by dividing the original sample (including both respondents and nonrespondents) into H mutually exclusive and nonoverlapping subsets, called *adjustment cells*, labeled using the subscript h, within which members are assumed to have similar values of the Y variable and all response probabilities are presumed to be equal. Since the entire sample is partitioned, it is apparent that information used to make assignments must be available for respondents and nonrespondents. The type of homogeneity desired for adjustment cells is similar though less restrictive than the homogeneity sought for the same cells if they were used for sampling. For this nonresponse adjustment we wish for $\bar{Y}_{1h} = \bar{Y}_h$ and $S_{1h}^2 = S_{0h}^2$, while for minimizing sampling error we try to form homogeneous cells to serve as sampling strata (we wish for $S_h = 0; h = 1, 2, \ldots, H$). The latter homogeneity criterion is, however, desirable for both hot-deck and cold-deck imputation methods of dealing with item nonresponse (Section 8.2.2).

The generalized estimator of p_i used in the weighting class adjustment method is a weighted within-cell response rate obtained as follows. For any probability sample where $W_{hi} = \pi_{hi}^{-1}$ is the unadjusted sample weight for the ith member of the hth adjustment cell, the estimator of p_{hi} (the subscript h must be added to the index adjustment cell) will be $\hat{p}_{hi}^{(2)} = \sum_{i=1}^{n_{1h}} W_{hi} / \sum_{i=1}^{n_h} W_{hi}$, and the nonresponse adjustment for the ith sampled unit in the hth cell is $1/\hat{p}_{hi}^{(2)} \geq 1$, where n_h is the number of sample members in the hth cell and n_{1h} is the number of respondents in the hth cell. The weighting-class-adjusted weight is thus $W_{hi}^{(2)} = W_{hi}/\hat{p}_{hi}^{(2)}$. When simple random sampling or any other equal probability design is used to choose the original sample, $\pi_{hi} = n/N$, for all $h = 1, 2, \ldots, H$ and $i = 1, 2, \ldots, n_h$, and $\hat{p}_{hi}^{(2)} = n_{1h}/n_h$.

A third approach to adjusting for nonresponse is the poststratification adjustment method. In the absence of nonresponse, poststratification enables one to reduce the sampling variance of estimates by adjusting sampling weights within each of H adjustment cells (Hansen et al., 1953a). The poststratification adjustment for each weight in the hth adjustment cell is computed here as

$$a_h = \frac{N_h \sum_{h=1}^{H} \sum_{i=1}^{n_h} W_{hi}}{N \sum_{i=1}^{n_h} W_{hi}}, \tag{8.11}$$

where measures of N_h, the number of population members in the hth adjustment cell or the size of N_h relative to $N = \sum_{h=1}^{H} N_h$, are obtained from the best data available (e.g., a recent census). For an equal probability sampling design the adjustment simplifies to $a_h = N_h n / (N n_h)$. We also note that a_h tends to be closer to unity as n_h increases and $\sum_{i=1}^{n_h} W_{hi} / \sum_{h=1}^{H} \sum_{i=1}^{n_h} W_{hi}$ improves as an estimator of N_h / N.

When used to deal jointly with sampling errors and nonresponse bias, poststratification and weighting class approaches can be combined to form the cell-level adjustment for W_{hi},

$$a_h^* = \frac{N_h \sum_{h=1}^{H} \sum_{i=1}^{n_h} W_{hi}}{N \sum_{i=1}^{n_{1h}} W_{hi}}. \tag{8.12}$$

For simple random sampling in the initial sample, $a_h^* = N_h n / (N n_{1h})$. We see from Eqs. (8.11) and (8.12) that $a_h^* \geq a_h$, with equality possible only if there is complete response within cells. The combined adjustment can be rewritten as $a_h^* = a_h / \hat{p}_{hi}^{(2)}$, from which we observe that the measure used to estimate response probabilities would be $\hat{p}_{hi}^{(3)} = \hat{p}_{hi}^{(2)} / a_h$, although a_h^{-1} may be more appropriately conceived as an adjustment to π_{hi} than as a modification to an estimate of p_{hi}. Thus the adjusted sample weight under the combined adjustment scheme is computed as $W_{hi}^{(3)} = a_h^* W_{hi}$, which, when equal probability sampling has been used, simplifies to $W_{hi}^{(3)} = N_h / n_{1h}$.

To this point we have assumed that the same adjustment cells are defined to use the weighting class and poststratification methods together. In practice, however, quite different cells may be used for each adjustment. When adjustment cells are defined separately for the two adjustments, perhaps one of the most commonly used approaches to dealing with unit nonresponse is first to compute a weighting-class-adjusted weight for a respondent in the h'th weighting class adjustment cell as $W_{h'i}^{(2)} = W_{h'i} / \hat{p}_{h'i}^{(2)}$, and then to compute the poststratification adjustment for the hth poststratification adjustment cell as

$$a_h^* = \frac{N_h \sum_{h=1}^{H} \sum_{i=1}^{n_{1h}} W_{hi}^{(2)}}{N \sum_{i=1}^{n_{1h}} W_{hi}^{(2)}}, \tag{8.13}$$

where $W_{hi}^{(2)}$ is the value of the weighting-class-adjusted weight for a respondent who is a member of the hth poststratification adjustment cell. The final combined weight used in analysis is $W_{hi}^{(4)} = a_h^* W_{hi}^{(2)}$, with a_h^* computed from Eq. (8.13).

From the description above of the procedure for computing weighting class and poststratification adjustments, we note one important difference in practical requirements. The weighting class adjustment requires two items of information for respondents and nonrespondents: unadjusted weights (W_{hi}) and the measures for each of the variables used to define the adjustment cells. The poststratification adjustment [a_h^* in Eq. (8.12)] requires an external measure of N_h and measures of the adjustment cell variables, but only for respondents.

Cohen and Kalsbeek (1981) give an example of the common practice of combining weighting class and poststratification adjustments for the National Medical Care Expenditure Survey. In this study, weighting class adjustment cells were census enumeration districts and block groups, which served as second-stage sampling units. Cells for the poststratification adjustment, computed using weighting-class-adjusted weights, were 32 age–race–sex cross-classification categories as adjustment cells. Bailar et al. (1978), as well as Rizvi (1983a) and Madow (1983), give other example of the joint application of the two adjustments.

A few other methods based on the idea of identifying reasonable measures of p_i are mentioned briefly. Chapman (1976) discusses one where in the p_i are estimated from regression analysis. Drew and Fuller (1980, 1981), in their strategy combining callbacks and adjustment cells, estimate p_{hi}, $\Delta_h = N_h/N$, and the proportion of hard-core nonrespondents from a fitted multinomial model. Thomsen and Siring (1983) likewise use weighting classes to fit a model relating p_i to the number of callbacks. As for all adjustment methods, the ability to reduce nonresponse bias is directly related to how well p_i can be estimated.

For simple estimators of means and totals the weighting class adjustment method can be viewed as an implicit form of imputation where the imputed value (Z_i) is inferred rather than a value of the Y variable for one of the respondents (Section 8.2.2). The linkage between these two types of compensation procedures is explored a bit further using Fig. 8.1, which illustrates the weighting class method within an adjustment cell consisting of six respondents and two nonrespondents. All unadjusted sample weights (W_{hi}) among units within the cell are assumed to be equal. In step 1, consider each of the eight units to consist of six equal-sized parts with the unit's Y variable associated with each part. The measure used by each unit in analysis is the simple average of the Y variable associated with the six parts. Since the Y variable is unavailable for nonrespondents U_7 and U_8, step 2 of the weighting class method calls for $Y_1, Y_2, \ldots, Y_6$ to be "weighted up" by a factor of 8/6 (the weighting class adjustment). This procedure is equivalent to dividing the nonrespondents' parts among the respondents and producing estimates with six rather than eight datum points coming from this cell. Since each unit consists of equal-sized parts, we see in step 3 that weighting up is equivalent to a redistribution of parts such that the simple average of the Y variable among respondents has effectively replaced its original value for the nonrespondents, U_7 and U_8. Thus we see in this case that applying the weighting class adjustment is the same as imputing the mean among respondents in the cell. We note also that the sample size for producing estimates from weighting-class-adjusted data would be the number of respondents (n_1) rather than the full sample size (n).

	Respondents						Nonrespondents	
Step	U1	U2	U3	U4	U5	U6	U7	U8

(1) The problem

Y_1	Y_2	Y_3	Y_4	Y_5	Y_6	Y_7	Y_8
Y_1	Y_2	Y_3	Y_4	Y_5	Y_6	Y_7	Y_8
Y_1	Y_2	Y_3	Y_4	Y_5	Y_6	Y_7	Y_8
Y_1	Y_2	Y_3	Y_4	Y_5	Y_6	Y_7	Y_8
Y_1	Y_2	Y_3	Y_4	Y_5	Y_6	Y_7	Y_8
Y_1	Y_2	Y_3	Y_4	Y_5	Y_6		

(2) The weighting-class adjustment applied

Y_1	Y_2	Y_3	Y_4	Y_5	Y_6
Y_1	Y_2	Y_3	Y_4	Y_5	Y_6
Y_1	Y_2	Y_3	Y_4	Y_5	Y_6
Y_1	Y_2	Y_3	Y_4	Y_5	Y_6
Y_1	Y_2	Y_3	Y_4	Y_5	Y_6
Y_1	Y_2	Y_3	Y_4	Y_5	Y_6
Y_1	Y_2	Y_3	Y_4	Y_5	Y_6
Y_1	Y_2	Y_3	Y_4	Y_5	Y_6

(3) Implicit imputation of the respondent mean

Y_1	Y_2	Y_3	Y_4	Y_5	Y_6	Y_1	Y_1
Y_1	Y_2	Y_3	Y_4	Y_5	Y_6	Y_2	Y_2
Y_1	Y_2	Y_3	Y_4	Y_5	Y_6	Y_3	Y_3
Y_1	Y_2	Y_3	Y_4	Y_5	Y_6	Y_4	Y_4
Y_1	Y_2	Y_3	Y_4	Y_5	Y_6	Y_5	Y_5
Y_1	Y_2	Y_3	Y_4	Y_5	Y_6	Y_6	Y_6

Figure 8.1 Relationship between weighting class adjustment and imputation (equal sample weights within adjustment cell).

The amount of survey error associated with both explicit and implicit imputation methods depends on how much imputed values differ from their corresponding true values (in Fig. 8.1 on how much the respondent mean differs from Y_7 and Y_8). The difference will be small when the adjustment cell is homogeneous with respect to the Y variable. The amount of homogeneity depends, in turn on the level of statistical association between the Y variable and the variables used to define the adjustment cell.

In the remainder of this section we discuss separately the Politz–Simmons weighting class and poststratification adjustment methods in the context of analyses where means and totals are being estimated. We will not consider the effectiveness of these adjustment procedures on other more complex estimates, although Santos (1981) has examined the effects of imputing adjustment cell means when estimating more complex parameters.

8.1.6.1 Politz–Simmons Adjustment

For the case where a total (Y) is to be estimated from a complete enumeration, Politz and Simmons (1949) demonstrate some properties for the estimates, $\hat{Y}_{PS} = \sum_{i=1}^{N_1} Y_i / \hat{p}_i^{(1)}$, where $\hat{p}_i^{(1)} \equiv p_i = (t_i + 1)/6$ and t_i is the number of times during the past five days that the ith respondent would have been available to be interviewed at the same time of day. The bias and variance of $\hat{Y}_{PS}$ are found to be

$$\text{Bias}\left(\hat{Y}_{PS}\right) = -\sum_{i=1}^{N}(1 - p_i)^6 Y_i \tag{8.14}$$

and

$$\text{Var}\left(\hat{Y}_{PS}\right) = \sum_{i=1}^{N} Y_i^2 \left\{ 6 \sum_{j=1}^{6} \frac{p_{ij}}{j} - \left[1 - (1 - p_i)^6\right]^2 \right\}, \tag{8.15}$$

respectively, where for the ith population member's p_i is a response probability (with *response* equated to availability), and p_{ij} is the probability that the member would be found at home j times if call attempts were made on six consecutive days at the same time each day.

We see from Eq. (8.14) that the Politz–Simmons estimator is not unbiased. $\hat{Y}_{PS}$ will always tend to underestimate Y unless $p_i = 1$ for all population members; however, the absolute value of the bias relative to Y will be small [i.e., less than $\bar{Y} - 1$, since $\sum_{i=1}^{N}(1_i - p_i)^6 \leq N$].

The estimator was originally designed to circumvent the cost of multiple call attempts. Adjusting the data with $\hat{p}_i^{(1)}$ was thought to be as useful as attempting to move response probabilities closer to unity by additional call attempts. One of the approach's limitations is that $\hat{p}_i^{(1)}$, while dealing with nonresponse from nonavailability, fails to account directly for nonresponse linked to refusal, although the value of t_i given by each respondent may to some extent reflect willingness and availability to participate. This point is relevant because when $\hat{p}_i^{(1)}$ is unrelated to the Y variable, the variance of $\hat{Y}_{PS}$ is needlessly high and its bias no lower than if nothing had been done to deal with the nonresponse. The approach also fails to account for those who would not have been available at all during the six-day period (those with $p_i = 0$). These hard-core nonrespondents have no chance of being represented in the sample and thus, given the schedule of call attempts, can never be represented in the survey. For additional discussion of the Politz–Simmons adjustment method, see Deming (1953), Zarkovich (1966), Kalton (1983), and Cochran (1977), Cochran presents properties of the Politz–Simmons estimator of a mean.

8.1.6.2 Weighting Class Adjustment

The cells defined to produce weighting class adjustments in multistage designs are often sampling units from one of the first stages of sampling. The rationale for defining cells in this way is that individuals living in geographic proximity are relatively more similar than the population as a whole, and that information on sampling unit affiliation is known for both respondents and nonrespondents. For example, the minor civil divisions selected in the first stage of a three-stage sample of households might be designated as the adjustment cells. Perhaps for this reason, Platek et al. (1977) refer to the criteria that define the cells for weighting class procedure as "design-dependent" balancing factors.

In unclustered list samples adjustment cells might be formed from one or more data items (variables) available on the sampling frame. For example, in a survey of physicians' record-keeping practices where the sample is selected from a membership listing of the American Medical Association, the variables used to form weighting class adjustment cells might be the physician's specialty and size of practice. In some instances these cells are defined by variables used earlier in the survey for stratified sampling. In other instances the sampling strata are also used as adjustment cells.

The ideal set of variables would be those strongly associated with the major study variables but mutually unrelated. It also seems preferable to define cells by a coarse division on several acceptable variables than to form the same number of cells by a finer division on the best single variable among those considered acceptable [see Kish and Anderson (1978)]. The choice of variables is often somewhat arbitrary, although more sophisticated cluster analysis techniques might be used [e.g., see Hartigan (1975)]. Since we hope for a high level of association between the variables used to form cells and the major study variables, which in turn are often related to the p_i, it is not uncommon and is even desirable to find observed within-cell response rates (l_{1h}) to vary considerably among cells. In picking the number (H) of cells to form, one is torn between two competing influences. On one hand, a large number of well-formed cells will more effectively reduce bias, but the wider variation in adjusted sample weights caused by more variable weighting class adjustments will cause the variance of estimates to increase. Thus, keeping H small is likely to reduce the variance, but increasing H is likely to reduce the bias.

To examine some of the statistical properties of estimates using the weighting class adjustment, we consider the case where the object is to estimate the population mean $\bar{Y}$. For any probability sample with known sample weights $W_{hi} = (\pi_{hi})^{-1}$, the estimator can be expressed as

$$\hat{\bar{Y}}_{WC} = \sum_{h=1}^{H} \hat{\Delta}_h \hat{\bar{Y}}_{1h}$$

$$= \frac{\sum_{h=1}^{H}\sum_{i=1}^{n_{1h}}W_{hi}^{(2)}Y_{hi}}{\sum_{h=1}^{H}\sum_{i=1}^{n_{1h}}W_{hi}^{(2)}}, \tag{8.16}$$

where $\hat{\Delta}_h = \sum_{i=1}^{n_h}W_{hi}/\sum_{h=1}^{H}\sum_{i=1}^{n_h}W_{hi}$ estimates $\Delta_h = N_h/N$, $\hat{\bar{Y}}_{1h} = \sum_{i=1}^{n_{1h}}W_{hi}Y_{hi}/\sum_{i=1}^{n_{1h}}W_{hi}$ estimates $\bar{Y}_{1h} = \sum_{i=1}^{N_{1h}}Y_{hi}/N_{1h}$, the mean among all respondents in the hth cell, and

$$W_{hi}^{(2)} = W_{hi}\frac{\sum_{i=1}^{n_h}W_{hi}}{\sum_{i=1}^{n_{1h}}W_{hi}}. \tag{8.17}$$

When $H = 1$, a single survey-wide adjustment is computed. Since this adjustment is a constant over n_1 respondents, $\hat{\bar{Y}}_{WC} = \sum_{i=1}^{n_1}W_iY_i/\sum_{i=1}^{n_1}W_i$, meaning that the unadjusted and weighting-class-adjusted estimators are equivalent.

For the special case where simple random sampling is used in choosing the initial sample, $\hat{\Delta}_h = \delta_h = n_h/n$, $\hat{\bar{Y}}_{1h} = \bar{y}_{1h} = \sum_{i=1}^{n_{1h}}Y_{hi}/n_{1h}$, and $W_{hi}^{(2)} = Nn_h/nn_{1h} \propto l_{1h}^{-1}$ with $l_{1h} = n_{1h}/n_h$. Thus $\hat{\bar{Y}}_{WC} = \sum_{h=1}^{H}\delta_h\bar{y}_{1h}$ and when $H = 1$, $\hat{\bar{Y}}_{WC} = \bar{y}_1 = \sum_{i=1}^{n_1}Y_i/n_1$. Kalton (1983) calls the weighting class adjustment in this setting a *sample weighting adjustment* and presents its basic properties. The bias of $\hat{\bar{Y}}_{WC}$ given by Kalton but shown earlier by Thomsen (1973) is

$$\text{Bias}\left(\hat{\bar{Y}}_{WC}\right) = \sum_{h=1}^{H}\Delta_h\lambda_{0h}\left(\bar{Y}_{1h} - \bar{Y}_{0h}\right) \tag{8.18}$$

and its variance is

$$\text{Var}\left(\hat{\bar{Y}}_{WC}\right)$$

$$\doteq \frac{\sum_{h=1}^{H}\Delta_hS_{1h}^2/\lambda_{1h} + \sum_{h=1}^{H}\lambda_{0h}S_{1h}^2/\lambda_{1h}^2n + \sum_{h=1}^{H}\Delta_h\left(\bar{Y}_{1h} - \bar{Y}_1^*\right)^2}{n}, \tag{8.19}$$

where S_{1h}^2 is the element variance among all respondents in the hth cell and $\bar{Y}_1^* = \sum_{h=1}^{H}\Delta_h\bar{Y}_{1h}$. We see from Eq. (8.18) why forming homogeneous adjustment cells is important. Complete homogeneity with respect to the Y variable among respondents and nonrespondents in each cell implies that $\bar{Y}_{1h} = \bar{Y}_{0h}$, which in turn implies that $\bar{Y}_{WC}$ will be unbiased.

In the event that adjustment cells are formed but that the unadjusted (no-compensation) estimator, $\bar{y}_1 = \sum_{i=1}^{n_1} Y_i/n_1$, is used to estimate $\hat{\bar{Y}}$, Kalton has also noted that the difference between the bias of $\bar{y}_1$ and $\hat{\bar{Y}}_{\mathrm{WC}}$ is

$$\mathrm{Bias}(\bar{y}_1) - \mathrm{Bias}(\hat{\bar{Y}}_{\mathrm{WC}}) = \sum_{h=1}^{H} \frac{\Delta_h(\bar{Y}_{1h} - \bar{Y}_1)(\lambda_{1h} - \lambda_1)}{\lambda_1}. \qquad (8.20)$$

The difference in bias therefore depends on properties of the adjustment cells. If the cells were formed so that $\lambda_{1h} = \lambda_1(h = 1, 2, \ldots, H), \bar{Y}_{1h} = \bar{Y}_1(h = 1, 2, \ldots, H)$, or more generally that the respondent means $(\bar{Y}_{1h})$ and expected response rates (λ_{1h}) were uncorrelated among cells, there will be no difference in the bias of $\hat{\bar{Y}}_{\mathrm{WC}}$ and $\bar{y}_1$. Moreover, when the cells are not completely homogeneous, we cannot be assured that $\hat{\bar{Y}}_{\mathrm{WC}}$ will have a smaller bias than $\bar{y}_1$. For example, we see from Eqs. (8.18) and (8.20) that when $\bar{Y}_{1h} > \bar{Y}_{0h}(h = 1, 2, \ldots, H)$ and λ_{1h} and $\bar{Y}_{1h}$ are inversely related among all cells, $\mathrm{Bias}(\bar{y}_1) < \mathrm{Bias}(\hat{\bar{Y}}_{\mathrm{WC}})$. In more practical terms these results warn us that it is possible to do more harm than good by using weighting class adjustments. A similar comparison of the variances of $\bar{y}_1$ and $\hat{\bar{Y}}_{\mathrm{WC}}$ by Kalton does not point to any clear preference between the two estimators, although when $\bar{Y}_{1h} = \bar{Y}_1$ and $S_{1h}^2 = \bar{S}_1^2(h = 1, 2, \ldots, H)$,

$$\mathrm{Var}(\hat{\bar{Y}}_{\mathrm{WC}}) - \mathrm{Var}(\bar{y}_1) \doteq \bar{S}_1^2 \sum_{h=1}^{H} \frac{\Delta_h/\lambda_{1h} - \Delta_{1h}/\lambda_1}{n}, \qquad (8.21)$$

where $\Delta_{1h} = N_{1h}/N_1$. The variance difference in Eq. (8.21) is nonnegative, with the increase in the variance of $\hat{\bar{Y}}_{\mathrm{WC}}$ over that of y_1 because of variation in the final weights $[W_{hi}^{(2)}]$ brought about by the variation among weighting class adjustments.

Comparisons of bias and variance for $\bar{y}_1$ and $\hat{\bar{Y}}_{\mathrm{WC}}$ lead to an interesting but troublesome paradox when useful adjustment cells can be formed. Cells with wide variation in λ_{1h} contribute to bias reduction in choosing $\hat{\bar{Y}}_{\mathrm{WC}}$ over $\bar{y}_1$ but a variance increase since variable λ_{1h} yields variable l_{1h}, which produce variable weights, since $W_{hi}^{(2)} \propto l_{1h}^{-1}$. Therefore when choosing to use $\hat{\bar{Y}}_{\mathrm{WC}}$, we must hope that our adjustment cells will lead to $\mathrm{MSE}(\hat{\bar{Y}}_{\mathrm{WC}}) \le \mathrm{MSE}(\bar{y}_1)$, where in general $\mathrm{MSE}(\cdot) = \mathrm{Var}(\cdot) + \mathrm{Bias}^2(\cdot)$.

Platek and Gray (1983) present an illustration of the weighting class adjustment procedure in the presence of errors from sampling, measurement, and nonresponse (from a stochastic viewpoint). Assuming that N is known, the estimator of $\bar{Y}$ would be

$$\hat{\bar{Y}}_{\mathrm{PG}} = \sum_{h=1}^{H} \sum_{i=1}^{n_h} \frac{R_{hi} W_{hi}^{(2)} V_{hi}}{\pi_{hi} N}, \qquad (8.22)$$

where for the ith member of the sample in the hth adjustment cell $W_{hi}^{(2)} = W_{hi}(n_h/n_{1h})$, $V_{hi} = Y_{hi} + \varepsilon_{1hi}$ is the value of Y_{hi} observed with measurement error equaling ε_{1hi}, and

$$R_{hi} = \begin{cases} 1 & \text{if the } hi\text{th sample member responds} \\ 0 & \text{if otherwise} \end{cases}$$

is the response status indicator. Assuming that adjustment cells have been predefined and that the adjustment is applied independently in each cell, the bias of $\hat{\bar{Y}}_{PG}$ is

$$\text{Bias}\left(\hat{\bar{Y}}_{PG}\right) = \sum_{h=1}^{H} \sum_{i=1}^{N_h} \frac{p_{hi} B_{1hi}}{\bar{p}_h} + \sum_{h=1}^{H} \sum_{i=1}^{N_h} \frac{(p_{hi} - \bar{p}_h) Y_{hi}}{\bar{p}_h}, \qquad (8.23)$$

where $B_{1hi} = E_t(\varepsilon_{1hi})$ is the expected individual measurement error (Section 7.3) and $\bar{p}_h = \sum_{i=1}^{n_h} p_{hi}/n_h$. The first component of Eq. (8.23) might be viewed as bias because of measurement error, while the second is bias attributed to within-cell correlation between response probabilities and the Y variable. The rather involved variance expression for $\hat{\bar{Y}}_{PG}$ consists of the following five additive components, attributed to:

1. Sampling
2. Measurement error arising from ε_{1hi}
3. The covariance between sampling and measurement error
4. The stochastic nature of nonresponse arising from R_{hi}
5. The covariance between R_{hi} and $R_{hj}(i \neq j)$ within adjustment cells

Although the weighting class adjustment method could conceivably be used to deal jointly with unit and item nonresponse, this option is seldom chosen since separate adjustments would be required for each data item. This can be seen by noting that the set of n_{1h} respondents at the item level would often differ in size and composition if "respondent at the item level" were defined as a sample member who participates in the survey and provides an answer to the item. When the sets of n_{1h} respondents within adjustment cells differ among items, the corresponding weighting class adjustments, $\sum_{i=1}^{n_h} W_{hi}/\sum_{i=1}^{n_{1h}} W_{hi}$, are likely to differ as well, thus requiring separate adjustments and final weights $[W_{hi}^{(2)}]$ for each item. This procedure is cumbersome and creates problems for any analysis involving combinations of items (e.g., cross-tabulations, regression modeling), which may explain why weighting class adjustments are used almost exclusively as a method to deal with unit nonresponse. The same point holds for the weighting class adjustment used in combination with poststratification.

Kalton (1983) investigates the properties of a version of the weighting class adjustment method for $H = 1$, which is implicitly hot-deck imputation (Section 8.2.2) applied at the unit level. This approach to dealing with nonresponse was first discussed by Hansen et al. (1953b) and Kish (1965) and will be called the randomized reweighting method, which when $n_1 \geq n_0 (l_1 \geq 0.5)$ works as follows. An equal probability subsample of size n_0 is selected with or without replacement from the n_1 respondents. Considering the case where there is one survey variable with missing data, each subsampled respondent is matched to a particular nonrespondent and becomes the donor unit for the nonrespondent, since the respondent's Y variable is imputed implicitly for the nonrespondent's Y variable. Imputation is implicit since application of the method calls for the sample weight of the nonrespondent to be added to the weight of its "donor," although properties of estimators when the imputation is explicit are the same [see Oh and Scheuren (1983)]. The respondent's (donor's) weight is incremented (or adjusted) by the amount equaling the nonrespondent's weight. Unlike the usual weighting class adjustment where (for $H = 1$) we assume that $\bar{y}_1 = \bar{y}_0$, in the randomized reweighting method we have $E_s(\bar{y}_0) = \bar{y}_1$, where expectation is over all samples of size n with n_1 respondents.

The distinction between the usual weighting class adjustment and randomized reweighting can be seen by examining the reweighting scheme more generally (by not requiring that $l_1 \geq 0.5$) when some equal-probability method is used to select the initial sample. Let $I = \mathrm{Tr}(l_1^{-1})$ denote $l_1^{-1} = n/n_1$ truncated to the nearest integer. Then $n_1^* = n \pmod{n_1}$ respondents in the hth adjustment cell are assigned the adjustment $I + 1$, and $(n_1 - n_1^*)$ respondents are assigned the adjustment I. By contrast, the adjustment assigned to all respondents under the weighting class procedure would be $l_1 - 1$ such that $I \leq l_1^{-1} < (I + 1)$.

Historically, randomized reweighting is a throwback to the earlier approach where computer punch cards for respondents were randomly duplicated to compensate for nonresponse and there was no reweighting of the data. This approach, conceptually identical to randomized reweighting, produces estimates with the same nonresponse bias as the survey-wide nonresponse adjustment (the weighting class adjustment with $H = 1$). Unfortunately, estimates using randomized reweighting have been shown to have less desirable variance properties than those of estimates using the survey-wide adjustment (Kalton, 1983). This may partially explain why duplication of respondent records is seldom used in present-day surveys.

The extent to which the variance of the estimator using the reweighting is greater than the variance of the estimator, $\bar{y}_1 = \sum_{i=1}^{n_1} Y_i / n_1$, which uses a survey-wide nonresponse adjustment, will depend on how the subsample of respondents is chosen in the randomized reweighting method. The more efficient the subsampling design (e.g., using stratification), the lower the amount of variance increase (Kalton and Kish, 1984).

Provided that $n_1 \geq n_0$ and that simple random sampling (without replacement) is used to choose the initial sample and the respondent subsample, the randomized reweighting estimator of $\bar{Y}$ is $\hat{\bar{Y}}_{RR} = (n_1 \bar{y}_1 + n_0 \bar{y}_d)/n$, where $\bar{y}_d$ is the mean of the n_0 respondent values assigned as donors. The proportionate increase in variance, conditioned on fixed n_1, is

$$\frac{\text{Var}\left(\hat{\bar{Y}}_{RR}\right) - \text{Var}(\bar{y}_1)}{\text{Var}(\bar{y}_1)} = (1 - l_1)(2l_1 - 1), \qquad (8.24)$$

where $l_1 = n_1/n \geq 0.5$ and the maximum proportionate increase of 0.125 occurs when $l_1 = 0.75$.

Platek and Gray (1983) present properties of a randomized reweighting scheme applied to adjustment cells and where errors associated with sampling, nonresponse (following a stochastic view), and measurement are considered. They assume that the population consists of H mutually exclusive and exhaustive adjustment cells, within which the cell total on the Y variable (Y_h) is estimated as

$$\hat{Y}_h = (I_h + 1) \sum_{i=1}^{n_{1h}^*} \frac{R_{hi} V_{hi}}{\pi_{hi}} + I_h \sum_{i=1}^{n_{1h} - n_{1h}^*} \frac{R_{hi} V_{hi}}{\pi_{hi}}, \qquad (8.25)$$

where $(n_{1h} - n_{1h}^*)$ respondents in the hth cell are assigned the adjustment $I_h = \text{Tr}(l_{1h}^{-1})$, and n_{1h}^* are assigned the adjustment $(I_h + 1)$. Assuming that the number of population members in the hth cell (N_h) is known, the population total and mean for the Y variable can be estimated as $\hat{Y}_{PG}^* = \sum_{h=1}^{H} \hat{Y}_h$ and $\hat{\bar{Y}}_{PG}^* = \hat{Y}_{PG}^*/N$, respectively. The expected value of $\hat{\bar{Y}}_{PG}^*$ from the random weighting method as described will be equivalent to the expected value of $\hat{\bar{Y}}_{PG}$ under the usual weighting class adjustment [see Eq. (8.22)], where the expectation for each estimator is taken over all stochastic sources of variation affecting each estimator. The variance of $\hat{\bar{Y}}_{PG}^*$ (or $\hat{Y}_{PG}^*$) consists of the same components as the variance of $\hat{\bar{Y}}_{PG}$ plus five similar components arising from the randomized subsampling of respondents to assign the adjustments, I_h or $I_h + 1$.

8.1.6.3 Poststratification Adjustment

Adjustment cells used for poststratification are formed in much the same way as strata for sample selection; however, they are defined by variables not available at the time the original sample was selected. They are always mutually exclusive and exhaustive subsets of the sample and one hopes that values of the Y variables in each cell are more similar than among all values in the sample. Since poststratification adjustment cells are formed after the survey has been done, the variables used to define the cells are often included in the survey questionnaire. The best poststratification variables are those strongly correlated to the Y variable. As a result, they are often correlated with individual response probabilities (if a stochastic view is taken), since the Y variable and the response probability for population members are usually correlated. Thus as with weighting class adjustment cells, we expect members of poststratification adjustment cells to have similar response probabilities.

Use of poststratification in the National Health Interview Survey (NHIS) conducted by the National Center for Health Statistics (Kovar and Poe, 1985) is fairly typical. In this survey of the civilian, noninstitutionalized population in the United States, each respondent is assigned to one of 60 age–race–sex cross-classification cells for which reliable current population figures $\Delta_h = N_h/N$ are available independent of the survey. A poststratification adjustment is computed for the hth cell ($h = 1, 2, \ldots, 60$) as

$$a_h^* = \Delta_h \frac{\sum_{h=1}^{H} \sum_{i=1}^{n_{1h}} W_{hi}^{(2)}}{\sum_{i=1}^{n_{1h}} W_{hi}^{(2)}},$$

where $W_{hi}^{(2)}$ is the raw sample weight ($W_{hi} = \pi_{hi}^{-1}$) times a weighting class adjustment. The final adjusted sample weight is $W_{hi}^{(3)} = a_h^* W_{hi}^{(2)}$, from which it can be shown that $\sum_{i=1}^{n_{1h}} W_{hi}^{(3)} / \sum_{h=1}^{H} \sum_{i=1}^{n_{1h}} W_{hi}^{(3)} = \Delta_h$. Thus we see in general that using a poststratification adjustment forces the weighted relative frequency distribution among cells to correspond precisely to the relative distribution among those same cells in the population. By using this adjustment the NHIS sample weights are in essence finally adjusted to bring the sample "into line" with the U.S. population, at least with respect to the joint distribution by age, race, and sex as defined in the 60 cells. A sample distorted by nonresponse, poor sample coverage, and sample variation now has weights allowing the weighted data more accurately to estimate parameters whose measurement (values of the Y variable) is correlated to the three poststratification variables.

Hansen et al. (1953a), Cochran (1977), and Oh and Scheuren (1983) make a number of practical suggestions regarding the formation of adjustment cells for poststratification. Several apply to the weighting class approach as well. One is that all cells should be large enough so that all n_{1h} are sufficiently great (say, 20 or more). Related to the importance of maintaining reasonable levels of n_{1h} is Thomsen and Siring's suggestion (1983) that gains from poststratification diminish as the number of adjustment cells (H) increases. The size constraints on n_{1h} and H point to the importance of considering cell-level estimates ($\bar{y}_{1h}$) when forming cells. Second, since the estimator is a function of $\Delta_h = N_h/N$, it is important for the measures of N_h to be reasonably accurate. Although perfect measures are seldom available, the measures used for N_h should be of higher quality than estimates of N_h that could be obtained directly from the survey data. Third, adjustment cells should be internally homogeneous and externally heterogeneous with respect to the Y variable. For that reason adjustment cells are sometimes determined by clustering algorithms which minimize the variance of the Y variable within cells [e.g., Morgan and Sonquist (1963) and Gillo and Shelley (1974)]. Fourth, the variables used for poststratification are always categorical (or categorized) and should all be highly correlated with the Y variable but not with each other. Adjustment cells that are internally homogeneous with respect to highly correlated poststratification variables will also be internally homogeneous with respect to the Y variable. Finally, in the case where one has several suitable poststratification variables from which to choose, it is usually preferable to form cells by crossing a few levels on all variables than to pick the best one and divide it into numerous levels.

Consider next the special case, where the initial sample is chosen by simple random sampling, the same adjustment cells are used for the weighting class and stratification adjustments, and the mean $(\bar{Y})$ is to be estimated. Kalton (1983) presents statistical properties of the estimator that uses the weighting class and poststratification adjustments, namely,

$$\hat{\bar{Y}}^*_{P-S} = \sum_{h=1}^{H} \Delta_h \bar{y}_{1h} = N^{-1} \sum_{h=1}^{H} N_h \frac{y_{1h}}{n_{1h}}$$

$$= \frac{\sum_{h=1}^{H} \sum_{i=1}^{n_{1h}} W_{hi}^{(3)} Y_{hi}}{\sum_{h=1}^{H} \sum_{i=1}^{n_{1h}} W_{hi}^{(3)}}, \tag{8.26}$$

where $\bar{y}_{1h} = \sum_{i=1}^{n_{1h}} Y_{hi}/n_{1h} = y_{1h}/n_{1h}$ and $W_{hi}^{(3)} = N_h/n_{1h}$. First note the distinction between $\hat{\bar{Y}}^*_{P-S}$ and two other estimators: (1) the poststratification estimator in the absence of nonresponse, $\bar{Y}_{P-S} = \sum_{h=1}^{H} \Delta_h \bar{y}_{1h}$, where $\bar{y}_h = \sum_{i=1}^{n_h} Y_{hi}/n_h$, and (2) the estimator $\hat{\bar{Y}}_{WC} = \sum_{h=1}^{H} \delta_h \bar{y}_{1h}$, using the weighting class adjustment only. We might reasonably expect that $\bar{Y}_{P-S}$ is the best of the three, although it makes the unrealistic assumption that there is complete response in the survey. In presuming the presence of nonresponse, the estimator $\hat{\bar{Y}}^*_{P-S}$ uses an adjustment that is the product of the weighting class and poststratification adjustments. It assumes that there reasonable measures for the Δ_h are available for the population and compensates for problems caused by an inadequate sampling frame (see Chapters 3 through 5) as well as nonresponse (e.g., in random-digit telephone surveys of the general population where some households do not have telephones and some of those that do would choose not to respond). Finally, $\hat{\bar{Y}}_{WC}$ uses data available from the sample alone to compensate for nonresponse. Since δ_h, a sample estimator of Δ_h, is used in $\hat{\bar{Y}}^*_{WC}$, we expect it to be subject to greater variation than $\bar{Y}_{P-S}$ and therefore less preferred.

One also notes from the first line of Eq. (8.26) the relationship between $\hat{\bar{Y}}_{\text{P}-\text{S}}$ and the separate ratio estimator of the mean presented by Hansen et al. (1953a) and discussed further by Cochran (1977, Sec. 6.10). The auxiliary X variables in the separate ratio estimator may be likened to a "count" variable which equals 1 for all members of the population. The sample data on the Y variable and X variable used to produce the estimate are limited to those who respond. To complete the analogy between $\hat{\bar{Y}}^*_{\text{P}-\text{S}}$ and the separate ratio estimator, we must assume for the former that respondents in each adjustment cell can be treated as a simple random subsample of those originally chosen.

Given the foregoing setting and that $E_s(\delta_h) = \Delta_h$, it is not surprising to find that

$$\text{Bias}\left(\hat{\bar{Y}}^*_{\text{P}-\text{S}}\right) = \text{Bias}\left(\hat{\bar{Y}}_{\text{WC}}\right) = \sum_{h=1}^{H} \Delta_h \lambda_{0h}\left(\bar{Y}_{1h} - \bar{Y}_{0h}\right), \qquad (8.27)$$

which implies that, as with $\hat{\bar{Y}}_{\text{WC}}$, the amount of nonresponse bias can be reduced to the extent that cells with equal respondent and nonrespondent means are formed. The variance of $\hat{\bar{Y}}^*_{\text{P}-\text{S}}$ can be expressed as

$$\text{Var}\left(\hat{\bar{Y}}^*_{\text{P}-\text{S}}\right) \doteq \text{Var}\left(\hat{\bar{Y}}_{\text{WC}}\right)$$
$$- \left[\frac{\sum_{h=1}^{H}(1 - \lambda_{1h})S_{1h}^2}{\lambda_{1h}^2 n^2} + \frac{\sum_{h=1}^{H}\Delta_h\left(\bar{Y}_{1h} - \bar{Y}_s\right)^2}{n} \right], \qquad (8.28)$$

where $\bar{Y}_s = \sum_{h=1}^{H}\Delta_h\bar{Y}_{1h}$. Since the biases of $\hat{\bar{Y}}^*_{\text{P}-\text{S}}$ and $\hat{\bar{Y}}_{\text{WC}}$ are equivalent and the bracketed term subtracted in Eq. (8.28) is greater than zero, we see that the variance and mean square error are always smaller for $\hat{\bar{Y}}^*_{\text{P}-\text{S}}$ than for $\hat{\bar{Y}}_{\text{WC}}$. This indicates that poststratification and the weighting class procedures applied to the same adjustment cells appear to complement each other in improving estimates, since when used in combination estimates are better than when the weighting class method is used alone. This finding then confirms our prior expectations. The reader is referred to Kalton (1983), the Panel on Incomplete Data (National Research Council, 1983), Oh and Scheuren (1983), and Thomsen and Siring (1983) for further discussion of the statistical and practical implications of the weighting class and poststratification approaches.

In general, we have seen that the adjustment methods operate by modifying the original sample weight ($W_i = 1/\pi_i$), which may vary because of the sampling design. The adjusted weight is computed as $W_i^* = W_i/\hat{p}_i$, where $\hat{p}_i$

as the estimator of the response probability for the ith sample member is likely to vary among survey respondents. Except for those instances where W_i and $\hat{p}_i$ are positively correlated, the variation in adjusted weights will exceed the variation of the original sample weights, since the $\hat{p}_i$ vary among respondents. The effect of added variation caused by the adjustments is to increase the variance of survey estimates.

One way to control the variation of adjusted weights is to smooth weights by setting upper and lower limits on the adjustments. For example, Banks et al. (1983) report setting an upper bound of 2 and a lower bound of 0.5 on the poststratification adjustments used in a 1970 national survey on health service use and expenditures in the United States.

In some instances, the excess from truncated weights will be spread over the sample, as seen in the following simple example. Suppose that for a set of n weights (W_i), we specify that all weights exceeding W' are to be trimmed and smoothed. Weights for the n'_1 respondents with $W_i > W'$ will have an excess $(\varepsilon_i = W_i - W')$ which must be spread over the sample so that the sum of weights over the n_1 respondents is the same before and after smoothing. One approach to spreading the excess would be to adjust each of the n_1 weights by the factor $\phi = \sum_{i=1}^{n_1} W_i / (\sum_{i=1}^{n_1} W_i - \sum_{i=1}^{n'_1} \varepsilon_i) \geq 1$, in which case the largest possible weight after smoothing will be $\phi W'$. If the smoothed weights must not exceed W', the excess can be smoothed iteratively, with the excess of each iteration smoothed over the remaining set of untruncated weights.

Truncating adjusted weights can also be done by pooling the data. This practice is followed in *The Current Population Survey* [see U.S. Bureau of the Census (1978)]. Weighting class adjustments exceeding 2.0 are smoothed by pooling counts of respondents and nonrespondents in the exceptional adjustment cell with counts in other cells with similar demographic characteristics. In the unlikely event that the pooled adjustment factor exceeds 3.0, it is set equal to 3.0. For example, an adjustment cell with 20 households and a response rate of 25 percent (implying a nonallowable adjustment of $a_h = 4.0$), when combined with a group of cells with 100 households and an overall response rate of 95 percent, would yield an allowable $a_h = 1.2$ for all respondents in the pooled cells.

Variation in sample weights can also be controlled by what has been called the *raking-ratio adjustment* or *iterative proportional-fitting method*. First suggested by Deming and Stephan (1940), the object of the raking method is to circumvent some of the instability of adjustments produced from cross-classifications of two or more adjustment variables by adjusting weights so that marginal distributions rather than joint distribution of the variables obtained external to the survey are maintained. The procedure assumes a log-linear model relating adjustment cell response rates to marginal effects of the adjustment variables.

As evident from this brief description, raking methods require only that the marginal frequencies for all raking variables be known. That joint frequencies are not required turns out to be a practical advantage of the method since marginal frequencies are often the only ones known or for which reasonable approximations are available.

Operationally, raking is similar to poststratification on two or more variables in that a single adjustment is the end product for each respondent. It differs from poststratification in that the process of computing each raking-ratio adjustment is iterative rather than a single computation. As with poststratification and the weighting class method, one presumes that probabilities within adjustment cells are equal. Moreover, rules for identifying adjustment variables and forming adjustment cells are the same as with the other two methods. In some instances raking might be used in combination with the other two methods. Oh and Scheuren (1978) present a bibliography of the use of raking methods.

Oh and Scheuren (1983) also describe a general protocol for raking when adjustment is over two variables, one with H categories, the other with K categories. If we let $n_{1hk}(h = 1, 2, \ldots, H; k = 1, 2, \ldots, K)$ denote the number of respondents in each cell of the cross-classification, and let W_{hki} be the weight to be adjusted for the ith respondent in each cell, then we start with a weighted sum, $\sum_{i=1}^{n_{1hk}} W_{hki}$, in each cell. If N_{hk} denotes the true population count for the hkth cell, the marginal frequencies $N_{h.} = \sum_k^K N_{hk}$ and $N_{hk} = \sum_h N_{.k}$ are assumed to be known (or at least well approximated) from auxiliary sources. Through an iterative process beginning with a seed value $(a_{hki}^{(0)})$, an adjustment (a_{hki}^*) is computed for each weight such that $\sum_{h=1}^H \sum_{i=1}^{n_{1hk}} a_{hki}^* W_{hki}$ and $\sum_{k=1}^K \sum_{i=1}^{n_{1hk}} a_{hki}^* W_{hki}$ are acceptably close to $N_{.k}$ and $N_{h.}$, respectively, for all $h = 1, 2, \ldots, H$ and $k = 1, 2, \ldots, K$. For example, starting with $a_{hki}^{(0)} = 1$ and $W_{hki}^{(0)} = W_{hki}$, we might compute for the vth iteration the raking-ratio adjustment,

$$
a_{hki}^{(v)} = \frac{N_{h.} N_{.k}}{\left[\sum_{k=1}^K \sum_{i=1}^{n_{1hk}} W_{hki}^{(v-1)}\right] \left[\sum_{h=1}^H \sum_{i=1}^{n_{1hk}} W_{hki}^{(v-1)}\right]}, \tag{8.29}
$$

where in general $W_{hki}^{(v)} = W_{hki}^{(v-1)} a_{hki}^{(v-1)} (v = 1, 2, \ldots)$. At each iteration the differences between the marginal weighted sums, $\sum_{h=1}^H \sum_{i=1}^{n_{1hk}} a_{hki}^{(v)} W_{hki}^{(v)}$ and $\sum_{k=1}^K \sum_{i=1}^{n_{1hk}} a_{hki}^{(v)} W_{hki}^{(v)}$, and the marginals for the auxiliary data, $N_{.k}$ and $N_{h.}$, respectively, would be computed to see whether adjusted weights have converged to the auxiliary marginals. If prespecified convergence criteria are eventually met on the Vth iteration, the final raking adjustment for W_{hki} is $a_{hki}^* = \prod_{v=1}^V a_{hki}^{(v)}$, and the final weight for analysis is $W_{hki}^* = a_{hki}^* W_{hki}$. As we see, then, raking is named for the r, the scaling of unadjusted weights, with the extent and direction of scaling (up or down) dependent on how the weighted relative frequency of each cell stacks up in producing some known marginal frequencies.

A cursory assessment of the raking-ratio adjustment technique reveals some limitations. One problem is that convergence of adjusted marginals to the auxiliary marginals is not assured and little is presently known about under what conditions convergence will occur. Much remains to be learned about this adjustment approach before it will be widely used, although several authors have already investigated the properties of estimates using raking-ratio adjustments [see, e.g., Brackstone and Rao (1981), Konijn (1981), and Kalton (1983)].

8.1.7 Bayesian and Other Model-Based Methods

Model-based methods are distinguished from classical methods in that true observable measures of the variable of interest (the Y variable) are considered to be the result of an assumed stochastic process, as opposed to being treated as fixed quantities. In effect, the vector $\mathbf{Y}' = (Y_1, Y_2, \ldots, Y_N)$ is a vector of random variables in model-based methods and a vector of fixed constants in classical methods.

In this section we focus our attention on ways in which the model-based approach to inference has been used in dealing with the nonresponse problem. We note in our discussion that the model in model-based approaches to nonresponses assumes a stochastic mechanism for nonresponse. The stochastic form specified for this mechanism is used to produce estimators that compensate for the nonresponse. We also noted that probability sampling, though not necessary for many model-based methods, is advisable, since it can produce samples suitably balanced in their representation of the population sampled. Quota sampling is thought by some to be sufficient because of the balance it produces through quota controls (King, 1983).

Stochastic models used to explain the outcome of $\mathbf{Y}$ may take several forms. Some models assume the Y_i to be random variables with common moments but make no specific assumptions concerning the shape of the distribution for Y_i. In other models the Y_i are assumed to follow some particular distribution function, $f(Y_i; \boldsymbol{\theta}_i)$, with an unknown but estimable L-dimensional parameter vector, $\boldsymbol{\theta}_i' = (\theta_{i1}, \theta_{i2}, \ldots, \theta_{iL})$. In most applications the Y_i are assumed to be mutually independent. Bayesian methods use these kinds of parametric models and add further prior stochastic assumptions concerning the parameters of the assumed distribution. In another type of model one assumes that the Y_i are related to a $(K + 1)$-dimensional vector of auxiliary measures $\mathbf{X}_i' = (X_{i0}, X_{i1}, \ldots, X_{ik})$ through some linear regression model such that $E_n(Y_i) = \boldsymbol{\beta}_i' \mathbf{X}_i$, where $E_m(\cdot)$ denotes expectation over outcomes of the model and $\boldsymbol{\beta}_i' = (\beta_{i0}, \beta_{i1}, \ldots, \beta_{ik})$ is a vector of unknown regression coefficients, thus implying zero expectation for random error attributable to the model. Any of the models above might be *exchangeable* if the joint distributional function for all possible permutations of $\mathbf{Y}$ is symmetric in its N terms.

As mentioned in Section 7.3, many model-based methods that accommodate the nonresponse problem use respondent data combined with assumptions about the joint distribution explaining the origin of $(\mathbf{Y})$, the vector of stochastic response outcomes $(\mathbf{R})$, and the vector of outcomes (D) from some sampling "mechanism" to make inference. By implication, then, these methods follow the stochastic view of nonresponse since the probability of outcomes to entries in **R** is tied to the vector of response probabilities, $\mathbf{p}' = (p_1, p_2, \ldots, p_N)$.

Fundamental to this perspective of statistical inference is the function of the sampling and response mechanisms. The sampling mechanism in some applications may be considered "ignorable," meaning that the probability of **D** conditioned on **Y**, **R**, and the sampling design depends only on features of the sampling design (e.g., stratification and clustering) and on **y** and **r**, the respective analogs of **Y** and **R** for the selected sample of size n. Because of the use of randomization in the selection process, the sampling mechanism is ignorable in model-based inference where probability sampling has been used. This situation implies that while probability sampling is not objectionable in this particular model-based setting, it does not play any statistical role in the analysis of data. The response mechanism leading to **R** may also be considered ignorable. This form of ignorability in a model with an ignorable sampling mechanism will occur if the probability of **R**, conditioned on Y and the sampling design, depends only on **Y**.

In nonignorable response mechanisms **R** and **Y** are generally assumed to be related in some specified manner. Whereas conceptualizing the relationship between the response outcome and measurement of interest in nonignorable response models has the advantage of formalizing subjective notations about the impact of nonresponse in surveys, its disadvantage is the added complexity it brings to inference. The need to consider nonignorable response models seriously seems clear in light of recent work by Greenlees et al. (1982), who present evidence that response probabilities are dependent on income, education, and other measures often the object of study.

Although all model-based methods rely on assumptions concerning the origins of **Y**, Bayesian and non-Bayesian methods are distinguished by how the assumed model is used. The theoretical basis for non-Bayesian methods are models that explain **Y**, in some instances using auxiliary information. These models in effect explain the process by which the population sampled is assumed to have been itself generated from some "superpopulation," associated with **Y**. Bayesian methods take things a step further by using the model to incorporate prior notions concerning parameters of the underlying distribution of **Y**. Speculation concerning the size of distributional parameters is allowed and incorporated into the process of making inference.

Despite their varying frameworks, both types of model-based approaches use the assumed model to predict values of the Y variable for nonresponding and unsampled population members. In Bayesian model-based methods, inference is made by revising stochastic prior information based on current information obtained from sample respondents. Respondent data thereby serve to confirm or modify information from the model which is then used to produce estimates. In non-Bayesian model-based methods, on the other hand, the respondent data are used to fit the assumed model, which is then used to make inferences about nonresponding sample members and other population members not sampled. This predictive nature of inference in both types of methods explains why some use the term *predictive approach* to describe these methods.

The use of model-based methods to deal with survey nonresponse in particular and for survey inference in general has evoked much debate. Most of the points raised in this debate have been summarized by Hansen et al. (1983), Rubin (1983), Little (1983), Cassel et al. (1983), and Little and Rubin (1987). Proponents of model-based methods argue that given an appropriate model their estimators have more desirable properties than do those developed by more classical methods, where the randomization used to draw the sample and assumptions concerning the relationship between respondents and nonrespondents form the basis for inference. For example, bias arising from the sampling design is less important in some model-based estimators with a correctly assumed model than in classical methods, where the type of sampling design affects the bias. Advocates of classical or traditional inference counter by arguing that the utility of model-based inference is heavily dependent on how closely assumed models reflect reality. Bias from an incorrectly specified model, they argue, may cancel out any gains, to the point that classical estimators would be preferable overall.

It has been noted that even classical approaches to the nonresponse problem are to some extent model dependent [see Norris (1983)]. Some adjustment methods, for example, implicitly treat respondents within an adjustment cell as a random subset of all sample members in that cell. Oh and Scheuren (1983) term this model of the response mechanism *quasi-randomization*, since it is assumed rather than supplied stochastic behavior. Moreover, some classical methods in adopting the stochastic view of nonresponse assume that response probabilities follow a certain pattern in the population. For example, a key supposition of the weighting class adjustment is that values of p_i collectively differ among adjustment cells but are equal within cells. Finally, classical inference also relies on assumptions concerning the distribution of the estimator among all possible samples. Normality, for example, must be assumed in completing inference (e.g., through hypothesis testing or interval estimation).

Having acknowledged the existence of models, classical theorists might claim that their estimators are more robust to assumptions that depart from reality. The model-based theorist could counter by arguing that chances of model misspecification can often be reduced. One way is to substantiate models through pilot studies conducted during the design phase. In some models the robustness of estimates against departures from the assumed model can be enhanced by obtained a "balanced" sample of respondents (Schaible, 1983). Moreover, one might apply several different models during analysis to test the sensitivity of findings to changes in the model.

We speculate that two reasons explain why the class of approaches popularly called *model-based methods* is not widely used in current survey practice. One (already discussed) is concern about the implication of misspecified models and an unwillingness to rely on often unverifiable assumption. Perhaps it seems safer to make inferences from random behavior that is created rather than assumed. A second important reason may be the greater practical difficulty in using the model-based methods. Perhaps it seems simpler to produce a single set of weighting adjustments that can be applied to several different Y variables than to specify and deal with the numerous separate models needed for each Y variable. Producing a large number of analyses involving many Y variables seems to be a less complex and more manageable task with most classical adjustment methods than with most model-based methods.

It is interesting and somewhat reassuring to note that in some instances classical and model-based methods lead to the same statements of inference when nonresponse is present. For example, Rubin (1983) shows that Bayesian inference with an ignorable response model and probability sampling leads to essentially the same 95 percent confidence interval for the population mean ($\bar{Y}$) as classical inference. In both cases the mean of sample respondents ($\bar{y}_1$) serves as the point estimate.

Because of the clear distinction in both approach and evolution between Bayesian and non-Bayesian model-based methods, contributions to the two approaches are discussed separately in the remainder of this section. We begin with a review of some of Bayesian methods for dealing with nonresponse, many of which are discussed by Singh (1983) and Rubin (1983). One Bayesian approach developed by Ericson (1967) for use in connection with nonrespondent subsampling has been discussed earlier (Section 8.1.4).

The end result of the predictive Bayesian approach Rubin (1977) suggests is an interval estimate of the overall sample mean ($\bar{y}$) based on the respondent data and an assumed multivariate linear regression model using auxiliary information available for respondents and nonrespondents. More specifically, the mean $\bar{y} = l_1 \bar{y}_1 + (1 - l_1)\bar{y}_0$, from a simple random sample of size n, is inferred by predicting the mean among sample nonrespondents ($\bar{y}_0$) by fitting the model

$$y_{gi} = \beta_{g0} + \boldsymbol{\beta}'_g \mathbf{x}_{gi} + \varepsilon_{gi} \qquad i = 1, 2, \ldots, n, \tag{8.30}$$

where lowercase x and y refer to measures associated with the sample; the subscript g indexes the case where the ith sample member is a member of the respondent stratum in the population ($g = 1$) or a member of the nonrespondent stratum ($g = 0$); β_{g0} and $\boldsymbol{\beta}'_g = (\beta_{g1}, \beta_{g2}, \ldots, \beta_{gK})$ are regression coefficients corresponding to the vector of auxiliary measures $\mathbf{x}'_{gi} = (x_{gi1}, x_{gi2}, \ldots, x_{giK})$; and the ε_{gi} are mutually independent error terms with $\varepsilon_{gi} \sim N(0, \sigma_g^2)$. The posterior distribution of $\eta_1 = (\boldsymbol{\beta}_{10}, \boldsymbol{\beta}_1, \sigma_1^2)$ is obtained from the prior distribution for η_1 and the likelihood expression for η_1 given the respondent data. This posterior distribution is used to predict or impute the vector of measures for sample nonrespondents $\mathbf{y}'_0 = (y_{01}, y_{02}, \ldots, y_{0n_0})$

by assuming that the parameters of the distribution of y_0 are centered on the comparable distribution for respondents, subject to some assumptions concerning the departure of these parameters about their expectation. The predicted mean for sample nonrespondents, $\bar{y}_0$, is then obtained directly from y_0.

Singh and Sedransk (1978a) give another approach where auxiliary data are involved in Bayesian inference. Here, however, the regression model itself is the object of study and Bayesian methods combined with two phases of sampling are used to make inference about the model's coefficients. Singh and Sedransk assume the linear regression model,

$$y_{gi} = \bar{y}_g + \beta'_g x^*_{gi} + \varepsilon_{gi}, \tag{8.31}$$

where the kth member of x^*_{gi} is $(x_{gki} - \bar{x}_{gk})$, with $\bar{x}_{gk} = \sum_{i=1}^{n_g} x_{gki}/n_g$, and the ε_{gi} are once again mutually independent but with $\varepsilon_{gi} \sim N(0, h_g)$. Assuming a joint prior distribution on β_g and h_g with separate marginal prior on λ_1, a first-phase sample, consisting of an initial sample followed by a random subsampling of nonrespondents, is selected. Conditioning on the resulting sample data from the first-phase sample, the posterior distribution of λ_1 and $\tau_g = (\beta, h_g)$ is obtained for $g = 0, 1$. This posterior distribution from the first-phase sample is treated as the prior distribution to which data from an independently chosen second-phase sample will be applied. As before, the initial sampling is follows by a subsampling of the initial nonrespondents. The likelihood function based on data from the second-phase sample is combined with the second phase prior to obtaining the posterior distribution of λ_1 and $\tau_g(g = 0, 1)$. Estimators of β_0 and β_1 are obtained as means of their respective marginal posterior distributions.

This same type of two-phase sampling has been used in other Bayesian approaches. In each case the first-phase sample serves as a preliminary to the main study. This technique is helpful when information used to formulate prior distributions for the main study, represented by the data collected in the second phase, is sketchy. Rao and Ghangurde (1972) use stratified two-phase sampling with nonrespondent subsampling as well as simple random sampling with poststratification to obtain an estimator of the population mean ($\bar{Y}$) subject to budget constraints. Singh and Sedransk (1978b) consider use of two-phase sampling combined with poststratification to estimate $\bar{Y}$. Kaufman and King (1973) suggest nonrespondent sampling but limit sampling to a single phase to establish the optimum initial sample size (n) and estimator for the sample proportion by minimizing assumed loss functions for each measure to be optimized. In addition to basing prior distributions on pilot samples, Rubin (1983) suggests that multiple imputation holds promise in simulating reasonable priors as well (see Section 8.2.2).

To summarize our discussion of Bayesian methods for dealing with nonresponse, we note first that most approaches assume some kind of probability sampling, whether from a finite or infinite population, although as Cassel et al. (1983) point out, neither the sampling distributions nor the bias associated with the designs are of relevance in the Bayesian logic of inference. Second, inference about the parameter of interest is often based on data from those who did respond and the posterior distribution of values for those in the population who did not respond, given some a priori assumptions. Unlike many classical methods that make assumptions about similarities between respondents and nonrespondents, Bayesian methods generally use assumptions concerning how they may differ, in some instances formalizing uncertainty about these differences [see Rubin (1977)]. Being model dependent, the assumptions needed in specifying prior distributions of important parameters are their main point of vulnerability, although some remedial suggestions have been made.

We move next to a discussion of some non-Bayesian model-based methods suggested to compensate for the problem of survey nonresponse. This class of methods is discussed separately from Bayesian methods, since the former, while once again making assumptions about the stochastic process that produces $\mathbf{Y}$, does not use a priori assumptions concerning parameters of the model describing that process. Inference is based on the properties of estimators of superpopulation parameters over repeat applications of the model that produces $\mathbf{Y}$, with $E_m(\cdot)$ used to denote expectation and $\text{Var}_m(\cdot)$, the variance over possible outcomes of this model. Inference at some point in these methods extends beyond the survey population to the superpopulation through the model on $\mathbf{Y}$, although estimated superpopulation parameters may be used to predict measures for those who fail to respond in the survey. Evaluation of the mean square error must thus include expectation over all possible outcomes of the model that produced the finite population.

Cassel et al. (1983) presented one non-Bayesian model-based approach to the nonresponse issue. Their general approach is similar to many classical adjustment procedures in that Horvitz–Thompson-like estimators are formulated by inflating the Y_i by $(\pi_i p_i)^{-1}$. This implies that information is needed concerning the N-dimensional vector of selection probabilities ($\boldsymbol{\pi}$) resulting from the indicator vector of sampling outcomes ($\mathbf{D}$) and the vector of response probabilities (p) resulting from the stochastic indicator vector of outcomes to the response process ($\mathbf{R}$). Inference involving the population mean ($\overline{Y}$) or total (Y) occurs in the following sequence:

1. Specify the stochastic nature of $\mathbf{Y}$ by assuming that it follows some distributional form (e.g., the normal distribution) or by relating it through some regression model to auxiliary data.
2. Formulate a reasonable prediction estimator ($\hat{\mathbf{Y}}$) of individual entries for $\mathbf{Y}$ using $\boldsymbol{\pi}$, $\mathbf{p}$, and estimated parameters for the model defining $\mathbf{Y}$.
3. Estimate $\overline{Y}$ and Y as the mean and sum of entries to $\hat{\mathbf{Y}}$, respectively.

When the object of analysis is the study of the relationship between the Y variable and the auxiliary variables, then Y is the result of a regression model with coefficients estimated using $\boldsymbol{\pi}$, $\boldsymbol{p}$, and the respondent data. The criterion for evaluating an estimator is its error evaluated jointly over $\mathbf{D}$, $\mathbf{R}$, and $\mathbf{Y}$.

The Cassel et al. approach is characterized by the multivariate linear model, assumed to have generated Y_i. In addition to assumptions concerning the stochastic nature of sampling and nonresponse, members of $\mathbf{Y}$ are assumed to be independent, and for all $i = 1,2,\ldots, N$,

$$E_m(Y_i) = \boldsymbol{\beta}'\mathbf{X}_i \qquad \mathrm{Var}_m(Y_i) = v_i\sigma^2, \tag{8.32}$$

where v_i and σ^2 are unknown constants, $\boldsymbol{\beta}' = (\beta_0, \beta_1, \ldots, \beta_k)$ is a vector of unknown regression coefficients corresponding to the vector of auxiliary variables $\mathbf{X}_i' = (X_{i0}, X_i, \ldots, X_{ik})$, $\mathrm{Var}_m(\cdot)$ denotes the variance of all outcomes of the assumed multivariate model, and $X_{i0} = 1$ if the model has an intercept (X_{i0} and β_0 do not exist otherwise). The Cassel et al. model also includes assumptions concerning randomization associated with the sampling design and a stochastic view of nonresponse. With this model as the framework, properties are presented for the estimator of $\overline{Y}$,

$$\hat{\overline{Y}}_{\mathrm{CSW}} = \sum_{i=1}^{N} \frac{\hat{\boldsymbol{\beta}}_{1i}'\mathbf{X}_i}{N}, \tag{8.33}$$

where

$$\hat{\boldsymbol{\beta}}_{1i} = \left(X_1'\Pi_1^{-1}P_1^{-1}V_1^{-1}X_1\right)^{-1} X_1'\Pi_1^{-1}P_1^{-1}V_1^{-1}X_1, \tag{8.34}$$

X_1 is an $n_1 \times (K + 1)$ matrix whose ith row (corresponding to the ith sample respondent) is X_i', and $\Pi_1 = \mathrm{diag}(N\pi_i)$, $P_1 = \mathrm{diag}(p_i)$, and $V_1 = \mathrm{diag}(v_i)$ are all $n_1 \times n_1$ diagonal matrices. It is apparent from Eq. (8.33) that $\hat{\overline{Y}}_{\mathrm{CSW}}$ resembles the classical regression estimator although, because of the model, its properties differ somewhat. The existence of P_1 in the formulation of the CSW estimator suggests that the p_i are known, although in practice this is seldom the case. Some approaches to estimating p_i are discussed in Sections 7.2 and 8.1.5. When $v_i = \sum_{k=0}^{K} z_k X_{ik}$, with the constants z_k not depending on the X variable, the expected value of $\hat{\overline{Y}}_{\mathrm{CSW}}$ over all possible outcomes of the stochastic response (or nonresponse) mechanism and all possible outcomes of the sampling design is approximately $\overline{Y}$, meaning that the estimator is nearly response and design unbiased. This property holds whether or not the model of Eq. (8.32) is true. When the model is true, $\hat{\overline{Y}}_{\mathrm{CSW}}$ will obviously also be model unbiased.

In the simple case where $K = 1$ with no intercept so that $v_i = X_i$ in the model and the sample is chosen by simple random sampling, $\hat{\overline{Y}}_{\mathrm{CSW}}$ resembles the well-known ratio estimator of the mean [see Cochran (1977, Chap. 6)]. When compared to the ratio estimator not adjusting p_i, $\hat{\overline{Y}}_{\mathrm{UR}} = \overline{X}\sum_{i=1}^{n_1} Y_i / \sum_{i=1}^{n_1} X_i$, it can be shown that the overall mean square error of

$\hat{\overline{Y}}_{\text{CSW}}$ over all outcomes of the sampling design, response mechanism, and model will be at least as large as for $\hat{\overline{Y}}_{\text{UR}}$ if the simplified model is realistic. When the model is unrealistic, $\hat{\overline{Y}}_{\text{UR}}$ will have a larger bias but a smaller variance than $\hat{\overline{Y}}_{\text{CSW}}$. The choice between estimators would then depend on the variability among the p_i and degree to which the model misrepresents reality. When the response probabilities are unknown and must be estimated, the error resulting from the estimated p_i would contribute further to components of the mean square error of $\hat{\overline{Y}}_{\text{CSW}}$ and thus affect our choice between estimators.

The model-based methods that Little (1983) discusses use the principle of maximum likelihood for inference. In addition to assumptions concerning the stochastic origin of the data vector (**Y**), the likelihood method requires assumptions about the random vectors **D** and **R**. The steps of inference for estimating the population total (Y) and mean $(\overline{Y})$ are as follows:

1. Specify a joint distribution for **D**, **R**, and **Y** with an associated vector of unknown parameters $\boldsymbol{\theta}$. As with the CSW approach, the ith member of **Y** may be related to a set of auxiliary variables through a regression model on the $(K + 1)$-dimensional vector $\mathbf{X}_i$.

2. The distribution in step 1 is used to determine the distribution of **D**, **R**, and **Y** in a sample from the population.

3. Using the observed data on **D**, **R**, and **Y** from the sample, the maximum likelihood principle is used to determine the value of $\boldsymbol{\theta}$ that is most likely to have produced the observed data.

4. The maximum likelihood estimator of $\boldsymbol{\theta}$ obtained in step 3 is then used to predict values for those either not sampled or sampled but not responding. This yields a predicted vector for **Y** which we might denote as $\hat{\mathbf{Y}}$.

5. The sum and average of the entries of $\hat{Y}$ are used to estimate Y and $\overline{Y}$, respectively. Methods following this likelihood approach become Bayesian when a prior distribution is attached to $\boldsymbol{\theta}$, and inference is based on the posterior distribution of **Y** given observed sample data.

Like the CSW model-based methods, the likelihood methods that Little describes use predictive modeling to obtain $\hat{\mathbf{Y}}$. The methods differ, however, in that the former uses adapted estimators from classical sampling theory, whereas the latter uses maximum likelihood as the basis for estimation. In addition to inference for descriptive measures such as Y and $\overline{Y}$, the maximum likelihood approach can be used to provide inference on the regression involving the Y variable and auxiliary variables as well as other parameters of the superpopulation assumed to have generated the population sampled. These parameters are part of $\boldsymbol{\theta}$, whose estimates are obtained as intermediate results to the obtained as intermediate results to the process described above.

Inference by maximum likelihood with nonresponse present is somewhat tedious because solutions to likelihood equations often cannot be solved directly, thus requiring iterative computational methods [e.g., using the expectation-maximization algorithm of Dempster et al. (1977)]. The need to use iterative methods may be determined by the pattern of nonresponse among items for the n units sampled. For "monotone" patterns of nonresponse among items (in which for the ith unit, nonresponse on the jth item, implies nonresponse on the kth item for all $k > j$), noniterative maximum likelihood solutions are often possible, however.

Although all models that Little presents assume probability sampling so that the sampling mechanism is ignorable (**D** can be excluded), both ignorable and nonignorable response mechanisms are considered. Likelihood methods with nonignorable response mechanisms are characterized by specific assumptions concerning the relationship between R_i and Y_i. One such model is given by Nelson (1977), who when dealing with interval-scaled measures, defines

$$C_i = \beta_0 + \beta_1 Q_i + \beta_2 Y_i + \varepsilon_i, \qquad (8.35)$$

where β_0, β_1, and β_2 are unknown parameters; ε_i a measure of random noise in the model is uncorrelated with Q_i and Y_i; and $E_m(\varepsilon_i) = 0$. One requires then that $R_i = 1$ if and only if $Y_i \geq C_i$, where C_i is a stochastic threshold point. Inference is made by estimating the βs from available data. Heckman (1976) presents a similar censoring approach. Greenlees et al. (1982) assume a logistic model relating Y_i and p_i. Most work with censored models of this type assumes C_i known, and we note that the condition for ignorability for this response mechanism is for $\beta_2 = 0$. For categorical measures Bayesian-type priors must often be attached to parameters in models of this type with nonignorable response mechanisms, since without the prior information the likelihood function of the observed data is flat.

8.2 METHODS DEALING WITH ITEM NONRESPONSE

Until now the methods presented in Chapter 8 have been those used to deal with unit nonresponse, although many, especially the Bayesian model-based methods just discussed, are specifically applicable to nonresponse at the unit and item levels. In the remaining sections of this chapter we consider methods typically thought of as remedies to item nonresponse. These are methods which, as we shall see, help the investigator by filling "gaps" caused by nonresponse among the data items obtained for the n_1 respondents from the original sample. One strategy calls for preventive measures which serve to reduce the chances that participants will fail to respond or provide useful data for individual questionnaire items. In the other strategy, gaps remaining after data collection are filled by some type of imputation in which a numerical measure replaces the missing item.

8.2.1 Preventive Measures

The preventive steps described here are those that would be implemented after location and solicitation has been completed and the sample member has agreed, in principle at least, to complete the survey questionnaire. These steps are taken to increase the likelihood that the participant will provide useful data for all questionnaire items.

To understand the motivation for the preventive measures taken in practice, one must understand what qualities of the survey contribute to increasing the incidence of missing data for some questionnaire items. One criterion is the type of question. It is well known, for example, that sensitive questions, such as those on income, produce higher item nonresponse rates than do nonsensitive questions. The mode of data collection may also be important. Groves and Kahn (1979) note that interview surveys conducted by telephone tend to have more missing data items than do those conducted in person.

Several other studies designed to better understand the reasons for item nonresponse have focused specifically on self-reporting questionnaries in mail surveys. Donald (1960), for example, found relatively low rates of item nonresponse on questions dealing with the participants' current behavior. Ferber (1966), in perhaps the first major study to investigate the effects of participant characteristics on item nonresponse in mail surveys, found participants' age, sex, occupation, and education to be important correlates of item nonresponse. Ferber's study also noted higher item nonresponse rates on questions requiring substantial thought or effort on the part of the respondent. Francis and Busch (1975) later confirmed most of Ferber's findings. A more recent study by Craig and McCann (1978) identifies a significant age and occupation effect while noting that the extent of item nonresponse did not seem to depend on questionnaire length. Messmer and Seymour (1982) also substantiate the effect of the participant's age. Ford (1968) found that item nonresponse in mail surveys tends to be less likely for easy questions and for those where adequate space was left on open-ended questions to record a response. Messmer and Seymour (1982) discovered that questions appearing after a branching question had notably higher item nonresponse rates than did other questions, undoubtedly because of confusion about the branching instructions.

The behavior and attitudes of survey interviewers may also affect item nonresponse. Rogers (1976) found interviewers who were more impersonal in conducting the interview to have lower item nonresponse rates than those of interviewers who had a more personable interviewing style. This differential was greater in personal interviews than in telephone interviews. Item nonresponse may also be affected by the interviewers' prior attitudes concerning the questionnaire. Sudman et al. (1977) discovered that interviewers who entered the interview thinking that the questionnaire would be difficult to administer had higher item nonresponse rates than did more optimistic interviewers. Bailar et al. (1977) reported that interviewers who thought it inappropriate to ask a sensitive question (e.g., on income) had higher item nonresponse rates on the question.

Several steps can be taken to prevent item nonresponse. Based on results of prior research and common sense where our intuition is unsubstantiated, several specific steps are suggested for dealing with item nonresponse. Since most research has been aimed at mail surveys, we assume with an accompanying note of caution that most of the conclusions of these research efforts will extend to interview surveys as well. First, expecting that nonresponse on some items will be higher for certain subgroups (e.g., the elderly, females, the less educated), steps might be taken to address any suspected concerns of these high-risk subgroups. For example, if an attitude question on balancing the federal budget is likely to cause some less educated respondents to skip over it since they do not understand the concept of a "balanced budget," a lead-in sentence explaining the concept might be used.

Another preventive step in interview surveys to reduce nonresponse rates to sensitive questions is to use the randomized response technique Warner (1965) first suggested. The participant uses a randomization device, whose outcome is unknown to the interviewer, to determine whether or not the answer given is to the sensitive question. At the expense of a loss in statistical precision, randomized response is designed to reduce item nonresponse rates while also reducing measurement error. Its usefulness presumes the willingness of the participant to believe that his or her anonymity is preserved by the device and that because of it, he or she can safely respond truthfully no matter which question is answered. Several adaptations of Warner's original idea have been published: use of a second unrelated question (Horvitz et al., 1967), avoiding the second innocuous question altogether (Greenberg et al., 1969), and the use of two alternative questions (Folsom et al., 1973). Emrich (1983) notes that despite its obvious advantages, the randomized response technique requires more sophisticated (costly) training of interviewers and longer interviews because of time spent in explaining the device. Some evidence is also given to raise some doubts about the participant's understanding or trust in the device, although improved interviewer training is seen as one way to overcome this problem.

Another remedy to the operation problems of the randomized response technique is to find a less complex approach to randomized questioning. One such approach, called the *block total response technique* (Smith et al., 1974; Raghavarao and Federer, 1979), avoids respondent involvement in choosing which questions are to be answered by randomizing (among respondents) questionnaires with alternative sets of questions. In the simplest case with binary responses to questions, there are two sets, one consisting of the sensitive question plus several innocuous questions and the other including only the latter. Response anonymity to the sensitive question is maintained by asking only for the sum of responses to the set of questions answered. The proportion of responses to the sensitive question can then be estimated as the difference in average totals from the two respondent subsamples. As with the randomized response technique, sampling variances are greater than if direct questioning were used, and negative estimates are possible especially when sample sizes are small.

A third preventive measure deals with item nonresponse occurring when the participant responds to the item but editing during the data processing finds the recorded data to be unusable. Steps are then taken to obtain the correct value of the edited item before analysis begins. For example, if a response of "93" is flagged for the age of a male who from the age of his 40-year-old wife and 10-year-old daughter should be younger, three steps might be taken to resolve this inconsistency: (1) the participant's age is verified against another reliable source (e.g., the sampling frame); (2) the interviewer is questioned as to whether the participant's age was actually "43" but entered by mistake as a "93"; and (3) the participant is recontacted to confirm his age.

Several other remedies have been successful in reducing the amount of item nonresponse. For example, the chances of obtaining an answer to *all* survey questions might be enhanced by offering the participant a special monetary incentive for completing the *entire* questionnaire. In mail surveys this incentive might be paid on receipt of a completed questionnaire. Another remedy for self-administered questionnaires is to leave plenty of writing space for open-ended questions and to avoid branching questions whenever possible. If branching questions are necessary, branching arrows or other formatting strategies should be used to clarify branching instructions [see Warwick and Lininger (1975)]. In general, every effort should be made to keep questions simple and understandable so that confusion about the question being asked will not cause a nonresponse. Item nonresponse may also be reduced by allowing only intended respondents to answer survey questions, especially when the proxy respondent is asked opinion or attitude questions that apply specifically to the intended respondent. Finally, the rate of data deletions because of unresolved inconsistencies among data items can be reduced by using portable computers or similar devices that enable editing of the data as they are collected. Lower deletion rates would be expected since inconsistencies can be more readily resolved at the moment they occur than later, during the final processing of the data.

8.2.2 Explicit Imputation Methods

Despite our best efforts to minimize item nonresponse through preventive methods, some missing items almost always appear in survey data, thus requiring us to find other ways to deal with the remaining nonresponse. As one might expect, the basic idea behind many of the methods used to handle unit nonresponse can be applied to item nonresponse as well, although as we shall see, only the notion of nonrespondent substitution (Section 8.1.3) has been widely adapted from the unit to item level. Other alternatives, such as weighting adjustment and model-based techniques, are not yet widely used because of practical limitations competing with their statistical advantages.

The idea of adjusting data for respondents to make up for the effect of nonrespondents on survey estimates, discussed in Section 8.1.5, can easily be extended to deal with nonresponse to individual items by computing adjusted sample weights for each variable with missing data. Computing item-level-adjusted weights might be done in a number of ways. One would be to compute a unit-level adjustment along with a set of separate adjustments for each missing data item and then to multiply each item-level adjustment by its unit-level adjustment. Another alternative would be to combine unit- and item-level nonresponse in producing an overall adjustment for each variable with missing data. Under either alternative, however, each missing data variable would have a corresponding set of adjusted weights to be used in analysis.

The fact that separate weights are needed for each missing data variable point to two important limitations of the method of weighting adjustment here. First, computing weights in this manner for surveys with many variables subject to nonresponse would be time consuming since the adjustment method chosen for each variable must be applied separately. Moreover, space must be allocated for each weight on analysis data sets. This problem can be minimized, of course, by computing adjusted item-level weights for only a few missing key data variables. Second, the analyst doing multivariate analysis involving more than one missing data variable faces the problem of deciding which set of item-level weights to use, although Little (1988) has proposed a single adjustment based on the regression of the response outcome, R_i, on the vector of variables with no item nonresponse. The problem of having multiple sets of weights does not, of course, apply to univariate estimates of means and totals, since the appropriate set of weights for the Y variable defining the parameter would be use.

The model-based methods of direct estimation discussed in Section 8.1.6 are easily extended to deal with item nonresponse by considering nonresponse for the Y variable to have arisen at either the item or unit levels. As with the weighting adjustment methods, however, applying separate models for each missing data item may be quite labor intensive and costly. Thus it would seem prudent to limit the use of these model-based methods to the most important survey variables.

Although the weighting adjustment and model-based methods have some desirable statistical properties (e.g., preserving statistical relationships with other variables involved in the method), a more widely used class of methods has emerged for dealing with item nonresponse and is therefore the major focus of our discussion of these methods. *Explicit imputation* is the term we use to describe those computation methods which, like substitution, use the idea of finding an actual replacement for the nonresponse. Using some protocol, a (we hope) reasonable proxy value for each missing data item is identified in this class of approaches. Although in the literature, this type of replacement process is generally called imputation, we consider the methods discussed later in this section to be explicit imputation strategies, since each replacement is expressed fully and clearly in the form of a numerical value in the survey data set. This qualifier is thought to be useful, since weighting class and other adjustment methods in effect impute means for each nonrespondent, without generating permanent replacement values, when the adjusted data are used to estimate means and totals. This imputation process is implicit rather than explicit since imputed values never physically replace the missing items in the data set.

We begin by describing several explicit imputation methods that have been suggested and used to some degree in practice. All methods described initially are univariate in the sense that missing values are imputed for a single Y variable at a time. We note later in this section that a univariate approach to imputation presents some problems in studying relationships, and we suggest some ways to deal with this issue.

All explicit imputation methods in current use have one important common objective: to pick a replacement value that is as similar as possible to the value of the missing item. This is generally done by defining similarity according to values of a set of *assignment variables* that are statistically related to the variable with the missing data and are available for respondents and nonrespondents. The auxiliary data obtained for the assignment variable may be used to (1) define imputation cells within which the replacement value is found, (2) define a regression model which when fitted becomes the basis for producing a computed alternate for the missing item, (3) quantify the degree of closeness so that the most similar substitute value can be identified to replace the missing item, (4) quantify a logical choice for the replacement value based on observed values of the auxiliary variables and their relationship to the Y variable, or (5) find the actual value of the missing item by linking into a set of administrative data containing a value for the missing item.

Kalton and Kasprzyk (1982) suggests that most explicit imputation methods can be expressed through a model linking the auxiliary data to the value of the missing item. If Z is the imputed value and $\mathbf{X}_i = (X_{1i}, X_{2i}, \ldots, X_{ki})$ is the k-dimensional vector of continuous or categorical auxiliary variables for the ith nonrespondent with actual Y_i, then the general *imputation model*,

$$Z_i = f(\mathbf{X}_i) + e_i, \tag{8.36}$$

can be used to describe most explicit methods, where $f(\cdot)$ is some function of the auxiliary data and e_i is a specified residual. This expression for Z_i enables us to distinguish imputation methods according to the form of f and whether e_i is randomized or fixed in identifying the imputed value.

Before proceeding further, the relationship between imputation and editing in surveys is worth mentioning since censoring as a result of data edits is, along with refusal and inability on the part of the respondent, among the reasons for missing item data. The analyst may choose to set a response to "missing" if, for example, it is perceived to be an outlier of an assumed underlying distribution and therefore likely to distort the planned analyses. These exceptional values may be the actual responses given by the respondent, but because of their extreme value, are truncated to make them more useful for analysis. Edits may also purge uncorrectable values which in view of responses to other items or the item itself are clearly impossible or at least highly questionable. Imputation of these improbable values represents a form of compensation for intended or unintended errors arising from the measurement process (see Chapter 9). Replacing a suspect value is often the preferred choice to the more costly option of retracing one's steps to find the correct value or to the undesirable option of using the existing data for analysis.

The linkage between imputation and editing also arises from the fact that imputed values, if they are to be useful to the analyst, should ideally be subjected to the same scrutiny as data actually collected. For that reason, many imputation systems call for the imputed value to be accepted only after it has passed the same edit checks as those the respondents have passed earlier (Sande, 1982). This requirement may be a problem for some imputation procedures—for example, those where passing the edit depends on variables other than those used to assign the imputed value. A reedit of the data after imputation, however, may help to preserve the intended relationship between the computed variable and those used to edit it. This will prove helpful in studies involving the interrelationship among these variables (e.g., regression analysis).

8.2.2.1 Hot-Deck Methods

The term *hot deck* is used to describe a family of imputation methods widely used in current survey practice. Although a universally acceptable definition for this approach to item imputation has not yet been suggested, a hot-deck method is generally one in which each missing value is replaced by the available value from a similar participant in the same survey. The choice of the imputed value is made within what we call *imputation cells*, which are mutually exclusive, exhaustive, and (intentionally) homogeneous subsets of the sample formed as a cross-classification of assignment variables. Algorithms developed to produce homogeneous groupings of categorical or categorized variables may be used in defining these cells [e.g., the Automatic Interaction Detector by Sonquist et al. (1971)]. Informative reviews of the hot-deck method have been presented by Chapman (1976), Rizvi (1983b), Ford (1983), and Sande (1983b), and Cox and Cohen (1985).

The word *deck* appearing in the name can be traced to the fact that when the approach was first used, the assigned value Z_i was made from a sorted deck of computer punch card. The deck was "hot" since the cards were taken from the current survey for which imputation of missing data was required. By contrast, in the *cold-deck method* the donor was identified from among population members in the assignment cell other than those responding in the current survey.

Cold-deck methods, although of interest historically, are rarely used in practice today and therefore are only mentioned in passing. Reviews by Nordbotten (1963) and Chapman (1976) point to the earliest uses of imputation to correct for unusable data discovered through editing as involving the cold deck. This approach was most useful in periodic surveys, where donors could be obtained from data in an earlier round of interviewing.

Oh and Scheuren (1980) trace the history and subsequent adaptation of the first form of hot-deck imputation as used by the U.S. Census Bureau. The original idea, although modified somewhat since it was first used to impute annual income in the 1962 Current Population Survey (CPS), is as follows. The traditional hot deck begins by identifying personal characteristics thought to predict income (age, sex, race, family status, amount of time employed, and major occupation group) and designating the chosen characteristics to be assignment variables. Seed values of the Y variable (those used to start the process) are then assigned to each imputation cell using data from prior surveys or some other logical source. The data set to which imputation is to be applied is then ordered in preparation for a single pass of the file required by the process. The object of this ordering step is to create a data set where consecutive records in each imputation cell are as similar as possible with respect to the Y variable (if there had been complete response). As the process begins the imputation cell of the first record is noted. If the record has a value for the Y variable, that value replaces the (seed) value currently in its cell. If, on the other hand, the value of the Y variable is missing, the seed value from its cell becomes the imputed value for the Y variable. The same assignment process is applied to subsequent records, with seed values (or their replacements) being imputed when missing data exist, or being replaced when there is a response for the Y variable.

It is useful to note that missing items on consecutive records in a cell will lead to the multiple use of some donor values, a feature that contributes to lowering the precision of survey estimates and understating variance estimates. We also observe that the imputed values (Z_i) are entirely dependent on the order of the data set and that variances of survey estimates are technically unattainable by classical methods since the assignment process is not randomized.

Starting in 1976, the concept of assignment in the traditional CPS hot deck was modified by increasing to 13 the number of assignment variables and by creating a hierarchical set of matching criteria (Welniak and Coder, 1980). Under this system a very similar donor, identified through an exceptionally detailed cross-classification of the assignment variables, is sought first. If a match cannot be made under these circumstances, the level of detail for the match is relaxed by combining certain imputation cells, and the search for a match continues. With each failure at matching, imputation cells are further combined until a match can be found. Although the complexity of this matching process is obviously greater than the original CPS hot deck, this version has the advantage of adding flexibility to the matching so that the donor is as similar to the nonrespondent as the data used for matching will allow. By increasing the number of assignment variables and imputing all missing information to a nonrespondent from the same donor, the 1976 revision also served to expand the set of interrelationships among variables that were preserved after imputation. This feature served partially to circumvent the problem of failing to maintain relationships that plagued the hot-deck and other imputation methods.

Evaluation of the hierarchical strategy seems to substantiate the utility of this modification. For example, one empirical study by Welniak and Coder indicates that mean levels of imputed values tend closely to resemble the values they replace, while an evaluation study by Oh et al. (1980) demonstrates some improvement in using the hierarchical hot deck (to impute total annual income) over using the version of the hot deck it replaced.

Another recent adaptation of the original CPS hot deck is called the weighted sequential hot deck (Cox, 1980). This version has two noted features, one being that individual selection probabilities for members of the data set are considered during imputation. The other novelty of this version of the hot deck is that a unique sequential sampling method developed by Chromy (1979) is applied to the ordered set of respondent data in selecting the donor value for the Y variable. The use of sequential sampling enables the analyst to deal with two drawbacks of the original CPS hot deck: that (1) some respondent records have no chance of donating an imputed value, and (2) some respondents donate values to more than one nonrespondent. The weighted sequential hot deck handles the first limitation by selecting each donor from a prespecified group of respondents which when merged with all other donor groups, constitutes the complete set of respondents. Problems with the second drawback are reduced by allowing multiple donations but controlling the number of times each respondent can be used as a donor. Although the weighted mean of imputed values will equal the weighted mean of respondents under repeated application of this revised hot deck, a practical limitation of this approach is that multiple passes of the data are required. Despite its cost, the weighted sequential hot deck has been applied successfully in two larger national surveys, the 1977 National Medical Care Expenditure Survey (Cox, 1980) and the 1980 National Medical Care Utilization and Expenditure Survey (Cox and Bonham, 1983).

A more theoretically understandable version of the traditional CPS hot deck is one where randomized assignment (either with or without replacement) is made within imputation cells. Randomization can be implemented by the sequential hot deck to a randomly sorted data set or by randomly picking from respondents in an appropriate imputation cell each time a nonresponse is encountered. Kalton and Kasprzyk (1982) call this version of the hot deck *random imputation within classes*, which they note can be expressed within the framework of their general imputation model as

$$Z_{ji} = b_0 + \sum_{j=1}^{H-1} b_j X_{ji} + e_{ji}, \qquad (8.37)$$

where b_0 and b_j are coefficients determined by the respondent sample, the X_{ji} are indicator variables, $X_{ji} = 1$ if the ith nonrespondent is a member of the jth imputation class (cell) and $X_{ji} = 0$ if otherwise, and the e_{ji} are respondent residuals chosen at random with the jth class. The e_{ji} under the traditional hot deck would be determined empirically. Precision losses from multiple uses of donor records are likely when donors are selected with replacement but can be avoided when without-replacement selection methods are used.

The randomized hot deck, while somewhat less practical to use than its sequential counterpart, has the advantage of providing a plausible basis (in classical inference) to formulate the mean square error of estimators where hot-deck methods have been applied. Bailar, Bailey, and Corby (1978), for example, compare statistical properties of the respondent sample mean ($\bar{y}_1$) with an estimator where a randomized hot deck with a single imputation cell was used. The hot deck is applied to the randomly sorted file of n_1 (unit and item) respondents from among the simple random sample of size $n + 1$ that was chosen initially (the extra sample member providing the seed value). Using the procedure outlined for the sequential hot deck in choosing imputed values for the $n_0 = n - n_1$ nonrespondents, an unbiased estimator of the respondent mean ($\bar{Y}_1$) can be obtained as

$$\hat{\bar{Y}}_{BBC} = \sum_{i=0}^{n} \frac{c_i Y_i}{n}, \qquad (8.38)$$

where Y_0 is the seed value taken from the set of respondents and c_i is the number of times the value Y_i is used ($c_i = 0$ if the ith sample member is a nonrespondent and $c_i \geq 1$ if the ith member is the seed or a respondent). The variance of $\hat{\bar{Y}}_{BBC}$ from this single-cell version of the hot deck can be expressed as

$$\text{Var}\left(\hat{\bar{Y}}_{BBC}\right) = \frac{\sigma_y^2\{1 + 2l_0[(n_1 + 1)n - 1]/(n_1 + 1)(n_1 + 2)\}}{n}, \qquad (8.39)$$

where σ_y^2 is the population unit variance for the Y variable and $l_0 = n_0/n$ is the nonresponse rate for the sample. The term in braces in Eq. (8.39) exceeds 1 and represents the increased variance (relative to σ_y^2/n, the variance of the complete response estimator) from this version of the hot deck. It can be shown that $\text{Var}(\hat{\bar{Y}}_{\text{BBC}}) > \text{Var}(\bar{y}_1) = \sigma_y^2(1 + l_0)/n$, which when combined with the fact that $\bar{y}_1$ is also an unbiased estimator for $\bar{Y}_1$, leads one to conclude that $\text{MSE}(\hat{\bar{Y}}_{\text{BBC}}) > \text{MSE}(\bar{y}_1)$, thus indicating that $\bar{y}_1$ is preferable to this single-cell hot deck. Even when the imputed value is without replacement (comparable to the random unit duplication approach of Section 8.1.5), the resulting unbiased estimator of $\bar{Y}_1$ still has greater variance (and therefore mean square error) than $\bar{y}_1$. In fairness to the hot deck, however, we must consider that the bias (and therefore the mean square error) of $\hat{\bar{Y}}_{\text{BBC}}$ would be smaller if a multicell hot deck were used, thus perhaps altering the seeming preference for $\bar{y}_1$. It is also worth noting that the variance of $\bar{y}_1$ (when we have in effect imputed the respondent mean for each nonrespondent) is greater than the variance of $\bar{y}$ (assuming complete response) by a factor or $1 + l_0$, thus implying that $\bar{y}_1$ affects both the bias and variance in estimating the population mean.

The statistical properties of estimators applied to data imputed from variously assumed versions of the hot deck have been addressed in several other articles. Bailey et al. (1978), applying similar assumptions to those used by Bailar et al. (1978), obtain the variances of $\hat{\bar{Y}}_{\text{BBC}}$ for two cases, one where the data set is randomly ordered, and the other for a serially correlated data set where for any two members (ith and jth), $\rho_{ij.} = \rho|i - j|$. Ernst (1980) has also considered both serially correlated and uncorrelated data in deriving the variances and asymptotic variances of estimators under several imputation procedures, including one that closely resembles the original CPS hot deck and another that mimics a randomized with-replacement hot deck. In comparing hot-deck variances of estimated means with the variance of respondent means (i.e., $\bar{y}_1$), both articles conclude that the randomized hot deck produces inferior variances to respondent means, although as before only a single imputation cell is assumed for all hot decks. With serial correlation present, Ernst (1980) has also shown, however, that for sufficiently large ρ the original CPS hot deck will have a smaller asymptotic variance than that of the respondent mean. These findings are further substantiated in work by Kalton (1983) on random imputation, which is conceptually akin to the randomized hot deck.

Bailar and Bailar (1983) and Kalton (1983) also compare the bias properties of respondent mean and the estimated means from a randomized hot deck. Kalton demonstrates that the bias of the hot deck estimator is the same as the bias of the weighting class estimator ($\hat{\bar{Y}}_{\text{WC}}$) discussed in Section 8.1.5. Both studies indicate that the relative sizes of bias will depend on the relationship between the Y variable and the response probabilities of individuals in the population.

8.2.2.2 Distance Function Matching

The donor in explicit imputation methods may also be identified in terms of a quantifiable measure of nearest distance to the nonrespondent, where distance is measured as a function of the assignment variables. This method is imputation by means of what is variously called *matching by distance function* or *nearest neighbor*. Like the hot-deck methods described earlier, this approach matches a missing value to a donor value that is identified from among those participating in the same survey. Some, in fact, consider distance function matching to be a "numerical" version of the hot deck, which, as described previously, is categorical because of its use of imputation cells (Sande, 1979).

Distance function matching has the advantage of seeking to find the donor who "best" matches the nonrespondent insofar as variables used to construct the distance measure are concerned. Finding the closest numerical match may therefore reduce the likelihood of the problem with the hot deck, where within imputation cell the values of the Y variable for the nonrespondent and its donor are still widely different. Nearest-neighbor matching does, however, require logical choices in measuring "nearness," leading us to review briefly some of the distance measures currently used in practice. Most logical measures of proximity are thought to be suitable, although the deterministic nature of matching used by most alternative measures makes it difficult for one to evaluate their efficiency (Sande, 1982).

Consider first the simplest case where a single continuous assignment variable (X_1) is used to identify the donor. Here nearness between the ith and jth sample members may be computed as

$$D_1 = |X_{1i} - X_{1j}|. \tag{8.40}$$

For assignment variables with skewed distribution, transforming the X variables [e.g., using $\log(X_{1i})$ or the rank of X_{1i} in place of X_{1i}] is often suggested for computing the distance measure. For example, imputing economic variables for businesses in Canada's Census of Construction is done using a log transformation of total business expenses in a distance function of the kind presented in Eq. (8.39) (Colledge et al., 1978). The value D_1 is used as part of an imputation system combining the concepts of categorical and numeric matching. Five potential donors are first identified on either side of the nonrespondent in a data set that has been ordered by gross business income within a set of imputation cells. The potential donor with the smallest value of D_1 is then considered to be "nearest" to the nonrespondent.

When there are $K > 1$ assignment variables and I_k measures the relative importance of the kth such variable, Sande (1979) suggests that

$$D_2 = \operatorname*{Sup}_{k} I_k |X_{ki} - X_{kj}|, \tag{8.41}$$

meaning that the ith and jth members are considered only as similar as their most dissimilar marginal difference among assignment variables. The Mahalanobis distance measure has also been suggested for matching by Vacek and Ashikaga (1980) and by Little and Smith (1983) for identifying outliers prior to regression imputation. In its simplest form the distance between any two sample members can be expressed as

$$D_3 = (\mathbf{X}_i - \mathbf{X}_j)' \sum (\mathbf{X}_i - \mathbf{X}_j), \tag{8.42}$$

where $\sum$ is the estimated covariance for the set of assignment variables and $\mathbf{X}_i$ and $\mathbf{X}_j$ are, respectively, the vectors of assignment variables for the ith and jth sample members.

Like the categorical hot deck, distance function matching allows donors to be used more than once. The chances of multiple usage can be controlled, however, by modifying the distance function so that nearness is inversely affected by the number of times the donor has been used previously (Colledge et al., 1978). Specifically, one might multiply D by a factor such as $1 + \lambda t$, where t is the number of times the donor has been used and λ is the assigned penalty for each usage (Kalton, 1983). Other considerations, such as concern about the quality of measurement for a potential donor, may be reflected by adding another distance increment, such as λt, to the distance function.

Although conceptually intended for use with continuous assignment variables, distance function matching can use categorical variables for imputation as well. One way is to use D, a function of the continuous variables, to identify the donor within imputation classes formed using the categorical variables. This hybrid version of the hot deck and distance function matching has been used successfully by Statistics Canada (e.g., in the Census of Construction). Categorical variables may also be incorporated into the distance function itself. Ordinal variables used in this way yield a measure of distance which is conceptually similar to one that is a function of continuous variables only. Using a nonordinal categorical variable with L categories would require expressing D as a function of $(L - 1)$ 0–1 indicator variables.

Relatively little theoretical work has been done to investigate the statistical properties of estimates produced when data have been imputed by a distance function match. One exception is the work by Ernst (1980), who considers an imputation method in which the donor is picked from observations on either side of the nonrespondent in a serially correlated list of sample members. When compared with the respondent mean and the estimated mean from the original CPS hot deck, this method is found to have a somewhat larger asymptotic variance, although computed relative differences were not great.

8.2.2.3 Mean-Value Imputation

Another form of explicit imputation in which the imputed value is determined by membership in imputation cells is called *mean-value imputation* (Kalton and Kasprzyk, 1982; Kalton, 1983). This method uses the estimated mean of respondents in the cell in which the missing item is a member as the imputed value for each missing item. Although mean-value imputation is relatively easy to do and reasonably effective in reducing the bias of point estimates for univariate parameters such as means and totals, it has two important disadvantages. First it distorts the distribution of the Y variable among sample members, since item nonrespondents can only take on values of the imputation cell means, causing the distribution to be spiked at these mean values. Another problem with mean-value imputation is that variances of means and totals will be understated, since variation among sample members of the same imputation cell will be too low by a factor approximately equal to the cell's response rate. Mean-value imputation is therefore most feasible for studies where analysis is limited to simple descriptive profiles (point estimates without accompanying variances) and the budget precludes the use of more complex analysis strategies where variance estimates are required.

8.2.2.4 Regression Methods

By specifying $f(\cdot)$ in the general imputation model of Eq. (8.36) to be some function of the assignment variables and by using the respondent data to fit the model, an imputed value can be predicted. Such is the premise of what we call the regression method of explicit imputation. The functional form of $f(\cdot)$ is almost always linear, so that in the framework of Kalton and Kasprzyk (1982) the ith imputed nonrespondent may be expressed as

$$Z_i = \hat{\beta}_0 + \sum_{k=1}^{K} \hat{\beta}_k X_{ki} + e_i, \qquad (8.43)$$

where the $\hat{\beta}$'s may have been estimated by standard methods such as ordinary least squares. Regression imputation differs from hot deck and distance function imputation in that Z_i is a predicted value rather than an actual value as taken from a designated respondent in the current survey.

Equation (8.42) presumes a continuous Y variable and that the K assignment variables are all quantitative, although qualitative variables could be used as well by defining appropriate indicator variables. We may specify that $e_i = 0$, in which case Z_i is a deterministic prediction given the respondent data. This unfortunately may distort the shape of the distribution of the Y variable or artificially inflate the degree of statistical association between the Y variable and the set of assignment variables. To deal with this issue, some have suggested that e_i be defined as a random residual with zero expectation.

Categorical Y variables can be accommodated by expressing Z_i as a log-linear function of the assignment variables. With a single assignment variable and $\beta_0 = e_i = 0$, regression imputation as defined in Eq. (8.42) simplifies a simple ratio adjustment method of imputation used in multiround panel surveys, where the response from a previous round is adjusted for time changes and used in place of the nonresponse in the current round (Ford et al., 1980).

Kalton (1983) lists several ways in which a random residual (e_i) can be identified. One is to take all residuals from the same (e.g., the normal) distribution with zero mean and residual variance as estimated in fitting the assumed model to respondent data. A second way is to choose e_i by applying the first approach separately within certain subclasses of the population. A third alternative is to use the residual (of the fitted model) from a randomly chosen respondent, and the fourth is the same as the third except that the donor residual is chosen from respondents with similar values of the assignment variables. Thus to varying degrees all four approaches assume that the relationship between the assignment variables and the Y variable are the same for respondents and nonrespondents.

In one application of regression imputation with continuous Y variables, Schaible (1983) investigates the properties of regression imputation using Eq. (8.42) with $e_i = 0$. Suppose that for $i = 1, 2, \ldots, N$,

$$Y_i = \beta_0 + \boldsymbol{\beta}'\mathbf{X}_i + \varepsilon_i, \qquad (8.44)$$

where β_0 and $\boldsymbol{\beta} = (\beta_1, \beta_2, \ldots, \beta_K)'$ are unknown but estimable coefficients and $\varepsilon_1, \varepsilon_2, \ldots, \varepsilon_N$ are mutually independent such that over the outcomes of the model $E_m(\varepsilon_i) = 0$ and $\mathrm{Var}_m(\varepsilon_i) = \sigma^2$, $i = 1, 2, \ldots, N$. The outcome of an assumed stochastic response mechanism is defined by the random indicator variable R_i, which, along with ε_i, is independent conditioned on the value of $\mathbf{X}_i$. Using a standard least squares unbiased estimator $(\hat{\boldsymbol{\beta}})$ of $\boldsymbol{\beta}$, one predictive estimator of $\overline{Y}$ [following Royall (1971)] by means of a fitted regression estimate of the Y_i for sample nonrespondents, given an initial without-replacement simple random sample of size n, is

$$\hat{\overline{Y}}_S = \frac{\sum_{i=1}^{n_1} Y_i + \sum_{i=1}^{n_0} Z_i}{n}. \qquad (8.45)$$

Given the assumptions of the model, the variance of $\hat{\overline{Y}}_S$ over the regression and response models is given as

$$\mathrm{Var}\left(\hat{\overline{Y}}_S\right) = \frac{(1-f)\sigma^2}{n} + \frac{(1-l_1)\sigma^2}{n_1\left[1 + \mathbf{d}'\boldsymbol{\Sigma}^{-1}\mathbf{d}/(1-l_1)\right]}, \qquad (8.46)$$

where $l_1 = n_1/n$ is the sample response rate; $(\overline{X}_{1k} - \overline{X}_k)$, the difference between respondent and full-sample means for the kth assignment variable, is the kth entry to the vector $\mathbf{d}$; and the entry on the kth row and k'th column of the $K \times K$ matrix Σ is $\sum_{i=1}^{n_1}(X_{ik} - \bar{X}_{1k})(X_{ik'} - \bar{X}_{1k'})/n_1$. The second component of $\mathrm{Var}(\hat{\bar{Y}}_s)$ is caused by sample nonresponse and the predictive imputation and is the price we must pay to deal with nonresponse in a way that avoids distortion in the Y distribution. The "incomplete data bias" of $\hat{\bar{Y}}_S$ (its bias from nonresponse in the sample alone) is zero assuming that the model is true.

Using the model in Eq. (8.38), Schaible also presents properties of a hot-deck estimator in which an imputed Y variable for each sample nonrespondent in the hth of H predesignated imputation cells is obtained as a with-replacement random selection from respondents in the same cell. The incomplete data bias and variance are determined, respectively, for the resulting estimator

$$\hat{\bar{Y}}_S^* = \sum_{h=1}^{H} \sum_{i=1}^{n_{1h}} (c_{hi} + 1) \frac{Y_{hi}}{n},$$ (8.47)

as

$$\mathrm{Bias}\left(\hat{\bar{Y}}_S^*\right) = \sum_{h=1}^{H} \sum_{k=1}^{K} \left(\sum_{i=1}^{n_{1h}} \frac{c_{hi} X_{hik}}{n_{0h}} - \bar{X}_{0hk} \right) \frac{\beta_k n_{0h}}{n}$$ (8.48)

and

$$\mathrm{Var}\left(\hat{\bar{Y}}_S^*\right) = \frac{(1-f)\sigma^2}{n} + \frac{(1-l_1)\sigma^2}{n_1} \left[l_1 \left\{ 1 + \frac{\sum_{h=1}^{H} \sum_{i=1}^{n_{1h}} c_{hi}}{n_0} \right\} \right],$$ (8.49)

where c_{hi} is the number of times the ith respondent in the hth cell is imputed and $\bar{X}_{0hk} = \sum_{i=1}^{n_{0h}} X_{hik}/n_{0h}$.

Scheiber (1978) applied a version of regression imputation to missing Social Security income data in a survey of the low-income aged and disabled. A two-stage procedure was needed in this approach since a large segment of the population studied has no Social Security income. The object of the first stage was to impute recipiency status. This procedure was done by fitting a logistic regression model using survey respondents with no missing data. The dependent variable in this first-stage model was an indicator of whether or not the respondent is a recipient. Item nonrespondents with a predicted value of Y_i exceeding 0.5 are subsequently treated as recipients, and those with predicted values less than or equal to 0.5 are considered to be nonrecipients (imputed a value of zero dollars for Social Security income). The second stage involved fitting a linear regression model with Social Security income as the dependent variable. This model was fitted using data from respondents who reported some Social Security income. The fitted model then was used to predict income for those item nonrespondents who were imputed to be recipients in the first stage. Herzog and Lancaster (1980) and Herzog (1980) presented another similar application of the regression approach for imputing Social Security recipiency status and benefits for persons surveyed during the March 1973 Current Population Survey.

Other illustrations of the use of regression imputation for dealing with earnings income missing from respondents to the Current Population Survey are given in comparison studies by Greenlees et al. (1982) and David et al. (1986). Comparison with the hot deck in the latter study showed the regression approach to produce somewhat more accurate imputation, at least when accuracy is determined by the mean absolute difference between imputed and comparison values.

Theoretically at least, the regression approach to predictive imputation is a reasonable alternative to consider in dealing with missing data. Given a well-fit and correctly specified model that applies equally to respondents and nonrespondents, Z_i and the missing Y_i may be very close. But this, of course, requires that all of the conditions above be met. Like hot-deck and distance function imputation, regression imputation has the attractive feature of maintaining the statistical association of the Y variable and the assignment variables. However, unlike the hot-deck and distance function methods, where appropriate donors may sometimes be difficult to find, regression imputation can always find a suitable replacement value for the nonrespondent (at least to the extent that the model itself is suitable). Using regression imputation also requires that occasionally predictive assignment variables be found and fitted through the model to the Y variable. The time and cost of doing this work with regression models may therefore limit the researcher to the number of variables imputed by regression methods to those where strong predictors are available.

8.2.2.5 Deductive Imputation

Sometimes the correct value for a missing Y variable can be determined with a high degree of certainty from values of the assignment variables. This approach might therefore be called *deductive imputation*. With imputation of this kind some known functional relationship exists among these variables so that $Z_i = f(\mathbf{X}_i)$, following the general imputation frameworks of Kalton and Kasprzyk (1982). Ignoring any measurement error on the assignment variables, this method of imputation is therefore completely deterministic since the functional relationship is known and there is no need for a randomized residual.

A few examples from a 1980 multiround panel study called the National Medical Care Utilization and Expenditure Survey (NMCUES) will serve to illustrate how the Research Triangle Institute used this relatively simple deductive approach to imputation (Cox et al., 1982). Before using the hot-deck method of imputation, an attempt was made to impute race for individuals from other members of their household. The race of predesignated individuals was used (e.g., race of the spouse if the race of the head of household was missing, the race of the head of household if the race of someone else in the household was missing, etc.). The functional relationship used for imputation in this case is simply $Z_i = X_{1i}$, where X_{1i} is the race of the predesignated individual. Missing values on the variables, age in years, for those born before the start of NMCUES were first imputed using the year of birth as X_{1i} so that $Z_i = 1979 - X_{1i}$. Finally, data on 1980 income from employment was obtained in a special employment supplement in the round of NMCUES. Missing income values for those dropping out of the study before the fifth round were imputed from the values of four assignment variables measured elsewhere in the study: weekly pay rate for primary job (X_{1i}) and secondary job (X_{2i}), as well as the number of weeks in 1980 worked in primary job (X_{3i}) and secondary job (X_{4i}). The estimate used as a replacement for employment income was computed as $Z_i = X_{1i}X_{3i} + X_{2i}X_{4i}$.

It is clear from these examples that $f(\mathbf{X}_i)$ can assume many different forms, and in some instances Z_i by deductive imputation only estimates Y_i (e.g., for employment income). We also see that deductive imputation fails for any nonrespondents with missing values on one or more donors, which is why it is often used as part of the editing process prior to other forms of imputation.

8.2.2.6 Exact Match Imputation

The donor value in another type of explicit imputation is the same measurement as the one missing, but is taken from the record for the same unit from some external source (e.g., administrative records). We call this approach *exact match imputation*, where unique identifying information for the unit with the missing item (e.g., Social Security number, driver's license number, etc.) is used to match nonrespondents to donor. Therefore, in most instances $Z_i = X_{1i}$, where X_{1i} is the corresponding value of the Y variable as obtained from the matched unit on the external data set.

Some noteworthy similarities with imputation methods described previously are mentioned briefly. First, exact match imputation resembles deductive imputation, since both produce Z_i without the use of a randomized residual (e_i), which is unnecessary since the actual value, as opposed to a predicted value, of Y_i is sought. These two methods differ in that deductive imputation tends to use data from the present survey to arrive at Z_i, whereas in unit matching Z_i is obtained from sources external to the survey. In this respect exact match imputation is more similar to cold-deck imputation, since both rely on matching and the use of external data sources to determine the donor value. The cold deck, however, uses less specific matching criteria by assigning nonrespondents to imputation cells, within which a donor is chosen from among all respondents in the cell. Of course, when the unique unit identifiers assumed by exact match imputation are not available for all nonrespondents, an approach similar to the cold deck, with imputation cells defined by certain demographic criteria, might be applied to the external source.

Several instances where exact match imputation has been successfully used, are mentioned briefly. For example, Schieber (1978) discusses the use of an administrative file of beneficiary payments to impute income from Social Security for a survey of the low-income aged and disabled. Social Security numbers are used to match nonrespondents to donors. An empirical comparison of Social Security income as imputed by exact match imputation, hot deck, and regression approaches in that study seems to indicate that the first two methods were of comparable accuracy, although the former tended to underestimate income, while the hot deck tended to overestimate it. Of the three approaches, regression imputation generally produced the least accurate imputations.

Cox and Bonham (1983) describe another application of exact match imputation for the 1980 National Medical Care Utilization and Expenditure Survey conducted by the Research Triangle Institute. In this study missing health care charges paid by Medicaid were imputed by matching survey records to Medicaid claims records. Availability of external data for imputing health costs was extremely important in this study since approximately 80 percent of charges for health care paid by Medicaid were missing among survey respondents. Other imputation strategies under these circumstances would in all likelihood have been ineffective with an item response rate this low.

Another study comparing the performance of exact matching imputation with other nonresponse compensation methods has been done by Platek and Gray (1983). Components of the mean square error of estimated totals are derived for the case where a matched donor from the external source is available for all nonrespondents and the case where a weighting adjustment is used for some nonrespondents for whom an external match does not exist. As expected, both of the bias variance components of the estimated total in the second case reflect the extent and quality of available external data for matching.

Despite its favorable performance in Schieber's comparison study, it is important to note some potential problems with exact match imputation. First, the quality of the value imputed by a unit match is no better than the quality of the external source. The conclusions in the Schieber study undoubtedly reflect the relatively high quality of the Social Security data used. More typically, there may be a significant proportion of nonrespondents for whom there is no match in the external data source and some matched records are subject to large measurement errors. Computational errors as well as differing definitions for comparable variables on the survey and external data sets may contribute to the latter. Second, the cost and complexity of unit matching may be prohibitive in some studies. A suitable external source of data must be found and the variables used in the imputation process made comparable between data sets. Then the survey nonrespondent and external data sets must be merged to identify donor values. The amount of time expended in completing this process may be great, especially in large studies, where the external data set is difficult to access and the process of matching records is complex.

8.2.2.7   Model-Based Methods

Although all imputation methods explicitly or implicitly involve an assumed model that defines a specific relationship between the Y variable and the set of assignment variables, we establish a separate class of imputation methods for those where assumptions concerning the stochastic origin of $\mathbf{Y}_n = (Y_1, Y_2, \ldots, Y_n)$, the vector of Y variables for the sample, are used to impute a value through prediction for each nonrespondent entry in $\mathbf{Y}_n$. Within the framework of the general imputation model, we might specify $Z_i = \hat{Y}_i$, where $\hat{Y}_i$ is the predicted value obtained through a distributional model for respondent and nonrespondent segments of $\mathbf{Y}_n$ among those sampled. Several articles discussed in Section 8.1.6 under methods for dealing with unit nonresponse would apply here [e.g., Rubin (1977) and Little (1983)].

We see from the preceding discussion that the approaches to explicit imputation for item nonresponse are many and varied but in some ways are conceptually similar. These approaches in most instances will reduce the size of nonresponse bias for univariate statistics and at the same time produce a survey data set "free" of item nonresponse, so that standard statistical software can be used (although perhaps inappropriately for estimating variances). Having a designated replacement value for all missing values also assures that analytical findings will be consistent, since only one set of values for the Y variable (including the original responses plus imputed values) is available to data users.

There are, however, several potentially important issues to consider in using imputed data. Most problems stem from fabricated data being the end result of the imputation process—data which to the unsuspecting analyst may be treated as if they were real. Imputed data are in fact a "delusion" as Sande (1983b:339) describes it, subject to imputation error whose effects on the bias and variance of estimates cannot be easily measured. Estimating the bias from imputation requires actual values of the Y variable for nonrespondents, which if available would eliminate the need for imputation in the first place. Clearly, however, the more accurately the donor value represents the missing value, on average, the lower the imputation bias for univariate measures such as means and totals. For multivariate measures, on the other hand, the relationship between individual imputation error and imputation bias is not as clear-cut.

Imputation may also increase the variance of an estimator in ways that are not easily detected. One cause of larger variances is the multiple use of donors in some versions of methods such as the hot deck and distance function matching. Multiple usage in effect creates additional random variability in weights for respondents, which in turn reduces the precision of estimates. Multiple-donor usage also causes variances estimated from the sample to be understated, since repeats of the same value in the sample gives the impression of greater homogeneity in the population than actually exists. A second source of added variance is imputation error, $Z_i - Y_i$, which usually varies among units. The variation in imputation error among individual units and the covariance of error among pairs of units adds to the overall variance of most univariate estimators (Platek and Gray, 1983).

Incorporating this imputation error into the evaluation of survey estimates can be accomplished in several ways. Variance estimates reflecting sampling and imputation error can be produced in multistage sampling designs by independently performing the imputation procedure within each PSU and then using standard methods for variance estimation in surveys with complex sample designs [e.g., Shah (1981)]. Unfortunately, facilitating variance estimation in this manner limits the effectiveness of the imputation procedure. When using the hot deck, for example, sample sites within PSU may be sufficiently small to cause many imputation cells to be empty, requiring that these cells be merged with others. The collapsed cells then will tend to produce fewer similar donor values for missing items.

Variances accounting for imputation may also be estimated by a strategy called *multiple imputation*, proposed by Rubin (1978, 1979) and discussed further by Herzog and Rubin (1983) and Rubin (1987). The basic idea of this strategy is to produce more than one imputed value for each missing value through a replication process. In the event that different methods are used to produce each set of imputed values, a comparison among estimates produced by each set allows one to observe how much using the various methods will affect analysis findings. On the other hand, to estimate variances that account for the imputation error produced by a given imputation method, the same method must be independently repeated in producing each set of imputed values. The multiple imputation data can then be used to simulate the distribution of the estimator under that particular method, assuming that the method involves some kind of random component so that the imputed values for a nonrespondent will differ.

The method of multiple imputation has been used to simulate the Bayesian posterior distribution of population parameters in the presence of nonresponse and to understand better and improve further imputation methods (Rubin, 1977; Rubin and Schenker, 1986). Oh and Scheuren (1980), for example, use two sets of hot-deck imputed values in estimating variances for the percentage income distribution of persons and families included in the March 1977 Current Population Survey. Findings of this study show the estimated standard error of estimates, ignoring imputation error, to be around 20 percent lower than standard errors using multiple imputation data. Another study coming to similar conclusions has been done by Herzog and Lancaster (1980) and Herzog (1980), who use multiple imputation to compare the estimated variance of benefits from Social Security under their regression imputation strategy with comparable estimates obtained by using the hot deck. Cox and Folsom (1981) use the idea of multiple imputation to isolate a component because of the use of a weighted hot deck in estimating the mean expenditure for health care visits in the National Medical Care Utilization and Expenditure Survey. Finally, Kalton (1983) and Kalton and Kish (1984) apply the principle of multiple imputations to the method of random reweighting (see Section 8.1.5) to reduce the component of variance caused by imputation in survey estimates.

The practical difficulty with multiple imputations, of course, is the need to produce and retain more than one donor value for each missing value. For imputation methods where a single set of imputed values is prohibitive, repeating the process may be out of the question except perhaps for a few key items in the questionnaire. Fortunately, there is some evidence that the number of sets need not be large for multiple imputation to be effective. A simulation study by Rubin and Schenker (1986) suggests, for example, that two sets of imputed values may be sufficient for variables with moderate rates of item nonresponse and that three sets would be satisfactory for larger item nonresponse rates.

The impact of imputation on the distribution of variables and on the interrelationship among variables offers two more issues with which one must contend in using imputed data for analysis. Distributions of variables whose missing values are imputed using respondent values will be distorted by the repeats appearing in the data set because of imputations. This problem is particularly applicable to the mean-value, hot-deck, distance function, and deterministic regression methods. It can be avoided for the most part by adding to the imputed value a randomized residual, which serves to smooth the distribution indicated by the respondent and imputed data combined, although selecting a suitable randomization scheme may not always be obvious.

In multivariate analysis of relationships among variables, where one or more have imputed missing values, the level of statistical association among the variables is usually underestimated. The tendency to underestimate the strength of interrelationships occurs because the added source of variation in imputed variables diffuses the variables affected, which in turn reduces the actual level of association. For the hot deck, the amount of added variation is directly related to the amount the imputed variable varies within the imputation cells from which donor values are assigned. Imputation may also cause some rather unusual relationships between individual values of the imputed variable and values of variables not used to define the imputation cells. For example, the infamous "pregnant male" may appear when gender is imputed and pregnancy status is not one of the assignment variables.

Relationships among variables can be retained when explicit imputation has been used. In some cases the method itself contributes toward this end. The fact that the hot-deck, regression, and distance function matching methods determine Z_i based on values of $\mathbf{X}_i$ assures that the relationship between the Y variable and the assignment variables will be maintained in the imputed data. Successful application of the logical-imputation and unit-matching methods will also often assure that the donor value Z_i is sufficiently close to Y_i to preserve its relationship with other variables. In the event that the imputation method creates unusual values of the Y variable, reediting the imputed data will help to assure that that relationship of the Y variable with the variables used in the reedit is maintained.

Another way to preserve interrelationships among variables in methods such as the hot deck and distance function matching methods (where another respondent is the donor) is to impute jointly for several variables. This imputation involves first defining a set of unit respondents with one or more missing values for a designated set of variables. A donor is found for each such record, and the missing values on it are replaced by the corresponding responses from the donor. The hot deck applied to the Current Population Survey has used multivariate imputation to preserve the relationship among income, work experience, and type of longest job (Oh and Scheuren, 1980). The problem with this strategy, of course, is that multivariate extensions to already complex imputation methods makes matters even more complex. With the hot deck, for example, it would be necessary to form imputation cells that are jointly suitable for all variables to be imputed. Multivariate imputation also reduces the chances of finding a suitable donor value and therefore tends to increase imputation error for individual variables. On one hand, the number of donors sought would be high since any unit respondent with one or more missing values would require imputed values. The supply of donors, on the other hand, would be limited to those appropriately similar unit respondents with values of all variables to be imputed for the recipient record.

8.3   CHOOSING AMONG METHODS

It is clear from this chapter and the extensive literature it represents that the problem of dealing with survey nonresponse has received much deserved attention. This focus is, of course, good for survey research in that intensive scrutiny helps us to understand nonresponse and its implications better. Knowledge builds on knowledge and new approaches can be devised and studied for dealing with the problem. An unfortunate consequence of this evolutionary process, however, is that the vast literature on research strategies for minimizing or measuring the effects of nonresponse may leave well-intended practitioners groping for a decision as to which approaches are best suited to meet their needs. The purpose of this concluding section is to draw from what we have learned in this chapter to suggest some things that may be important in formulating a plan for addressing the nonresponse problem in a survey.

We begin by recalling that methods for dealing with nonresponse are basically of two types, preventive and compensatory. Preventive methods (see Sections 8.1.1 and 8.2.1) are designed to reduce the rate of nonresponse, while compensatory methods are those which, once all feasible preventive steps have been taken, serve to reduce the effect of the remaining nonresponse.

Deciding on a suitable preventive strategy for survey nonresponse is made relatively straightforward for the researcher because the sole objective of all preventive methods is to improve response rates. Unfortunately, it is seldom easy to disentangle the relative effectiveness of individual preventive methods in reducing survey response since several are often used together in surveys. Thus we might best search for that combination of methods (e.g., incentives plus multiple callbacks plus endorsements) which, based on prior experience, is likely to be most effective given the qualities and constraints of the present study.

Assessing the utility of nonpreventive methods in deciding on a strategy for dealing with nonresponse may be even more difficult than finding ways to maximize response rates. Building from some ideas given by Bailar (1978), the statistical implications of the alternative methods are one of the first things to consider, since nonpreventive methods designed to deal with survey nonresponse will in one way or another affect the statistical behavior of survey estimates. Thus we wish to find that method (or methods) which lets us make the statistical inference we had intended while minimizing the effect of nonresponse on inference.

In traditional statistical inference this method has involved investigating components of the mean square error of survey estimators. While observing, for example, that the nonresponse bias of estimators of univariate parameters such as means and totals will be reduced, the variance may be affected as well, thus making it important to identify those methods with the smallest mean square error in evaluating nonpreventive strategies.

Statistical considerations other than the mean square error may also be important. One example is for cross-tabulations, regression, and other analyses where the object is to investigate relationships among variables. Here the mean square of univariate statistics may be less important than the effect of the compensation method on the relationship being studied. Obviously, one wishes to pick the method that least alters the relationship being studied. In the model-based view of inference, one may be concerned primarily about finding approaches that minimize the bias and variance arising from the assumed model and whose estimators are most robust to departures from the assumed model. Fast rates of convergence for iterative methods would also be desirable.

Since many nonpreventive methods require auxiliary data, this information must be available for the method to work. Some methods use these data to compute adjustments (e.g., weighting class method), while others use the auxiliary data to assign imputed values (e.g., hot-deck method). In most methods these data are needed for all sample members, although some model-based methods require that these data be available for the entire population.

Evaluating the various alternatives for dealing with nonresponse requires that we consider their statistical effectiveness as well as the practical implications of using them. Doing so, however, usually forces the practitioner to make difficult decisions. In surveys of health providers, for example, the increase in provider response rate that comes about by obtaining professional endorsements may not be justified in view of the cost of getting those endorsements. Then again, researchers doing mail surveys must consider the increased cost of nonrespondent subsampling against its known effectiveness.

Another aspect of the cost-effectiveness issue that one must consider in choosing among approaches to dealing with nonresponse is the complexity of implementing one. More sophisticated approaches such as multiple imputation applied to the hot-deck method may not be practical when staff are unavailable to apply the method and interpret its findings. Other factors affecting the complexity of a compensation method are the site and scope of the survey, the extent of nonresponse, and the availability of computer software required by the approach. The first two issues tend to amplify the complexity of all approaches, while the third tends to be more important for nonpreventive methods.

A review of recent comparison studies may also aid in deciding among nonpreventive methods. A brief synopsis of some recent comparison studies is presented in Table 8.3. Ford (1983) reviews many of the studies included in Table 8.3. Some are "analytical" in that they compare the theory of prevention methods, some are real or simulated data to make "empirical" comparisons, and some make analytic and empirical comparisons. Methods for dealing with unit and item nonresponse are compared, sometimes together, because of conceptual similarities. The criteria for comparison on analytical studies are usually the bias or variance of estimators, while many of the empirical studies evaluate methods by comparing imputed values against known values for the missing data. Others use multiple imputation to compare the variances of estimates obtained from data that have been subject to the imputation methods being studied.

Despite the impressive amount of research represented by these and other studies of methods for dealing with survey nonresponse, we are left with the unalterable conclusion that hard and fast rules for choosing among methods presently evades us and probably always will. Varying surveys and assumptions made in evaluation often lead us to differing conclusions concerning the relative merits of existing methods.

Despite the uncertainties faced in choosing among methods, two points clearly emerge. One is that there is no substitute for complete response because estimates produced by using adjustment, imputation, subsampling, or an assumed model to compensate for existing nonresponse will rarely be superior to comparable estimates produced in the absence of nonresponse. Another is that it is better when attempting to reduce nonresponse bias to use a well-chosen compensation method than to do nothing at all, unless the

Table 8.3 Summary of Recent Comparison Studies on Methods to Deal with Nonresponse

Author(s)	Methods Compared[a]	Type of Comparison: Analytical (A) or Empirical (E)	Source of Empirical Data
Ford (1976)	RR, HD, DF	E	Simulation[b]
Colledge et al. (1978)	NC, RR, HD	E	Simulation[b]
Cox and Folsom (1978)	HD, WC	E	National Medical Care Expenditure Survey
Bailar and Bailar (1978)	HD, WC	A	—
Bailar et al. (1978)	HD, WC	A	—
Ernst (1978)	HD, WC	E	CPS[c]
Pare (1978)	NC, RR, HD	E	Simulation[b]
Schieber (1978)	RR, EM, HD	E	CPS[c]
Ernst (1980)	NC, HD, RD	A	—
Herzog and Lancaster (1980) and Herzog (1980)	RR, HD	E	CPS[c]
Santos (1981)	RR, WC, RD	A, E	Simulation[b]
Little and Samuhel (1983) and David and Triest (1983)	RR, HD	E	CPS[c]
Kalton (1983)	NC, WC, PS, RD, MV	A, E	Simulation[b]
Rizvi (1983a)	NC, HD, WC	A	National Longitudinal Survey
Platek and Gray (1983)	NC, HD, EM, WC	A	—

[a]Key: NC, no compensation (ignoring missing data); RR, ratio or regression prediction method; HD, hot-deck method; EM, exact match imputation; WC, weighting class adjustment method; PS, poststratification; RD, random duplication; MV, mean-value imputation within-class imputation; DF, distance function matching.

[b]Data used in this simulation study came from an existing survey data set.

[c]Current Population Survey conducted by the U.S. Bureau of the Census.

rate of nonresponse is low because all such methods will be of some benefit in dealing with this effect of nonresponse. The challenge in making the final choice, however, is recognizing the relative strengths and weaknesses of competing alternatives for the survey in this area to focus more intently on finding ways to give the practitioner a rational basis for choosing among methods.

Finally, model-based compensation methods, although not widely used in survey practice, merit further study by methods researchers and consideration by practitioners. General acceptance of model-based methods by the survey research community, however, would seem to be tied to finding a convincing answer to the question of robustness to departures from the assumed models and to making easy-to-use computer packages generally available to those electing to apply these methods.

CHAPTER 9

Measurement: Survey Measurement and Measurement Error

This chapter and the next two are devoted to survey measurement and assessment of measurement error. In Chapter 9 we focus on the general nature of survey measurement and efforts to establish a conceptual framework for the study of measurement error. Our focus in Chapters 10 and 11 is on assessment of bias and variability in survey measurements.

We first discuss the general nature of measurement and measurement error and then focus specifically on surveys. Measurement in a survey involves deciding what characteristics or aspects of the target population should be measured, developing a method for measuring each characteristic, carrying out the measurement processes, and evaluating the quality of the measurements. As indicated in earlier chapters, the survey can give erroneous results because of an incorrect choice of the characteristics to be measured. Little methodology is available for assessing the usefulness of statistical information, and although this is a "shameful gap" (Kruskal and Mosteller, 1980), we do not discuss methods for making such an evaluation.

What does it mean to make a measurement? Stevens (1959:19) defines *measurement* as "the assignment of numerals to objects according to a rule—any rule." This very broad definition is not accepted by everyone. Ellis (1967) restricts measurement to the study of quantitative relationships where a quantitative relationship is expressed as greater than, equal to, or less than. Ellis focuses on the philosophy of measurement and gives very explicit definitions of a quantity, measurement scales, and measurement. Measurement is always made on some scale, and we have "a measurement scale for a quantity q if and only if we have a determimitive rule for making the numerical assignments to things possessing q, such that if those things are arranged in the order of the numerals that are assigned to them according to this rule, then they are also arranged in the order of q." This definition is more restrictive than that of Stevens because it does not allow for the use of nominal scales (see below).

Because of this fundamental quality of scales, understanding the nature of the different types of scales is important to understanding what we have learned in making measurements and using them. Methods of classifying scales include classification in terms of the kind of arithmetic the scale supports, an approach Coombs (1952) describes; N. R. Campbell's classification in terms of the kinds of operations by which the scales are set up; and Stevens' approach, which classifies scales in terms of the types of mathematical relationships that make the levels the scale forms invariant (Ellis, 1967).

For illustration consider Steven's approach and Ellis's modification of Campbell's approach. The Campbell approach identifies three types of measurement, including *fundamental measurement*, in which the measurement operations have the formal characteristics of the arithmetical operation of addition of real positive numbers. A fundamental scale is established as follows:

1. Some object, say $s(0)$, possessing q, is chosen to act as an initial standard and is assigned a positive number $a(0)$.
2. The following rules for making numerical assignments are then adopted:
   a. If $s(1)$ and $s(2)$ are any two systems possessing q and the system $s(1)$ has already been assigned the number $a(1)$, the system $s(2)$ is to be assigned a number $a(2)$ such that $a(1) \langle, = , \rangle$, $a(2)$ according to whether the quantity q of $s(1)$ is $\langle, = , \rangle$ quantity q possessed by $s(2)$.
   b. If $s(1)$ and $s(2)$ are any two systems possessing q to which the numbers $a(1)$ and $a(2)$ have already been assigned, the composite system $o[s(1), s(2)]$ is to be assigned the number $a(1) + a(2)$.

Examples of properties with fundamental measurement are length and weight.

The second kind of measurement is *associative measurement*, in which one independently measurable quantity is taken under precisely defined conditions to be the measurement of another quantity. The scales used in psychology and sociology for intelligence, attitudes, opinions, and so on, are examples. In this case, a quantity q that is fundamentally measurable is taken as the measurement of the target quantity p under certain defined conditions. The two quantities p and q are not the same thing because under some other set of conditions q will not be a measure of p.

The third type of measurement is *derived measurement*, in which measurements are defined by certain numerical laws that relate the measurements of other fundamentally measured quantities. Many measurements in physics are derived, such as the fundamental gas laws that relate temperature, pressure, and volume.

Stevens's classification (1959) is widely used by statisticians because he relates measurement scales and appropriate statistics, pointing out that some types of statistics are appropriate for the scale and others are not. A statistic is considered appropriate for a scale if it is invariant under the transformations permitted by the scale, either because its numerical value remains the same or because the item "designated by the statistic remains the same." The latter type of invariance is well illustrated by the order statistics for a sample, which remain the same under all transformations that maintain order. The types of measurement scales Stevens gives are:

1. *Nominal*, in which each object is classified as to whether or not it possesses a certain aspect or belongs to a certain category. Race, political party affiliation, and sex are examples. Any one-to-one transformation of the numbers is allowed.

2. *Ordinal*, in which the numbers indicate whether the object has more or less of the characteristic. Monotonic transformations are allowed.

3. *Interval*, in which the differences between the numbers assigned indicate the amount of difference in the aspect being measured. There is no absolute zero on such scales. Linear transformations of the form $y = ax + b$, $a > 0$ are permitted. Time is an example.

4. *Ratio* scales are like interval scales, with the additional property that an absolute zero can be defined, where zero indicates that the object does not possess the aspect being measured. Transformations of the form $y = cx$, $c > 0$ are permitted. Income from wages is an example.

Eisenhart (1963) distinguishes between a measurement method and a measurement process. The *measurement method* is the specification of how the measurements are to be made. A measurement method includes specification of the measurement scale, the actions to be executed, and the conditions under which the measurements are to be made. The *measurement process* is the "realization of the measurement method," which then becomes the basis for defining error. The *error* for any particular measurement is defined as the difference between the true value of the measurement and the value obtained during the measurement process.

There are two problems with this definition: the difficulties inherent in defining and in determining a true value. The degree of difficulty in defining and determining a true value will vary according to the nature of the characteristic in question. At least three cases can be distinguished:

Case 1. The characteristic to be measured has a clear operational definition, and the measurement method adopted for the survey is that implied by the operational definition. A characteristic is defined operationally if the definition of the characteristic defines or identifies the process to be carried out in obtaining values of the characteristic. Determining weight of the respondents in a survey by asking them to step on a scale and recording the results is an example of adopting a measurement method implied by the operational definition.

Case 2.   In this case the characteristic to be measured has an operational definition, but the process one must go through to obtain the values is so difficult, cumbersome, or expensive that it is not practical to carry it out in the context of the survey. In this case a measurement method is developed for the survey that is believed to give results that are close to those that would be obtained using the operational definition. Asking the respondents to report their weight to an interviewer is an example.

Case 3.   In this case the characteristic to be measured has no clear agreed-upon operational definition. This case encompasses subjective phenomena such as attitudes, beliefs, and opinions. These phenomena are directly observable only by the subjects who possess these characteristics. This case also includes certain concepts, such as intelligence, marital satisfaction, and kindness, which are concerned with complexes of behaviors and emotions. Asking a person whether or not he is satisfied with his weight is an example of such a characteristic.

Our willingness to accept a particular measurement method as the operational definition of the characteristic in question depends on the purposes of the survey. For example, consider measuring people's weight. If we are interested in a general description of the distribution of weight among the adult population, asking people to report their weight may be an adequate operational definition. However, if we are studying a group of infants in which weight gain is to be used to assess the general heartiness of the child or the success of a treatment program, exact specifications for weighing the infant may be required. This procedure might include specifying the type of scale, the amount of clothing, the timing of the weighing in relation to feeding, and so on.

Eisenhart (1963) describes what one might do when developing a measurement method for the length of a board. At first this seems like an easy task, but closer reflection reveals some of the difficulties in defining or developing a measurement method that will give the true value. Can we measure the longest axis that will pass through the board? What will we do about uneven ends? A long list of such questions can be constructed.

We are willing to accept different measurement methods as being the operational definition of a characteristic of different times because of the role measurement plays in science. Ellis (1967) points out that a major contribution to the development of modern science has been the development of measurement techniques that are "sufficiently refined to allow distinction between rival hypotheses and theories." Similarly, Eisenhart (1963:162) states that "the 'true value' of a quantity is intimately linked to the purposes of which knowledge of the magnitude of the quantity is needed, and cannot, in the final analysis be meaningfully and usefully defined in isolation from these needs." Carmines and Zeller (1979) use these ideas in defining measurement for the social sciences. Much of social science measurement is concerned with measuring abstract concepts (case 3 above), and Stevens's definition of measurement as the assignment of numbers to objects or events is not appropriate. The phenomena to be measured are too abstract to be considered either events or objects. Examples cited by Carmines and Zeller are political efficacy, alienation, gross national product, and cognitive dissonance. Stevens's definition is also faulted as ignoring the theoretical components that guided the development of measurement methods. The authors define measurement as "the process of linking abstract concepts to empirical indicants" (1979:10). These abstract concepts can be measured in a large variety of ways.

This criticism of Stevens's definition of measurement may actually have more to do with how it has subsequently been used by operationalists than with the way that Stevens conceived of measurement. His defense of the nominal scale as constituting a measurement scale is made by pointing out that classification into common categories is the first step toward conceptualizing and understanding what goes on in the world.

In attempting to determine the extent of error associated with a particular measurement method, Eisenhart makes a distinction between precision and accuracy. *Precision* refers to the degree to which repeated measurements yield consistent values. In fact, if the measurements are not reasonably consistent with one another, one could question whether a well-defined measurement method existed, because without consistency, no "rule" is being followed in the assignment of values. *Accuracy* is concerned with the net difference between the obtained measurement and the true value. A measurement method can be precise without being accurate.

The terms *precision* and *accuracy* are akin to the terms *reliability* and *validity* used by psychologists and sociologists. However, validity often entails an examination of the larger question of meaning, whereas accuracy tends to be confined to assessing differences between measured values and more-well-defined criterion values.

To understand this difference, consider again the three general cases described above. Our example for case 1, in which the measurement method implied by the operational definition of the characteristic is being used in the survey, involved having respondents step on a scale to determine their weight. We might ask whether the scales being used in the survey are precise (or reliable), meaning to inquire whether repeated measurements on the same person would give sufficiently similar results. In addition, we might ask whether or not the scale is accurate, or: "Is the scale giving a result close enough to that which I could achieve if such and such more rigorous measurement method was used?" The results under this more rigorous method are a criterion value I am willing to accept as a "true value" within the context of my particular research. Thus the criterion value has become the yardstick for judging the accuracy of the measurement method used in the survey. Of course, in making that statement, the researcher has ignored the fact that even the more rigorous measurement method will not always give exactly the same results. This discovery has led some to state either that the "true value" does not exist or that it is unknowable.

In case 2, where we are not using a method implied by the operational definition but simply query the respondents, we are still concerned with precision and accuracy. However, in this case we might have a tendency to speak of validity of the measurement process rather than accuracy. Are respondents' self-reports a valid measure of their weight? No matter how we phrase it, we are, however, relatively certain that we could, in principle at least, determine the accuracy or validity of a respondent's weight report. We would compare the report to the criterion value.

There is no consistent use of the terms *accuracy*, *reliability*, *validity*, and *precision* in the scientific literature. Sometimes the terms are given specific definitions and thus become technical jargon useful for distinguishing among related concepts. In other cases, the words are used as they are generally understood and no precise technical distinctions are implied. Readers must stay alert.

Where attempting to assess the validity (or accuracy) of conceptual characteristics such as those described in case 3, the researcher is severely hampered because often no agreed-upon criterion values exist. Social scientists have developed several methods for assessing the validity of these types of measurements. Basically, these methods are concerned with whether or not the measurements are consistent with observable phenomenon implied by the broader definition of the characteristic or its role in theoretical structures, and whether or not the measurements for a particular characteristic can be distinguished from those of another.

Generally, three types of validity are distinguished: *content validity*— whether or not the results make sense; *predictive* or *criterion validity*—how well the measurements are related to subsequent or concurrent measurements on other characteristics; and *construct validity*—how well people and characteristics are distinguished one from another in the context of theoretical constructs about the nature of social or psychological characteristics. There is considerable literature in the social sciences on statistical methods for assessing the reliability and validity of survey measurements of abstract concepts and subjective phenomena. These methods are not discussed in this book. The interested reader can consult Carmines and Zeller (1979) for an elementary introduction to methods used in assessing reliability and validity. Zeller and Carmines (1980) provide a fuller discussion. Additional references include Blalock and Blalock (1968). Results of an investigation of reliability and validity for surveys of subjective phenomena are presented in a two-volume work edited by Turner and Martin (1984). Numerous references to additional books and articles can be found in these works. These procedures are not discussed further in this book, and in what follows, we largely confine ourselves to characteristics or situations where a criterion value exists conceptually.

We have stated that scientific measurement entails the development, execution, and evaluation of measurement methods. We discuss below the nature in three steps in the context of surveys.

9.1   SURVEY MEASUREMENT PROCESSES

Survey data are collected in a variety of ways, and the type of data collection method determines the sources and types of errors that can occur. There is, however, no well-developed theory or even a consistent classification of the types of measurement methods. Despite this fact, it is widely recognized that the measurement method holds a key position in determining the quality of the final results.

In the preceding section we examined the general nature of measurement by considering types of measurement scales and characteristics studied. Attempts to formalize the notion of how the measurement method influences quality have involved the allocation of survey measurement methodologies into broad classes. However, there is no generally agreed-upon classification of survey measurement methods. We present two as an illustration of how such classifications may be useful in organizing one's thinking about sources of error. Moser and Kalton (1972) list four methods for collecting data: (1) use of documentary sources, (2) direct observation, (3) interviewing, and (4) use of mail questionnaires. When data are collected from documents, the terms used in the documents may vary in meaning from document to document, thus introducing error. With direct observation, there are problems of the representativeness of the observed behavior and with assessing its relative frequency; observer bias may occur; or the observer may alter the normal process and produce a *control effect*. During personal interviews the quality of the data will be affected by respondents' understanding, their accessibility to the information desired, their motivation to answer the questions, and the interaction between interviewer and respondent. In addition, the interviewer may record responses incorrectly and introduce biases when interpreting the respondent's verbal responses. Similarly, mail questionnaires have special problems, including the interaction of the question wording with the respondent's degree of literacy, lack of control over who completes the questionnaire, absence of an opportunity for probing, and so on.

In contrast, Dalenius (1974) gives three classes of measurement methods: (1) using special recording instruments, (2) using instruments handled by professional technicians, and (3) asking questions. Direct observation is considered a subset of instruments handled by professionals. Asking questions includes two main approaches: the questionnaire approach, in which respondents fill out forms, and the schedule approach, in which interviewers fill out forms.

These classifications are not well developed in the literature; however, they have great potential for evolving into a theory of survey methodology that will provide a conceptual framework for controlling the quality of survey measurements. Much of the detailed discussion in the survey methodology literature is devoted to the interview (or schedule) approach. Consider the conceptual framework Sudman and Bradburn (1974) developed. The interview is considered to consist of three components: the task to be accomplished, an interviewer role, and a respondent role.

The task is seen as having an impact on the error by means of its structure, the problems of self-presentation it induces in the respondent, and the saliency to the respondent of the information requested. Some of the elements of the task structure are the type of questions—open ended or closed; the supplementary devices used such as aided-recall devices; and the mode of administration—telephone or personal. The task may raise problems of self-presentation within the respondent because some of the questions may be threatening, may require the respondent to admit ignorance in order to tell the truth, may have socially desirable responses, and so on. The saliency of the requested information, including the complexity of recall tasks or clarity of attitudes, will also affect the results.

Sudman and Bradburn hypothesize that measurement errors will be smaller with greater structure, with fewer evoked problems of self-representation, and with greater saliency. They state, also, that the task sources of error are determined primarily by the questionnaire and "that questionnaire construction and question formulation lie at the heart of the problem of response effects" (p. 13).

The role of the interviewer is the second major component of this framework. Surveys vary in the extent to which they prescribe the interviewer's behavior, some requiring interviewers to follow detailed instructions, others giving them considerable leeway. Sudman and Bradburn call these the interviewer role demands. The interviewers' behavior and extra role characteristics, including age, sex, race, social class, religious or ethnic affiliation, and so on, are all elements influencing the quality of the data.

The authors hypothesize that the greater structure in the interviewer's role, the lower the measurement errors; and the greater the adherence to the role demands, the lower the errors. They also hypothesize that the greater the saliency of the interviewer's extra role characteristics for the questions, the greater the measurement errors.

The last component involves the respondents' role and their motivation to respond well. Concomitantly, Sudman and Bradburn state that the better the motivation, the fewer the errors. They review many studies which investigate the size of response effects associated with various sources of error.

The usefulness of this type of framework lies in its ability to systematize our knowledge of what affects survey results. A more extensive discussion of these methods is beyond the scope of this book, and the interested reader should consult the survey methodology literature. Sources of information include Bradburn et al. (1979), Sudman and Bradburn (1982), Turner and Martin (1984), Tourangeau (1984), Schuman and Presser (1981), and Moser and Kalton (1972). This list is by no means exhaustive, and each source has an extensive bibliography.

9.2 DEFINITIONS OF MEASUREMENT ERROR

The approaches used to define measurement error in surveys vary according to a particular researcher's position on true values. One approach considers true values to exist independent of the survey conditions; the other takes a strict operationalist approach and defines true or preferred values only in relation to the survey conditions. These two approaches reflect different positions on how to deal with the difficulty inherent in defining a true value. As an example of independently existing true values, consider a series of definitions given by Hansen et al. (1951). This article and its subsequent revision (Hansen et al., (1953b) have had a strong influence on subsequent work: for example, Kish (1965), Raj (1968), O'Muircheartaigh (1977), and Moser and Kalton (1972). Hansen et al. (1951:149) had this to say:

> The individual true value will be conceived of as a characteristic of the individual quite independent of the survey conditions.... Three criteria for definition of the true value are (the first two essential, the third useful but not essential):
> (1) The true value must be uniquely defined.
> (2) The true value must be defined in such a manner that the purposes of the survey are met. For example, in a study of school children's intelligence, we would ordinarily not define the true value as the score assigned by the child's teacher on a given day although this might be perfectly satisfactory for some studies (if, for example, our purpose was to study intelligence as measured by teacher's ratings).
> (3) Where it is possible to do so consistently with the first two criteria, the true value should be defined in terms of operations which can actually be carried through (even though it might be difficult or expensive to perform the operations).

Thus the authors prefer the true value to be operationally defined but do not require it. They go on to define the *individual response error* as "the difference between an individual [survey] observation and the true value for the individual" (p. 152).

Sukhatme and Sukhatme (1970) adopt a similar definition; however, they speak of the discrepancy between the survey values and the true values, rather than the difference.

It may be preferable to use the word *discrepancy*, as Sukhatme and Sukhatme do, rather than the word *difference*, because in some instances using the mathematical difference to describe the degree of the discrepancy does not make sense. Such would be the case for characteristics measured on nominal scales, because the numbers assigned to the categories are arbitrary. Ordinal variables also exhibit this difficulty to some extent. For example, suppose that parents in a certain city were asked to report on which of five elementary schools their children attended. A nominal scale might be established with the following categories:

1. Silkwood Elementary
2. Crabtree Elementary
3. Center Elementary
4. Mann Road Elementary
5. Villa Heights Elementary

A positive difference between the survey value and the true value is no different in meaning than is a negative difference. The only relevant question to ask is whether the child was classified correctly.

Now suppose that the schools have been ordered according to perceived prestige within the community, with Silkwood being the least prestigous and Villa Heights the most. The scale is merely ordinal, however, and there is no claim that Villa Heights is five times more prestigous than Silkwood. Now, a positive difference between the survey value and true value has a different meaning than that of a negative difference. An average positive difference would indicate that respondents tended to report that their children attended more prestigous schools than is actually the case. Even more information is conveyed about the extent of error if the scale is numeric (interval or ratio).

A later article by Hansen et al. (1967) recognized that it is sometimes not feasible in a survey to adopt a technique for obtaining the true values. They define three sets of measurements: (1) those produced if the *ideal goals* had been met, (2) those specified under a more operationally feasible set of survey specifications which were executed without error, and (3) *expected values of survey procedures*, the set of statistics that are the expected values of the survey statistics over a large number of independent replicates.[1]

In contrast to this approach, which defines true values separate from the survey conditions, Zarkovich (1966) defines true values in the context of an *adopted system of work*, which consists of the chosen measurement methods, concepts and definitions, tabulation plans, data collection instructions, and so on. The true values are those that would result if the adopted system of work could be carried out without error. The results actually obtained are the *survey values*, and error is defined as the difference between the survey and true values.

Deming (1960) also takes this approach, stating that there is no such thing as a true value but that "we do have the liberty to define and to accept a

---

[1]For the terms in italics, the Compendium of Nonsampling Error Terminology gives the exact wording used in the reference.

specific set of operations as preferred" (p. 62). Subject matter experts may provide a *preferred survey technique* for the material to be studied. In some cases this technique cannot be used because of expense or other difficulties, and it is supplanted by a *working survey technique* which will give different results from the preferred technique. He defines the *equal complete coverage* of a survey as the result that would be obtained from examining all the units in the sampling frame. The difference between the preferred and working technique under the equal complete coverage is called the *bias* of the working technique. Accuracy then, refers to the size of this bias. Deming uses the term *validity* to refer to the presence or absence of sampling bias, not to assessment of the quality of measurements.

In many practical cases the distinction between a true value and a preferred value will be unimportant. Preferred values can take the place of true values in definitions of measurement error. However, in other cases the issue of true values will be important because of its implication for the responsibilities that must be considered when assessing the quality of the survey measurements. If true values do not exist apart from the survey specifications, the researcher is only responsible for determining how the survey values differ from these preferred values. If an independently existing true value is defined, some assessment of its relationship to the survey value is needed. The extent to which one needs to consider true values that exist apart from the survey specifications will depend on the nature of the research, as was illustrated early in this chapter when we considered the difficulties in defining a true value. In every case, however, it must be remembered that someone must ultimately take responsibility for determining whether or not our surveys measure what they claim to measure.

In the next two chapters we consider the methods used to assess the presence of measurement error, the magnitude of its impact on the survey estimates, and the methods that can be used to adjust for measurement errors. In Chapter 10 we focus on measurement bias, in Chapter 11 on measurement variability.

CHAPTER 10

Measurement: Quantifying Measurement Error

10.1 HISTORICAL PERSPECTIVES

It is well recognized that repeated measurements on the same material will not give the same results. In many cases the existence of measurement variability can be thought of as arising from errors in the measurement. Studies of measurement variability have had a long history. An early article by Karl Pearson (1902) reported on an experiment that he conducted concerning errors in judgment. He was particularly concerned with estimating parameters of the "personal equation."

The personal equation for a human measurer is the average over a large series of judgments or measurements of the errors that an observer makes in measuring of a fixed quantity. Using Pearson's terminology, if ξ is the actual value of a physical quantity, and x_1 its value according to the judgment of an observer, then the mean value of $x_1 - \xi$ over a larger series of judgments, P_{01}, is the personal equation of the observer.

Briefly, the theory or mathematical model for measurement error that Pearson used was as follows. Let ξ be the true value of the quantity being measured and x_j the value obtained by the jth observer. Then for every observer the mean value of $x_j - \xi = P_{0j}$ is the personal equation of the jth observer, and the variability of the jth observer's judgment is σ_{0j}.

It is assumed that the x_j are independently normally distributed. The goodness of the observer is then measured by two characteristics (Pearson, 1902:238):

1. The smallness of his personal equation, p_{0j}
2. The smallness of the variability of his judgment, σ_{0j}

Now, in most of the measurement situations Pearson was concerned with, ξ was not known, so it was only possible to measure $x_j - x_{j'}$, the relative error

245

of judgment of two observers. However, Pearson noted that if we had three observers, σ_{0j}^2 could be determined for each of them. Let $\sigma_{jj'}^2$, be the variability of the relative error of judgment. Assuming the independence of measurements among observers, then with three observers we can measure each one's stability by using the following:

$$\sigma_{21}^2 = \sigma_{01}^2 + \sigma_{02}^2$$

$$\sigma_{32}^2 = \sigma_{02}^2 + \sigma_{03}^2$$

$$\sigma_{13}^2 = \sigma_{03}^2 + \sigma_{01}^2.$$

These give the following as estimates for the variability in judgments:

$$\sigma_{01}^2 = \frac{\sigma_{21}^2 + \sigma_{13}^2 - \sigma_{23}^2}{2}$$

$$\sigma_{02}^2 = \frac{\sigma_{32} + \sigma_{21}^2 - \sigma_{13}^2}{2}$$

$$\sigma_{03}^2 = \frac{\sigma_{13}^2 + \sigma_{32}^2 - \sigma_{21}^2}{2}.$$

Based on this theoretical development, Pearson designed two experiments to measure the variability in judgments. In the first experiment, he had three observers each bisect 500 line segments by eye estimate; in the second, over 500 estimates of the distance transversed by a bright light were made. In each case the true values were known, and Pearson and his colleagues calculated the absolute and relative personal equations of the three people. They calculated the means and variances by hand, and it took the *entire summer* to make the calculations, aided by something called an "American Comptometer, which for some years past we have found to be of great aid in statistical investigations" (Pearson 1902:250).

Much to Pearson's surprise, he found a positive correlation between the errors of judgment made by the three observers. Instead of zero correlation, the correlations $r_{jj'}$ for the bright line series of experiments were $r_{12} = 0.0132$, $r_{23} = 0.3819$, and $r_{13} = 0.1571$. Pearson was not able to explain this correlation, since the judgments were made independently and spoke of "some particular source of mental or physical likeness which leads to this correlation in judgments" (p. 236).

Pearson's work on the personal equation is a prototype of subsequent investigations that try to quantify measurement error—a statistical model is developed, experiments are designed, data are gathered, and estimates of the model components are made.

Cochran (1968) discusses several simple mathematical models used in the study of measurement errors. In describing a simple model for numeric variables, Cochran (1977) asks one to imagine that a large number of independent measurements could be made for each member of the target population. The measurement value for a particular instance t is Y_{it}, where i indicates a specific member of the target population. The measurement is thought of as follows:

$$Y_{it} = X_i + e_{it},$$

where X_i is the true or correct value for the ith population unit and e_{it} is an error of measurement that has a frequency distribution with a mean B_i and a variance of σ_i^2.

Models of this nature can be used in various ways: The effects of measurement errors on various statistical techniques can be studied analytically. Cochran (1968) reviews the impact on analysis of variance, regression, and so on, under various assumptions and elaborations of the foregoing model. Similar models can also be used to study the effects of alternative survey methods, such as comparing telephone and face-to-face interviews. A third approach is to use mathematical models of this nature to investigate the impact of specific sources of error, such as context effects in questionnaires, recall biases, interviewer errors, and so on.

In this book we focus on the use of these models to evaluate the overall quality of the survey, the overall quality of some particular data collection methodology, or on their use for investigating the impact of specific sources of error connected with a particular methodology. Therefore, much of the discussion is directed at looking at the impact of measurement errors on estimates of fairly simple population statistics, such as means and totals. We recognize the extreme importance of investigating impact of measurement error on complex statistical techniques; however, such studies are beyond the scope of this book. The interested reader can consult the following sources: Chai (1971), Koch (1969), Bross (1954), Giesbrect (1967), Korn (1982), Chua and Fuller (1987).

To organize the discussion, we first ignore measurement variability and focus on the bias or fixed errors. This approach is taken when one is willing either to assume that there is no variability in the measurements or that, for the purposes at hand, the results of a particular implementation of the measurement method are to be investigated. This type is the simplest approach to the study of measurement errors, and most researchers recognize that response variability exists, although they are choosing to ignore it for the moment.

In this chapter we study this fixed-bias case. We begin with a section on modeling in which we first discuss the mean square error of estimates. We consider different approaches for the study of numeric and categorical data. We discuss the need to look beyond simple mean-square-error models or overall error and present several statistical techniques for doing so.

Following the section on modeling, we discuss experimental methods for quantifying the models in general and then give some examples of the large variety of issues to which they have been applied. Refinements of these general models/experimental approaches for the study of specific sources of error are also discussed. In the final section of the chapter we discuss how the information from an error study can be used to adjust for bias.

Chapter 11 is devoted to measurement variability. If focuses on some of the controversies with the concept and some of the alternative approaches used to model measurement variability in surveys.

10.2 ERROR MODELS FOR NUMERIC DATA

The simplest models for the effect of measurement errors on survey estimates are those that consider fixed biases in the measurements and no random variations. Zarkovich (1966) discusses models of this type. These models are expressed either for a particular element of the population or in terms of net effects on population parameters. We first consider models for numeric (ratio and interval) data.

10.2.1 Model

First consider models for the individual. Specifically, let

$Y_i =$ measurement for the ith target element

$X_i =$ true value (or preferred value) for the ith element

$B_i = Y_i - X_i =$ individual error of measurement for the ith element.

Then

$$Y_i = X_i + B_i.$$

The B_i are called by some authors the individual biases or element biases. These individual errors may or may not have net effect on the survey estimates. Net effects are usually discussed for estimates of totals and means; however, it is easy to generalize the discussion. Let

$\mathbf{X} =$ vector of population true values

$\mathbf{Y} =$ vector of measurements for the target population

$\mathbf{B} =$ vector of element biases

$\mathbf{Y} = \mathbf{X} + \mathbf{B}.$

Now suppose that the goal of the survey is to estimate a function of the population of true values. This value could be their total or their mean. Let $f(\mathbf{X})$ be the quantity to be estimated from the survey. Suppose that $g(\mathbf{y})$ is the sample estimate of $f(\mathbf{X})$, where $\mathbf{y}$ is the vector of measurements on the sample and g is some function of them, such as the sample mean. Then the mean square error of $g(\mathbf{y})$, MSE$[g(\mathbf{y})]$, is given by

$$\mathrm{MSE}[g(\mathbf{y})] = E_s[g(\mathbf{y}) - f(\mathbf{X})]^2,$$

where E_s denotes the expected value over all possible samples:

$$\mathrm{MSE}[g(\mathbf{y})] = E_s[g(\mathbf{y}) - E_s g(\mathbf{y})]^2 + [E_s g(\mathbf{y}) - f(\mathbf{X})]^2.$$

Assuming that the function $f(y)$ is one for which an asymptotically unbiased estimation procedure exists, that is, $E_s[g(\mathbf{y})] \doteq f(\mathbf{y})$,

$$\mathrm{MSE}[g(\mathbf{y})] \doteq \text{sampling variance of } g(\mathbf{y}) + \text{squared bias in } g(\mathbf{y})$$
$$= E_s[g(\mathbf{y}) - f(\mathbf{Y})]^2 + [f(Y) - f(X)]^2.$$

The adjective *asymptotically unbiased* is used to remove from consideration those components of sampling estimation bias that tend to zero as the sample size gets large: for example, the bias in combined ratio estimators. This simple model can be applied in the study of many sources of error, such as respondent reporting errors, systematic errors in measuring devices, quality of results obtained in telephone versus face-to-face interviews, and so on.

10.2.2 Methods for Quantifying Error

A numerical study of biased measurements can be carried out by collecting additional accurate or unbiased measurements for the sample members. This is usually done on a subsample basis, because if the researcher could afford to obtain accurate measurements for the entire sample, he or she would not bother with the faulty ones. Two methods that are used are (1) record check studies, in which a source of records is checked for a subsample of units included in the survey and (2) resurveys, in which a subsample of the units is resurveyed or remeasured using more accurate methods. One then assumes that the record check values or more accurate values are the true ones. Of course, it is usually recognized that these more accurate measurements are not "true values" in any ultimate sense, but rather, are the preferred values for the current issue under study.

In the context of the simple aggregate model given above, let

$\mathbf{y}_{ss}$ = original measurements for members of the resurvey or record check subsample

$\mathbf{x}_{ss}$ = true or preferred values obtained from the record checks or the resurvey

and the net bias may by estimated by

$$\hat{\mathrm{Bias}}[g(\mathbf{y})] = g(\mathbf{y}_{ss}) - g(\mathbf{x}_{ss}).$$

If we again assume that the sample design and subsample are ones for which an asymptotically unbiased estimation procedure exists in the sense that $E_s\{E_{s|ss}[g(\mathbf{y}_{ss})]\} \doteq f(\mathbf{Y})$, then

$$E_s\Big\{E_{ss|s}\hat{\text{Bias}}[g(\mathbf{y})]\Big\} = f(\mathbf{Y}) - f(\mathbf{X}).$$

A variety of statistics based on this simple model can be computed to make statements about the magnitude of error. To illustrate some of these, consider simple random sampling and estimation of the population mean

$$\bar{X} = \frac{1}{N}\sum_{i=1}^{N}X_i.$$

If a sample of size n is drawn from a population of size N and the sample mean is used to estimate $\bar{X}$, the mean square error of

$$\bar{y} = \frac{1}{n}\sum_{i=1}^{N}Y_i$$

as an estimate for $\bar{X}$ is

$$\text{MSE}(\bar{y}) = \frac{N-n}{N}\frac{1}{n}S_y^2 + \bar{B}^2,$$

where

$$S_y^2 = \frac{1}{N-1}\sum_{i=1}^{N}(Y_i - \bar{Y})^2$$

$$\bar{B} = \frac{1}{N}\sum_{i=1}^{N}B_i.$$

Ignoring the finite population correction factor, $\text{MSE}(\bar{y})$ can be written as

$$\text{MSE}(\bar{y}) = \frac{1}{n}\Big[S_X^2 + 2\,\text{Cov}(X_i, B_i) + S_B^2\Big] + \bar{B}^2,$$

where

$$S_X^2 = \frac{1}{N-1}\sum_{i=1}^{N}(X_i - \bar{X})^2$$

$$S_B^2 = \frac{1}{N-1}\sum_{i=1}^{N}(B_i - \bar{B})^2.$$

Table 10.1   Error Measures Under a Fixed Bias Model for Error

Error Measure	Definition	Sample Estimator	What It Tells You
Bias	$\bar{B}$	$\bar{b}$	Difference between true value and measured values
Mean square error (MSE)	$\mathrm{Var}(\bar{y}) + (\bar{B})^2$	$\dfrac{s_y^2}{n_s} + b^2 - \dfrac{s_b^2}{n_s}$	Overall measure of quality of survey data, including sampling error and measurement bias
Coefficient variation of biases terms	$\dfrac{S_B}{|\bar{B}|}$	$\dfrac{s_b}{|\bar{b}|}$	Measure of dispersion of biases
Relative bias	$\dfrac{\bar{B}}{\bar{X}}$	$\dfrac{\bar{b}}{\bar{x}}$	Net bias measure—proportionate error
Increase in MSE from using faulty measurements	$\mathrm{MSE}(\bar{y}) - \mathrm{Var}(\bar{x})$	$2s_{xb} + \bar{b}^2$	Loss from using faulty measurements
Bias ratio	$\dfrac{\bar{B}}{[\mathrm{Var}(x)]^{1/2}}$	$\dfrac{\bar{b}}{\left(s_x^2/n_s\right)^{1/2}}$	Impact of bias in units of standard error

Some of the error measures used in addition to the aggregate bias and mean square error and the type of information each provides are summarized in Table 10.1. In the table,

$$s_y^2 = \frac{1}{n_s - 1} \sum_{i=1}^{n_s} (Y_i - \bar{y}_s)^2,$$

s_b^2 is defined similarly, and

$$s_{xb}^2 = \frac{1}{n_s - 1} \sum_{i=1}^{n_s} (X_i - \bar{x}_s)(B_i - \bar{b}_s).$$

Faulty and accurate measurements can be used to study a variety of issues. For example, Fleischer et al. (1958) used the method to compare different ways of obtaining acreage estimates of cotton fields. They compared direct measurement (chain measurements), farmers' estimates during an interview, planimeter measurements using aerial photographs, and rotometer measurements on aerial photographs. For the purposes of the study, direct measurements were considered to be the accurate or unbiased technique. Their results are summarized in Table 10.2. The study was done in two years, and data for both years are shown.

These figures reveal some interesting results concerning the different measurement processes used. Although the farmers' estimates are practically unbiased in the aggregate, there is greater variability in their errors than for either the planimeter or rotometer estimators. The negative covariance between the bias terms and true values indicates that farmers tend to underestimate acreage in large fields and overestimate acreage in small fields. In contrast, the planimeter and rotometer methods tend to have net overall biases but less variability in these biases. The positive covariance between biased and true values indicates that these methods tend to under-measure acreage in small fields and overmeasure it in large fields.

Although this study indicated that farmers do a fairly good job in the aggregate, the authors of this article were unwilling to recommend using only farmers' reports of field sizes because farmers may be inaccurate in other years or for different types of crops. The larger variance of the farmers' errors tends to support the feeling that the interview data are unstable.

Kish and Lansing (1954) carried out a similar study on response errors in estimating the value of homes. Homeowners' and appraisers' estimates of the market value of houses were compared. Their model explicitly recognized

Table 10.2   Errors Associated with Alternative Ways of Measuring Cotton Acreage[a]

Measurement Method	Bias	Relative Bias	Variance of Bias Terms	MSE	Coefficient of Variance of Bias Terms	Increase in MSE	Covariance of Bias and True Values
				1954			
Farmers' estimates	−0.118	−0.01	1.73	67.83	11.15	0.27	−2.00
Planimeter	0.806	0.074	0.99	79.71	1.23	11.61	10.02
				1955			
Farmers' estimates	−0.001	0	2.07	92.18	—	−3.37	−5.44
Planimeter	0.342	0.033	0.21	102.14	1.33	6.48	6.27
Rotometer[b]	0.520	0.050	0.33	104.89	1.10	9.34	8.75

Source: Adapted from Table 2 of Fleischer et al. (1958).

[a]The accurate measurements are

$$1954: \quad \bar{x} = 10.825, \quad s_x^2/n = 68.10$$

$$1955: \quad x = 10.410, \quad s_x^2/n = 95.55.$$

[b]Only done in 1955.

that the appraisers' estimates may also be in error. Some of the error measures the authors used were:

- A comparison of the frequency distribution of the two measurements
- Cross-tabs of owner versus appraiser values (which revealed some coding errors)
- Frequency distribution of the ratio of the homeowner to the appraiser values
- Estimates of overall mean square error and relative contribution of bias and variance terms for different sample sizes
- The root-mean-square difference between pairs of values for various subgroups where (using the simple random sampling formulation above)

$$\text{Mean-square difference} = \frac{1}{n_s} \sum_{i=1}^{n_s} b_i^2.$$

Many of the analyses were directed at determining what variables might be associated with errors that homeowners made. The hope was that if accurate respondents could be identified, methods of analysis that focused on these accurate respondents could be developed.

Measures other than direct estimates of the overall bias are, in general, aimed at looking beyond the net bias in some overall statistic to detect any biases that may differentially affect subgroups of the target population. In general, the aim of these analyses is to determine whether analytic statements are likely to be adversely affected by measurement bias, as illustrated by the following study description.

Borus (1966) describes a record check study in which respondent and employer-reported earnings are compared. He calls this research a *response error study*. He used a variety of measures to compare the "true" (employer) to the measured (respondent) values. These include:

1. Mean absolute deviation of measured values and true values.
2. Mean percentage deviation measured as a percentage of the respondent values.
3. Pearson product moment correlation between respondent values and true values.
4. Distribution of absolute deviations.
5. Distribution of the percentage deviations measured as a percentage of the average of the two values, that is,

$$\frac{X_i - Y_i}{1/2(X_i + Y_i)}.$$

6. Distributions of positive and negative deviations. Chi-squared tests of the equivalence of these distributions were done.

Finally, trying to discover if certain subpopulations have more misreporting than others, he regressed the

$$\text{average response error per week} = \frac{\text{actual value} - \text{respondent value}}{\text{number of weeks}}$$

on various characteristics of the respondents. Borus does not discuss the relative usefulness of these measures but does say that his finding that response error regressed significantly on certain characteristics (age, sex, education, etc.) indicates that the data require adjustment to remove the bias in the survey estimates. Presumably, this need would be because subgroup comparisons would be done in the survey data analysis.

In comparing the foregoing measures of the effect of bias, we note that the net bias can be zero even if the mean absolute deviation is not, so that the latter statistic may be preferable to net bias when subgroup comparisons are anticipated. The percentage deviation is a relative bias measure; however, the one in which the average of the true and response values is used in the denominator understates the effect of bias on the estimates. The correlation coefficient can be very high even though the net bias is high. This situation will occur if some relatively constant bias affects the results of each sample member. The distribution of the absolute errors indicates whether or not there is stability in the response error: Do they tend to cluster around some average error, or are they widely dispersed? The distribution of positive and negative deviations and the test of significance indicates whether or not there is a net bias and a difference in the tendencies to under- or overestimate.

The type of bias analysis that will be most useful in a particular study depends on the goals of the survey; however, in many studies it is necessary to go beyond an assessment of overall bias by using some of the techniques described above. In the absence of any hard data on bias, the argument is often made that although the measurements may be biased, we can assume that the bias is the same for each subgroup, so that subgroup comparisons remain valid. This assumption can be very wrong. Consider, for example, the results of the Woltman et al. study (1980) of the effect of interviewing mode in the U.S. National Crime Survey. An interviewing method that made maximum possible telephone use (about 80 percent interviewed by telephone) for obtaining data was compared to the usual data collection procedure (about 80 percent interviewed in person). The victimization rates per 1000 persons aged 12 or older were compared. Overall, some 8 percent fewer personal victimizations were reported in the maximum telephone group. However, there was a dramatic difference between males and females in this difference. For males the difference between the usual procedure and the maximum telephone was 12 percent lower for the telephone group. In contrast, this difference for females was only 3 percent. Thus the subgroup comparison of males versus females is adversely affected by the differential bias experienced by the two groups. If the usual survey procedure is used, the male rate is 38 percent greater than the female rate. If the telephone interview procedure is used, the male rate is only 25 percent greater than the female rate. In this study it was assumed that the interview results were more accurate.

Sudman and Bradburn (1974) compared a large number of surveys using a statistic they call the *relative response effect*. They define this statistic as (in the context of our previous notation) follows:

$$\text{Relative response effect} = \frac{\bar{b}}{s_x}.$$

They compared a very large number of studies using this measure. Their study was directed at verifying certain hypotheses concerning the nature of the response task and interviewer and respondent characteristics (see Chapter 9 for a brief discussion of their theory). Bradburn (1983) summarizes fundings from a number of studies.

10.2.3 Effect of Errors in the Error Study

The record check and resurvey procedures discussed above for evaluating the impact of biased measurements assume some criterion value that is free of error and can be compared to the survey value either on an element-by-element basis or by using aggregate statistics. These are very restrictive assumptions, and it is possible to study the existence of bias even when the record check study or the resurvey is also subject to error.

One source of error in a record check study is mismatching—the failure to associate correctly survey values with the record check or resurvey values for the same element. this situation can occur because the identifying information is not adequate to make an unambigious match or because of errors made during the resurvey. For example, one might make a record check of household respondent hospitalization reports. The household respondent may be identified as Jane Smith of 125 Main Street. The hospital may have records for

Mrs. George Smith	Main Street
Jane Smith	125 Glenn Avenue
Jane Smith	416 West Brule Avenue

It is not clear which record should be matched with the survey respondent.

Neter et al. (1965) studied the effect of mismatching on the measurement of response errors in a record check study. They present an unrestricted matching error model and a model in which mismatches are restricted to subsets of the population.

In summary, for the unrestricted model, we have $Y_i - X_i = B_i$, the true response error. The sample element, however, may be incorrectly matched with element X_k, $k \neq i$. Let

$$Z_i = \text{record check value matched with element } Y_i$$

$$Z_i = \begin{cases} X_i \text{ with probability } p \\ X_k \text{ with probability } q \end{cases}$$

where for a population of N units, $p + (N - 1)q = 1$.

Thus the measured response error M_i is given by the following:

$$M_i = \begin{cases} Y_i - X_i = B_i \text{ with probability } p \\ Y_i - X_k \text{ with probability } q; \, k \neq i. \end{cases}$$

The authors show that under this model the estimates of the average bias, $\bar{B}$, are unbiased and the variance of the measured response errors will be greater than the variance of the actual response errors if the correlation between the true values and reported values is positive. Algebraically, this is expressed as

$$\text{Var}(M) = \frac{1}{N} \sum_{i=1}^{N} (M_i - \bar{M})^2$$

$$\text{Var}(B) = \frac{1}{N} \sum_{i=1}^{N} (B_i - \bar{B})^2$$

$$\text{Cov}(Y, X) = \frac{1}{N} \sum_{i=1}^{N} (Y_i - \bar{Y})(X_i - \bar{X}).$$

The variance of the measured response errors is shown to be equal to $\text{Var}(M) = \text{Var}(B) + 2Nq\,\text{Cov}(Y, X)$.

In addition, the authors note that if the measured response errors are regressed on the values from the records, the slope of the regression line will be understated if there is a positive correlation between the true and reported values. In the restricted model, matching errors are constrained to occur only within subsets of the population: for example, only for people with the same last names. Results similar to those for the unrestricted case are demonstrated.

10.3   ERROR MODELS FOR CATEGORICAL DATA

The procedures above are most appropriate when the scale of measurement for a characteristic is numeric (see Chapter 9 for a discussion of measurement scales). When data are categorical, the misclassification approach is

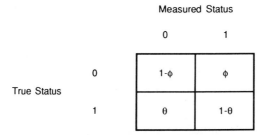

Measured Status

		0	1
True Status	0	$1-\phi$	ϕ
	1	θ	$1-\theta$

Figure 10.1 Misclassification matrix for presence or absence of a characteristic.

preferable for the study of error. To illustrate this approach, consider a population of size N whose members are classified as possessing or not possessing a particular characteristic, using a particular survey technique that is subject to error. A misclassification matrix for the proportion of the elements that are incorrectly classified is shown in Fig.10.1. A 1 indicates the presence of the characteristic; 0, its absence. In the figure, θ represents the proportion of elements that have the characteristic and are classified as not having the characteristic, and ϕ is the proportion of elements that do not have the characteristic and are classified as having the characteristic.

In some settings θ is called the proportion of false negatives, and ϕ is termed the proportion of false positives. In medical studies the terms sensitivity and specificity are often used, where the sensitivity of the measurement process is the proportion with the characteristic (usually a disease) who are correctly classified as having the characteristic. Sensitivity is $1 - \theta$; specificity is $1 - \phi$. Figure 10.2 expresses the same idea in terms of the number of population elements that fall into the cells of the matrix.

The true proportion p possessing the characteristic is

$$p = \frac{N_{1+}}{N}.$$

Measured Status

		0	1	
True Status	0	N_{00}	N_{01}	N_{0+}
	1	N_{10}	N_{11}	N_{1+}
		N_{+0}	N_{+1}	N

Figure 10.2 Classification of a population of N elements by a fallible survey method and true status.

In addition,

$$\theta = \frac{N_{10}}{N_{1+}}$$

$$\phi = \frac{N_{01}}{N_{0+}}.$$

If a simple random sample of size n is selected from this population and measured by the faulty survey measurement method such that

$$Y_i = \begin{cases} 1 & \text{if the element is classified in the survey as having} \\ & \text{the characteristic} \\ 0 & \text{otherwise,} \end{cases}$$

the expected value of $\bar{y}$ as an estimate of p is

$$E(\bar{y}) = p(1 - \theta) + q\phi.$$

The net bias $\bar{B}$ in the measurement method is

$$\text{Bias}(\bar{B}) = -p\theta + q\phi.$$

The net bias is not zero even when $\theta = \phi$, except when the proportion of false positives equals the proportion of false negatives and p is $1/2$. However, if the *number* of false positives equals the *number* of false negatives, the net bias is zero; however, this requires that $p\theta = q\phi$ because (see Fig. 10.3)

$$N_{01} = N\phi q = N\frac{N_{01}}{N_{0+}}\frac{N_{0+}}{N}$$

$$N_{10} = N\theta p = N\frac{N_{10}}{N_{1+}}\frac{N_{1+}}{N}.$$

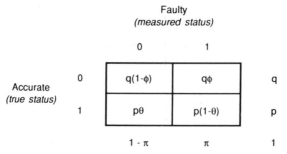

Figure 10.3 Cell probabilities for joint classification of an element for a faulty and an accurate measurement process.

Because of these facts, an error study examining the measurement of a dicotomous variable should look beyond the net bias in the measurement method and consider other measures of error. This step is because analytic comparisons of subgroups may be subject to large errors even if the net estimates for the entire population are subject only to small errors, as was illustrated in the preceding section with results from a study of the National Crime Survey.

A numeric study of the errors can be carried out in a manner analogous to that described in the preceding section. A subsample is selected, and accurate measurements (either through record checks or remeasurement) are obtained for members of the subsample. Estimates of the net bias, ϕ and θ then can be obtained. Tenebein (1970) suggests an additional summary measure of the quality of the survey measurements. This measure is the *coefficient of reliability*, which Tenebein describes as "a measure of the strength of the relationship between the true and fallible classification, i.e., a measurement of how well the true classification can be predicted from the fallible one on a particular element" (p. 1356). This measure is defined as the square of the correlation coefficient of the true and faulty measurements and can be derived using the joint distribution in Fig. 10.3.

$$\text{Reliability coefficient} = \rho^2(X,Y) = \left[\frac{\text{Cov}(X,Y)}{S_X S_Y} \right]^2$$

$$= \frac{pq(1 - \theta - \phi)^2}{\pi(1 - \pi)}.$$

The quantity π is the proportion of population with the characteristic as measured by the fallible method.

Analogous techniques can also be used for the study of multinomial data in which the elements are measured on a scale that classifies them into one of r categories. Figure 10.4 illustrates this general case. In this situation $\theta_{kk'}$ is the proportion of population members in category k that are classified into category k' by the fallible survey measurement method. Then $\Sigma_{k'}\theta_{kk'} = 1$. The elements in each row of this misclassification matrix are similar to the binomial case. The expected value of the estimated proportion in the kth class using the fallible survey measurement method is

$$\pi_{k'} = \sum_{k=1}^{r} \theta_{kk'} p_k$$

where p_k is the true proportion in the kth class (Cochran, 1968). The impact of misclassification on measures of association has been studied by numerous authors, including Giesbrect (1967), Mote and Anderson (1965), Cochran (1968), Goldberg (1975), and Korn (1982). Summary discussions are given by

Measured Status

		1	2	... k' ...	r
	1	θ_{11}	θ_{12}	. . . . . .	θ_{1r}
	2	θ_{21}	θ_{22}		θ_{2r}
True Status					
	k			$\theta_{kk'}$	
	r	θ_{r1}	θ_{r2}	. . . . . .	θ_{rr}

Figure 10.4   Misclassification matrix for general multinomial case.

Fleiss (1973) and Landis and Koch (1975a, b). Numerous additional references can be found in these works.

Bishop et al. (1975) discuss measures of association and agreement for two-way tables. They point out that measures of agreement are special cases of measures of association where the contingency tables have the same number of categories in each marginal, the categories on each marginal are in the same order, and our main interest lies in the proportion of measurements that lie on the main diagonal.

10.4   METHODS USED IN THE ABSENCE OF SPECIAL DATA COLLECTION FOR THE STUDY OF ERROR

In the preceding two sections we described methods developed for error studies that entail the collection of additional data on the sample or a subsample as a part of (during or soon after) the survey activities. For at least a portion of the sample elements, the investigator has measurements by an inaccurate and a presumed-more-accurate survey method. The accurate method becomes the standard or criterion measure by which the quality of the fallible survey method is judged. We described methods for numeric data that are based on the numeric difference between the inaccurate and accurate measurement. Various summary statistics (bias, MSE, covariance of accurate and faulty values, and so on) used to evaluate survey measurement methods were described. For categorical data, we described the misclassification approach to the study of error. Statistics based on the proportion of elements misclassified by the fallible measurement method were presented.

All the methods above require measurement of the same elements by an accurate and inaccurate measurement method. In many cases, the survey researcher will not have such data available and will still want to make some assessment of the quality of the results.

A simple method for detecting bias is to compare estimates from the survey to estimates or results from another source. In general, comparison of data from two independent sources which should agree can indicate that bias is present; however, the measurement of bias requires the assumption that one source is accurate. This situation is illustrated as follows. Let

$$\bar{Y}_1 = \text{mean from source 1}$$
$$\bar{Y}_2 = \text{mean from source 2.}$$

Under the simple model for numeric data,

$$\bar{Y}_1 = \bar{X} + \bar{B}_1$$
$$Y_2 = \bar{X} + \bar{B}_2.$$

The difference,

$$\bar{Y}_1 - \bar{Y}_2 = \bar{B}_1 - \bar{B}_2,$$

is often called the *differential bias*. If either $\bar{B}_1$ or $\bar{B}_2$ can be assumed to be zero, the difference above is an estimate of the bias in the source with possible nonzero bias. Note that even when the differential bias is zero, bias can still exist—both sources may be equally biased.

Often, when results from a survey are compared to those from another data source that is assumed to be accurate, the real interest is not in the specific data items being compared. After all, if accurate data concerning some characteristic are already available, there is no reason to do the survey. Usually, this type of comparison is done as a means of determining how well the survey has performed in general. It is assumed or implied by the persons making the comparison that if there is good agreement between the survey and an accurate independent source for some set A of characteristics, the survey operated well. Thus one can place more confidence in the measurements for some set B of characteristics, for which no independent external data exist.

This situation of course may not be true. Ability to measure set A does not necessarily imply ability to measure set B, particularly when you consider the set of characteristics for which comparison data are likely to be available. This set is likely to encompass the more commonly measured demographic items, such as age, race, sex, education, and so on. This set of characteristics may be much better defind than set *B*. Of course, the finding that the survey is not in agreement with the presumed accurate source for set *A* (the more easily and widely measured characteristics) is not likely to excite confidence in measurements for set *B*.

A second method for detecting bias is through the use of consistency studies in which the consistency of the data with known relationships is examined. For example, Zarkovich (1966) describes some consistency studies for demographic data. Examples he gives are:

1. Certain indices, such as Myer's, Whipple's, and Bachi's, which are aimed at detecting the tendency to misreport age by rounding or other digit preferences
2. Age and sex ratios, which may indicate that people of certain ages and sexes have been missed in a demographic survey
3. Use of balancing equations, in which the total of some characteristics in the survey should equal sums and differences of data from other sources
4. Cohort survival techniques used in censuses, in which the actual number in a later census is compared to the expected number surviving from a previous census.

Sometimes the consistency study is not based on presumed relationships among the survey characteristics, but rather a theory about missreporting is used to decide on the magnitude of the bias. In many cases, it is assumed that the major component of the net bias in a survey estimate is because of underreporting of various events. For example, people may underreport episodes of illness, fail to report having had an illegal abortion, or not report a criminal victimization. This may be because the respondent forgot the event, did not wish to reveal it, or did not want to talk about it with an interviewer. In any case, in this situation when data from two surveys or sources are compared, the source having the higher estimate of the number of events is assumed to be the more correct one. For example, Woltman et al. (1980) compared three data collection procedures for the U.S. National Crime Survey in the later 1970s. This survey asks household respondents about the number of criminal victimizations (thefts, assaults, and robberies) experienced by the household or its members. The validity criterion for judging the preferred or least biased interview procedure was the victimization rate, so that the method having the highest number of reported criminal victimization was considered the most accurate.

Internal consistency studies are also used to check for biased reports. In this case, data for a single element are examined to see if the results of several measurements are logically consistent. This method is often used during the editing phase of the survey checking, for example, to see that the same person is not reported as male and hospitalized for a gynecological problem.

Split-ballot experiments are another method used in the study of error. Schuman and Presser (1981) report on a large number of experiments concerned with question form, wording, and context. In this method the sample is randomly assigned to alternative versions of the questionnaire. Errors are examined by comparing results obtained under the two versions and are not compared to a standard. In fact, most of the characteristics that Schuman and Preser study are attitudinal-type variables for which no independently existing "true values" are defined. Evidence of an effect due to the question form, wording, or context is inferred from an assessment of differences in distributions of responses to the alternative forms. Statistical methods used to study experimental effects include the likelihood ratio chi-squared tests of significance, log-linear analysis of measures of association (Goodman, 1978), and odds ratios (Fleiss, 1973).

10.5 STUDIES OF SPECIFIC SOURCES OF ERROR

The procedures discussed above are very general and may be adapted to investigate a variety of bias sources. Some examples are respondents' reports of the value of their homes, accuracy of bank balance reports by different groups of respondents, effects of different measurement methods, and impact of question form and effect of interview mode on the survey measurements. In this section, some procedures used to treat specific sources of bias are presented.

Neter and Waksberg (1964) developed an experimental design for estimating bias in which data that can reasonably be assumed to have smaller net bias are compared to data with presumably larger biases. Three sources of error in an expenditure survey were considered:

1. *Telescoping*, the response effect caused by the tendency of respondents to allocate events to an earlier or later time period than the one in which they actually occurred.

2. *Conditioning*, the response effect in a panel survey in which the number of events reported decreases as the number of times interviewed increases.

3. *Recall loss effect*, owing to two sources:
   a. Losses because of the respondent's inability to remember expenditures.
   b. Losses because of a "report loading effect," in that there is a reduction in the number of events reported because the longer recall period entails a greater reporting burden, causing the respondents to omit events to shorten the interview or to believe that with a longer reporting period there is less interest in smaller jobs.

The authors make use of bounded and unbounded recall periods of different lengths and staggered starting dates for subsamples of the entire sample to obtain estimates of the various effects. The estimates are arrived at by assuming that certain subsamples of the respondents are less subject to one of the effects than are other subsamples. The entire sample was divided into 14 panels, each of which was interviewed four times using a combination of four types of recall periods:

1. Unbounded, one-month recall period
2. Unbounded, six-month recall period
3. Bounded, one-month recall period
4. Bounded, three-month recall period

In unbounded recall, respondents were asked to report events occurring since a previous date (housing alteration and repair expenditures, in this case). Bounded recall consisted of providing the respondent in a particular interview with his reports from a previous interview to remind him of what he reported previously. By staggering the times the panels entered the survey over the entire year of data collection, several estimates for a particular month or other time period were available. These different estimates varied as to the type of recall task involved.

The following models and estimates are given by Neter and Waksberg (1964):

Telescoping effect: Let

u = estimated events from an unbounded one-month recall period
i = estimated events from a bounded one-month recall period
B = telescoping effect,

where B satisfies the relationship

$$E(u) = (1 + B)E(i).$$

The estimate of B is

$$\hat{B} = \frac{u}{i} - 1.$$

Conditioning effect: Let

i_2 = estimated number of events from a bounded one-month recall period and the second interview

i_3 = estimated number of events from a bounded one-month recall period and the third interview

α = conditioning effect,

where α satisfies the relationship

$$E(i_3) = (1 - \alpha)E(i_2).$$

Thus the estimate of α is

$$\hat{\alpha} = 1 - \frac{i_3}{i_2}.$$

The *recall loss effect* is estimated by comparing a three-month bounded recall period to a one-month bounded recall period. Assuming no conditioning or telescoping effects, we have

t = estimated events for a particular one-month period from a bounded three-month recall period

i = estimated events for a particular one-month period from a bounded one-month recall period

P = recall loss effect,

where P satisfies the relationship

$$E(t) = (1 - P)E(i).$$

The estimate of P is

$$\hat{P} = 1 - \frac{t}{i}.$$

The authors also give expressions for internal telescoping and the joint effect of conditioning and telescoping. These estimates required additional assumptions as to the relationships of various error effects.

The conditioning effect is an important source of bias in panel surveys. Panel surveys are those in which the same sample is measured repeatedly. This is done for two reasons: either to increase the precision of a series of statistical estimates such as unemployment rates, or to track particular individuals through time so that one can study not only changes from one time to the next but also the components of that change. In the latter case the panel survey is often called a *longitudinal survey*.

Panel biases arise because of the panel nature of the survey, that is, because of trying to survey the sampling units repeatedly. There are several possible causes of panel bias. One, as we have seen, may be a conditioning effect that may arise in one of two ways:

1. **Reporting Bias.** Reporting bias arises when sample units in the survey are more likely to report in certain ways the longer they have been in the survey. The actual measurement does not change, just the reporting behavior. Thus reporting bias will occur when measurement errors are a function of the number of times interviewed.

2. **Reactive Effect.** The reactive effect arises when presence in the sample and exposure to certain measurement techniques causes the characteristics to change within the sample units. Thus the units in the sample are no longer representative of those in the population. An example is when people are asked if they used a public service of which they were previously unaware. The proportion of the sample using a service may increase over time because of the increased knowledge of the respondents. However, the proportion using the service in the rest of the population may not be changing.

Neter and Waksberg (1964) assumed that the conditioning effect is always positive or that the number of events reported decreases with the number of times interviewed. This assumption is consistent with the hypothesis that the conditioning effect is related to respondent burden, in that to shorten the interview, respondents fail to report events.

A different hypothesis as to the source of panel bias is that it is due to differential nonresponse. The reasoning is as follows: If the probability of response is related to the values of the characteristics being measured, changes over time in the probability of response will cause estimates to change even when no real changes occur. This situation can happen even when the overall response rate remains the same.

Williams and Mallows (1970) considered the case of differential nonresponse and presented a model for systematic biases in panel surveys due to nonresponse.

Sometimes, panel bias is called *rotation group bias* because several large-scale U.S. surveys were designed to take advantage of the correlation between repeat measurements on the same individuals to estimate change in a statistical series.

Bailar (1975) examines the effect of panel bias on estimates from panel surveys designed to estimate a statistical series. Her approach is different from that of Neter and Waksberg. Bailar defines *rotation group bias* as "the effect of variations between responses at different times with repeated interviewing" (p. 13).

This bias is detected by dividing the entire sample into subsamples called *rotation groups* and inducting these groups into the sample over time. Thus, for any particular time period, a reporting month, for example, some people will be in their first interview, some their second, and so on, depending on the total number of rotation groups.

Existence of the rotation group bias is detected by calculating the rotation group index. Bailar defines *rotation group* index as "computed by dividing the total number of persons in a given rotation group having the characteristic of interest by the average number of persons having the characteristic over all rotation groups, and then multiplying by 100" (1975: 23).

If all the rotation groups had the same number of people with the characteristic, the index would be 100.0 for each group. (In Bailar's study these indices ranged from 120.0 to 91.0.)

Bailar's approach may be defined mathematically as follows. Let

Y_{hi} = measurement for time period h for rotation group i
($i = 1, 2, \ldots, R$), R = total number of rotation groups

X_h = population value to be estimated.

Then

$$\hat{X}_h = \sum_{i=1}^{R} Y_{hi}$$

with

$$E(\hat{X}_h) = \sum_{i=1}^{R} E(Y_{hi})$$

$$= \sum_{i=1}^{R} \left(\frac{1}{R} X_h + a_i \right),$$

giving

$$a_i = \text{rotation group bias}$$

$$= E(Y_{hi}) - \frac{1}{R} X_h.$$

The difference between the Neter and Waksberg approach and Bailar's is that the former examines the effect of conditioning in terms of the ratio of the true and reported values; the latter focuses on the difference.

10.6 ADJUSTING FOR BIAS IN SURVEY MEASUREMENTS

If data are available to estimate the bias in a survey estimate, they can also be used to correct the estimate for the bias. Several methods have been developed for adjusting for bias in a survey measurement process. One entire class of procedures may be termed *double sampling methods*. In these procedures, more accurate data are collected for a subsample of the original sample. Instead of being used to estimate the bias and/or mean square error of the estimates from the original survey, these data are used to correct the estimates from the entire sample. The more accurate data are generally more expensive and more difficult to collect, so that the measurement process providing the accurate data is more expensive to use than that which provides the inaccurate data. Several of the procedures provide methods for determining the optimum allocation of the entire survey effort to the original sample and subsample given the cost of the accurate and inaccurate measurement methods.

10.6.1 Methods for Numeric Measurements

Frankel (1979) gives three models for use of what he calls verification information. His basic model assumes that a survey instrument (questionnaire) will be administered to a sample of n population elements. The data for some portion k of these sample elements may be subject to verification from outside sources. Let

Y_i = the survey value obtained for the ith element

X_i = verification or accurate measurement for the ith element

n = sample size

kn = subsample size

C_1 = cost of survey data

C_2 = cost of verification data

C = total cost of study

$C = C_1 n + C_2 kn$

S_x^2 = variance of the accurate values

S_y^2 = variance of the survey values.

Using the basic model, three methods Frankel described for using the verification or accurate data are (1) substitution, (2) ratio adjustment, and (3) regression adjustment.

We assume simple random sampling and ignore finite population corrections in all cases. The goal in each case is to estimate $\overline{X}$, the population mean.

1. The *substitution method* consists of substituting accurate measurements when they are available for less accurate survey values.

$$\hat{\bar{X}} = \hat{y}_s = \frac{1}{n} \sum_{i=1}^{n} Z_i,$$

where

$$Z_i = \begin{cases} X_i & \text{if available} \\ Y_i & \text{otherwise.} \end{cases}$$

This estimate is not unbiased unless $k = 1$. The mean square error is given by

$$\text{MSE}(\bar{y}_s) = \frac{(1 - k)S_y^2 + kS_x^2}{n} + (1 - k)\bar{B}^2.$$

For fixed survey cost, C, Frankel gives the optimum value of k for the case when $S_y^2 = S_x^2 = S^2$. In that case

$$k(\text{optimum}) = 1 - \frac{C_2 S^2}{2C\bar{B}^2}.$$

2. The *ratio adjustment method* entails multiplying the sample mean by the ratio of the accurate-to-survey values for the subsample. Let

$$\bar{y} = \frac{1}{n} \sum_{i=1}^{n} Y_i$$

$$\bar{x}' = \frac{1}{kn} \sum_{i=1}^{kn} X_i$$

$$\bar{y}' = \frac{1}{kn} \sum_{i=1}^{kn} Y_i$$

$$r = \frac{\bar{x}'}{\bar{y}'}.$$

Then the ratio adjustment estimate of the mean is

$$\hat{\bar{X}} = \bar{y}_r = r\bar{y} = \frac{\bar{x}'}{\bar{y}'}\bar{y}.$$

Ratio estimates are not strictly unbiased; however, assuming that kn is sufficiently large so that this bias is negligible, we have

$$\text{MSE}(\bar{y}_r) = \text{Var}(\bar{y}_r) = \frac{S_r^2}{kn} + \frac{S_x^2 - S_r^2}{n}$$

where $S_r^2 = R^2 S_y^2 + S_x^2 - 2R\,\text{Cov}(x, y)$. The optimum value of k is

$$k = \frac{S_r(C_1)^{1/2}}{\left[C_2(S_x^2 - S_r^2)\right]^{1/2}}.$$

3. For the *regression adjustment method*, assume that

$$X_i = A + \beta Y_i + e_i.$$

Least squares is used to estimate A and β, giving

$$\bar{y}_{rg} = \bar{x}' + b(\bar{y} - \bar{y}'),$$

where b is the least squares estimate of β. This estimate is asymptotically unbiased and

$$\text{MSE}(\bar{y}_{rg}) = \text{Var}(\bar{y}_{rg}) = \frac{S_x^2(1 - \rho^2)}{kn} + \frac{S_x^2 \rho^2}{n},$$

where ρ is the correlation between x and y.

The optimum value of k that minimizes the variance for fixed total cost is

$$k = \left[\frac{C_1(1 - \rho^2)}{C_2 \rho^2} \right]^{1/2}. \tag{10.1}$$

Assuming that the same values of k and n are used for the ratio and regression estimator, Kish (1965) gave an approximate expression for the efficiency of the ratio estimator relative to the regression estimator:

$$\frac{\text{Var}(\bar{y}_r)}{\text{Var}(\bar{y}_{rg})} = 1 + \frac{\left[CV(\bar{x})/CV(\bar{y}) - \rho \right]^2}{1 - \rho^2},$$

where

$$CV^2(\bar{x}) = \frac{\text{Var}(\bar{x})}{\bar{X}^2}$$

$$CV^2(\bar{y}) = \frac{\text{Var}(\bar{y})}{\bar{Y}^2}.$$

This formula implies that the regression estimator will always be better than the ratio estimator. Perhaps a fairer comparison would use optimum values of k and n for each method. Assuming fixed total cost C, we have the following:

$$\text{MSE}(\bar{y}_r) = \frac{S_r^2}{kn} + \frac{S_x^2 - S_r^2}{n}$$

$$C = C_1 n + C_2 kn$$

$$\text{Optimum } k = \left(\frac{S_r^2}{S_x^2 - S_r^2}\right)^{1/2} \left(\frac{C_1}{C_2}\right)^{1/2}$$

$$C = nC_1 + nC_2 \left(\frac{S_r^2}{S_x^2 - S_r^2}\right)^{1/2} \left(\frac{C_1}{C_2}\right)^{1/2}$$

$$= n\left[C_1 + (C_1 C_2)^{1/2}\left(\frac{S_r^2}{S_x^2 - S_r^2}\right)^{1/2}\right]$$

$$n = \frac{C}{C_1 + (C_1 C_2)^{1/2}\left(\dfrac{S_r^2}{S_x^2 - S_r^2}\right)^{1/2}} \, .$$

Substituting the optimum values for n and k, we have

$$\text{MSE}(\bar{y}_r) = \frac{\left\{(S_r^2 C_2)^{1/2} + \left[C_1(S_x^2 - S_r^2)\right]^{1/2}\right\}^2}{C} \, .$$

Using the same method to solve for the optimum values of k and n for the regression estimator, we have

$$\text{MSE}(\bar{y}_{rg}) = \frac{\left\{\left[S_x^2 C_2(1 - \rho^2)\right]^{1/2} + \left(S_x^2 C_1 \rho^2\right)^{1/2}\right\}^2}{C} \, .$$

Using these values, we have

$$\left[\frac{\text{MSE}(\bar{y}_r)}{\text{MSE}(\bar{y}_{rg})}\right]^{1/2} = \frac{(S_r^2/S_x^2)^{1/2}C_2^{1/2} + (1 - S_r^2/S_x^2)^{1/2}C_1^{1/2}}{\rho C_1^{1/2} + (1 - \rho^2)^{1/2}C_2^{1/2}}$$

Now

$$\frac{S_r^2}{S_x^2} = R^2 \frac{S_y^2}{S_x^2} + 1 - 2R\beta.$$

If the regression is through the origin, then

$$\frac{S_r^2}{S_x^2} = 1 + R^2\frac{S_y^2}{S_x^2} - 2R\frac{S_y}{S_x}\rho.$$

If the faulty and accurate measurements are equally variable, that is, if $S_x^2 = S_y^2$, the expression for the ratio of the MSE becomes

$$\left[\frac{\text{MSE}(\bar{y}_r)}{\text{MSE}(\bar{y}_{rg})}\right]^{1/2} = \frac{(1 - 2R\rho + R^2)(C_2/C_1)^{1/2} + [R(2\rho - R)]^{1/2}}{(1 - \rho^2)^{1/2} + (C_2/C_1)^{1/2} + \rho}.$$

Suppose that $R = 1.2$, $\rho = 0.85$, and $(C_2/C_1)^{1/2} = 2$; then the ratio estimator is preferred because

$$\left[\frac{\text{MSE}(\bar{y}_r)}{\text{MSE}(\bar{y}_{rg})}\right]^2 = 0.82.$$

4. A fourth method not discussed by Frankel is the *difference adjustment method*. This is a simple method that would probably be preferred over the substitution method if one did not wish to use one of the two nonlinear methods. The difference adjustment method consists of subtracting the subsample estimate of the bias from the sample mean. Let

$$B_i = Y_i - X_i$$

$S_B^2 = $ variance of the individual bias terms.

Then

$$\hat{\bar{x}}(\text{difference}) = \bar{y}_d = \bar{y} - (\bar{y}' - \bar{x}').$$

This is unbiased with

$$\text{Var}(\bar{y}_d) = \frac{S_x^2}{n} + \frac{1 - k}{kn}S_B^2$$

$$= \frac{S_x^2 - S_B^2}{n} + \frac{S_B^2}{kn}.$$

Here S_B^2 is the population variance of the individual bias terms. The optimum value of k is

$$k = \frac{(S_B^2 C_1)^{1/2}}{[C_2(S_x^2 - S_B^2)]^{1/2}}.$$

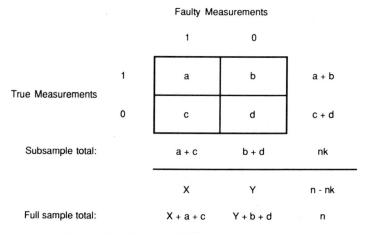

Figure 10.5 Tenebein (1970) subsampling scheme.

10.6.2 Categorical Data

The procedures described above are most appropriate for numeric data. When data are categorical, the adjustments are based on estimates of the probabilities of misclassification. Tenebein (1970) describes the use of such a subsampling procedure. A sample size of n and a subsample of size nk are drawn and classified as shown in Fig. 10.5. The maximum likelihood estimate for the true proportion having the characteristic is

$$\hat{p}(\text{true}) = \frac{a}{a+c} \frac{X+a+c}{n} + \frac{b}{b+d} \frac{Y+b+d}{n}.$$

Tenebein derives the asymptotic variance of $\hat{p}$, assuming the following model. Recall that θ is the probability of misclassifying an element that has the characteristic and ϕ is the probability of misclassifying an element that does not have the characteristic. The cell probabilities of joint classification of an element by the accurate and faulty measuring devices were shown in Fig. 10.3. Recalling that π is the proportion said to have the characteristic when measured by the faulty method, the expression Tenebein gives for the variance is

$$V(\hat{p}) = \frac{pq}{n}\left[1 - \frac{pq(1-\theta-\phi)^2}{\pi(1-\pi)}\right] + \frac{p^2q^2(1-\theta-\phi)^2}{n\pi(1-\pi)}.$$

Using the expression defined above for the coefficient for reliability, the variance of $\hat{p}$ becomes

$$V(\hat{p}) = \frac{pq}{nk}(1 - \rho^2) + \frac{pq\rho^2}{n}.$$

This has the same form of expression as the double sampling regression estimator. Tenebein derives the values for n and nk. His results are the same as expression (10.1), that is,

$$k = \left[\frac{C_1(1 - \rho^2)}{C_2\rho^2}\right]^{1/2}.$$

The value for n under fixed costs is obtained by substituting for k in

$$C = nC_1 + nkC_2.$$

All of these optimum values can also be derived for fixed variance and variable cost.

Elton and Duffy (1983) use such a double sampling procedure to adjust estimates of relative risk for misclassification bias. These procedures have been extended for multinomial data by Hochberg (1977) and Chen (1979b). Hochberg uses the least squares approach to the analyses of categorical data, and Chen uses the log-linear models approach.

10.6.3   Reduction of Bias in Responses to Sensitive Questions

One source of bias in surveys is respondents' reluctance to answer sensitive questions truthfully because they wish to conceal from the interviewer their true status on certain characteristics. *Randomized response* is a method developed to compensate for the reluctance to answer sensitive questions truthfully. Warner (1965) introduced the technique, which involves using a randomizing device to conceal from the interviewer the exact meaning of the respondent's answer.

Suppose that one wishes to estimate the proportion of people who had characteristic A, where A might possibly be "have committed a crime during the past year." Two statements are provided to the respondent:

1. I have committed a crime in the past year.
2. I have not committed a crime in the past year.

The respondent is presented with a randomizing device that selects one of the two statements and the respondent is to answer either "Yes" or "No".

As observed by Chen (1979a), randomized response is a special case of a misclassification problem in which the data are purposely misclassified to

Response

No: 0   Yes: 1

	No: 0	Yes: 1	
$\tilde{A}$: 0	$(1-\phi)(1-p)$	$\phi(1-p)$	$1-p$
A: 1	ϕp	$(1-\phi)p$	p
	$1 - \pi$	π	1

True Status

Figure 10.6 Misclassification matrix for randomized response.

create a series of faulty responses for which the misclassification matrix is known. The respondent is presented with two statements.

Statement 1: I have A.
Statement 2: I do not have A.

If we let ϕ be the probability that the randomizing device picks statement 2, then

ϕ = Probability of selecting statement 2

  = false positive rate, that is, the proportion of "Yes" answers that do not mean that the respondent has A.

The false negative rate also equals ϕ and we have the misclassification matrix shown in Fig. 10.6. Thus, in the entire population, p is the true proportion with the characteristic, and

$$\pi = \phi(1 - p) + (1 - \phi)p.$$

The MLE of p is

$$\hat{p} = \frac{\hat{\pi} - \phi}{1 - 2\phi} \qquad \phi \neq \frac{1}{2}.$$

The variance of $\hat{p}$ is given by

$$\text{Var}(\hat{p}) = \frac{p(1 - p)}{n} + \frac{\phi(1 - \phi)}{n(1 - 2\phi)^2}.$$

This expression for the variance (Greenberg et al., 1969) illustrates the effect of uncertainty introduced by the randomizing device. The first term is the expression for the variance for a direct question about possessing characteristic A. The second component represents the increase in variance due to the randomizing device. However, the advantage of using the randomizing device comes through reduction of bias and accompanying reduction in the mean square error of the estimate. Whether or not the mean square error of the randomized device is less than that of a direct question depends on the values of ϕ and the willingness of the respondent to answer truthfully under the two survey procedures: the direct question procedure or under the randomized response procedure.

There has been considerable theoretical development of the randomized response procedures since Warner's article (1965). In general, these methods have focused on ways to increase the statistical efficiency of the technique, new methods for constructing randomizing devices, and extensions of the methods to include the study of quantitative and multinomial data. Of particular usefulness is the unrelated question randomized response model [see Greenberg et al., (1969) and Folsom et al. (1973)], which has been shown always to be more efficient than the original Warner procedure (Dowling and Shactman, 1975). Chen (1979a) also shows how the unrelated question randomized response methods can be considered in the context of purposely misclassified data.

If the pancake does not decide to enter the Olympics as a disk, causing the athletes to fall far below records set previously by Russian athletes because of the sticky maple syrup, we will tell you next, dear reader, about measurement variability.

Measurement: Quantifying Measurement Error, Variability in Measurement

In Chapter 10 we noted that attempts to understand measurement error have focused on measurement variability and net error of measurement or bias. We restricted our attention to studies of what was termed fixed bias with the recognition that this was a simplification. In this chapter we discuss various approaches to the study of measurement variability. We first discuss the general nature of measurement variability and then focus on the mathematical models developed for its study. In examining these models, we point out differences between various models, some of which are not obvious on first inspection. Experimental methods for quantifying the components of the models are also discussed.

11.1 NATURE OF MEASUREMENT VARIABILITY

What causes response variability? What causes errors in measurement to be correlated? Is it a mental and physical likeness of the persons making the measurements as Pearson (1902) speculated? Do measurements behave like random variables? What effect do respondents have on measurement variability? What about interviewers, coders, phrasing of questions, and mode of interview? What is their impact? Answers to these and similar questions are the goal of studies of measurement variability.

There is disagreement in the literature as to the nature of measurement or response variability. This disagreement centers on the nature of the process that generates measurement variability. Consider a survey that involves personal interviews. The survey measurements take place in a milieu composed of many factors. Some of these factors we have controlled to a high degree, such as the design of the sample and the questionnaire; others we have controlled to a lesser degree, such as interviewer actions, which have been prescribed by the training and instructions; still others we have very little control over, such as the respondent's mood, the political climate at the time of the interview, and the respondent's reaction to characteristics of the interviewer such as his age, race, sex, or social class. Hansen et al. (1961) have called this whole complex of factors the *general conditions* of the survey. Hansen et al. (1953b) call the subset of the conditions that are subject to the control of the researcher the *essential survey conditions*.

One point of view assumes that the uncontrollable factors in the survey milieu cause the measurements for a particular element of the target population to behave like a random variable. Hansen et al. (1951:152) discuss the concept of response variability. We could "consider interviewing each individual a large number of times under exactly the same conditions. ... This would yield a population of responses for all individuals." The *response error* is defined as the difference between an individual observation and the true value for the individual. The *individual response bias* is the expected value of this response error for a particular individual. Hansen et al. state that there will be a random component of variation around this expected value with the essential survey conditions determining the expected value and the random component of variation.

Other researchers are opposed to modeling measurements as random variables because they believe that these models define quantities that are unmeasurable. Zarkovich (1966) presents a case for this point of view. He argues that measurements are unique. Zarkovich accepts the fact that some factors are under the control of the researchers, and some have a random character: for example, the mood of the respondent and interviewer, time of calling, interviewer–respondent interaction, respondent interest in the survey, and so on. Other factors are seen as having a variable effect but as not behaving in random fashion. The examples that Zarkovich gives are "(1) the quality of the enumerators selected, (2) the effects of the publicity campaign, and (3) special temporary features of the external circumstances, such as housing problems, psychological tensions created as a result of wars, a difficult economic or political situation, etc." (p. 44). Zarkovich calls these occasional factors.

In addition, Zarkovich delineates a group of factors that produce a trend over time in responses, including (1) the respondent's memory of previous survey responses, especially to those questions that required work by the respondent; (2) the educational value of the survey for the respondent; and (3) the reluctance to respond to another survey soon after participating in one survey.

Thus Zarkovich reasons that because of the joint action of these factors, we must conclude that successive responses are not random in character. "Each response must therefore be considered a product of particular circumstances and cannot be regarded as representative of the same population as the responses resulting from other trials. [It is the] product of some response generating system that disappears after the trial is carried out" (p. 46). Basically, Zarkovich is saying that although each response may have a random component, it is impossible ever to measure that component, so we should not consider it in our models.

There is considerable merit to Zarkovich's point of view. The major barrier to using the concept of individual response variability is the inability to design experiments that give valid measures of this variability. Despite this and similar objections to the concept, there has been considerable interest in it. There is, however, no general agreement as to how the concepts should be formulated.

Three main points of view are prevalent. Some authors simply state that the measurements are a random variable with a mean and a finite variance. Others speak of the measurements as a random variable generated by a conceptual sequence of repeated independent trials of the measurement process. In some cases the sample is allowed to vary from trial to trial; others define the trials only for fixed samples. Hansen et al. (1953b) described a discrete distribution for the measurements over a distinct number of possible responses each occurring with certain probabilities. Later articles by Hansen and Hurwitz used the concept of independent trials. A third point of view does not allow for response variability at the element level but assumes that the variability results from interviewers and subsequent handling of the data. The compendium of terminology summarizes the views on response variability expressed in various writings. In the following sections we describe the details of several of these different approaches to modeling response variability.

In Chapter 10 we distinguished between approaches for numeric data and categorical data. The models described in the next section have been applied to numeric and dichotomous measurements. Because several very similar models have employed similar terminology but somewhat different approaches, in the next section we discuss in some detail the similarities and differences among the various models. We do this assuming that the measurements are numerical. We first discuss general models and examine how estimates for the error components can be obtained. Next we focus on two general approaches to modeling the effect of interviewers (or other survey personnel) in measurement variability—the variance decomposition approach and a linear models approach. Estimators are given for the general case and in connection with several more complex models. The final section focuses explicitly on application of the methods to categorical data.

11.2 GENERAL MODELS FOR MEASUREMENT VARIABILITY

The models in the literature for the effect of measurement error are expressed either as mean-square-error decomposition models or as mixed linear models. Sometimes the net bias is assumed to be zero, so that the models deal only with variability; however, this assumption is usually a simplifying one because the author(s) wishes to focus on the response variability in a particular paper. The mean-square-error decomposition formulation and the linear model formulation tend to merge as the various models focus on a particular source of error—most usually the interviewer. In such a specific case, both approaches can be expressed in terms of a linear model, although the assumptions are different under different approaches. The major difference between the two approaches is that the decomposition approach often has a component attributable to the interaction between sampling and measurement errors (not always defined the same) and the linear models approach generally omits this component. In addition, the linear model approach usually defines response variability in terms of variation around a true value.

11.2.1 Variance Decomposition Models

In this section we discuss general variance decomposition models that have been formulated without reference to a specific survey measurement design (such as personal interviews). We consider in the next section models which explicitly consider the effect of interviewers. The U.S. Census Bureau has been the leader in the development of models of this type. Many of the models have been developed by staff at the U.S. Census and by other researchers working with them.

Probably the most influential writing on the variance decomposition approach is by Hansen et al. (1961). That article and Mahalanobi's technique of replication or interpenetrating samples appear to have been the major influences on succeeding work. Other authors do not always have the exact same model as in the Hansen–Hurwitz–Bershad model, but they often have components that have the same names and more or less similar definitions. In other cases, corresponding components have different names. This situation has undoubtedly caused some confusion among readers. Because of these differences we include some derivations in the following that delineate some of these similarities and differences.

11.2.1.1 Hansen–Hurwitz–Bershad Model
The Hansen–Hurwitz–Bershad model was originally explicated for dichotomous variables; however, it can be used for continuous data also. The more general approach is used in the following description of the method. Let

X_i = true value for the ith element

Y_{it} = measurement for the ith element at the tth trial

$$E(Y_{it|i}) = E_{s \supset i} E_{t|s \supset i}(Y_{it}) = Y_i. \tag{11.1}$$

Element Number	Trial Number			
	1	2	· · · · · ·	t
1	0	1	· · · · · ·	0
2	1	0	· · · · · ·	1
.	.	.		
.	.	.		
.	.	.		
j	1	0	· · · · · ·	1
.	.	.		
.	.	.		
.	.	.		
N	0	1	· · · · · ·	0

Figure 11.1   Index matrix.

Hansen, Hurwitz, and Bershad define this expected value over all possible samples and all possible trials. The exact quote is: "We can conceive of the possible repetitions of the measurements over all possible samples and trials under the general conditions represented by G, on one unit of the population, say the ith unit. Then the conditional expected value over all such possible measurements on this particular unit is $E_i(Y_{it}) = Y_i$" (p. 361).

Hansen, Hurwitz, and Bershad do not specify the structure of $E(Y_{it|i})$ as is done in (11.1). Equation (11.1) says that the expected value of Y_{it} given i over repetitions of the measurement process is the expected value over all possible samples that contain element i of the expected value over trials of the measurements given a sample that contains i. Thus the conceptual sequence of repeated trials occurs for a fixed sample.

Another formulation of this model by Bailar and Dalenius (1969) allows for new samples to be drawn for each trial. Bailar and Dalenius state that "the survey procedure is such that the process of recording a response for any unit is repeatable; it gives rise to a random variable whose value at trial t is not correlated with its value at any other trial" (p. 342). The authors then show the above index matrix (Fig. 11.1). Note that the same sample is not necessarily used for every trial. Under such a model one would not want to use the structure in (11.1), which implies that the trials occur for a fixed sample.

Hansen, Hurwitz, and Bershad define the conditional expected value for the ith unit over all possible repetitions of the survey for a fixed sample, s, as $E_{is}Y_{it} = Y_{is}$. The distinctions may seem immaterial; however, they have been important in the various attempts to understand the nature of certain covariance or interaction variance components that arise in the variance decomposition models.

The notion that the survey measurements are random variables does not necessarily depend on defining a conceptual sequence of independent trials. One may simply state that the measurements for each element are random variables with a finite mean and variance. This approach is the one Raj (1968) takes in his model, discussed in detail later.

Returning to the Hansen–Hurwitz–Bershad model, assume a simple random sample of size n. The authors characterize the mean square error of $\bar{y}_t = (1/n)\sum_{i=1}^{n} Y_{it}$ as follows:

$$\text{MSE}(\bar{y}_t) = E\left[\left(\bar{y}_t - \bar{X}\right)^2\right]$$

$$= E\left[\left(\bar{y}_t - \bar{Y}\right)^2\right] + \left(\bar{Y} - \bar{X}\right)^2$$

$$= V(\bar{y}_t) + B^2.$$

The variance of $\bar{y}_t$ is decomposed into components using

$$\bar{y} = \frac{1}{n}\sum_{i=1}^{n} Y_i$$

$$V(\bar{y}_t) = E\left\{\left[(\bar{y}_t - \bar{y}) + (\bar{y} - \bar{Y})\right]^2\right\}$$

$$= E\left[(\bar{y}_t - \bar{y})^2 + (\bar{y} - \bar{Y})^2 + 2(\bar{y}_t - \bar{y})(\bar{y} - \bar{Y})\right]. \quad (11.2)$$

The three terms in (11.2) are, respectively, the *response variance*, the *sampling variance*, and the *covariance between response and sampling deviations*. The covariance component is not further characterized; however, the authors state that it will be zero for a complete census and for repetitions using a fixed sample.

The response variance term is considered in more detail and is decomposed using the individual response deviation

$$d_{it} = Y_{it} - Y_i.$$

Then, letting $\sigma_d^2 = E(d_{it}^2)$ and $\sigma_d^2\rho = E(d_{it}d_{i't})$, we have

$$E(\bar{y}_t - \bar{y})^2 = \text{Var}(\bar{d}_t) = \frac{1}{n}\left[\sigma_d^2 + (n-1)\rho\sigma_d^2\right].$$

The authors also elaborate the definition of σ_d^2 (the simple response variance) in terms of the conditional expectation (E_t) of the squared individual response deviations (d_{it}^2) averaged over all population units, that is,

$$\sigma_d^2 = \frac{1}{N}\sum_{i=1}^{N} E_t(d_{it})^2.$$

Similarly, the correlated component is

$$\rho\sigma_d^2 = \frac{1}{N(N-1)} \sum_{i \neq i'}^{N} E_t(d_{it}d_{i't}),$$

although this is not explicitly defined this way by the authors.

To summarize, the variance of the sample mean in the Hansen–Hurwitz–Bershad model consists of a sampling variance component, a measurement (response) variance component, and a component due to the covariance of response and sampling deviations, that is,

$$\mathrm{Var}(\bar{y}_t) = \mathrm{SV} + \mathrm{MV} + \mathrm{CRS}$$

$$= \mathrm{SV} + \frac{1}{n}\left[\sigma_d^2 + (n-1)\rho\sigma_d^2\right] + \mathrm{CRS}.$$

The last component is not further characterized except by the statement that it is zero for complete censuses and for fixed samples.

11.2.1.2 Other Models

We now consider several other general variance decomposition models that are available in the literature and illustrate the existing similarities and differences. Koch (1973) presents a response error model for the multivariate unequal probability case. To simplify the discussion and the comparison between models, Koch's model has been expressed in the univariate case. Following Cornfield (1944), Koch uses indicator random variables U_i to describe the nature of the samples, where

$$U_i = \begin{cases} 1 & \text{if population element } i \text{ is in the sample} \\ 0 & \text{otherwise.} \end{cases}$$

The random variables U_i are used to reflect the sampling errors and random variables Y_{it} are defined which reflect response errors.

Koch then defines $E_t(Y_{it}|U_i = 1)$ as a conditional expected value over repeated trials with respect to the ith individual in the population. To compare Koch's model to the Hansen–Hurwitz–Bershad model, we will expand the various terms of the Koch model and assume simple random sampling. We have

$$\bar{y}_t = \frac{1}{n}\sum_{i=1}^{N} U_i Y_{it}$$

$$E_t(Y_{it}|U_i = 1) = Y_i$$

$$\mathrm{Var}(\bar{y}_t) = E(\bar{y}_t - \bar{Y})^2$$

$$= E\left[(\bar{y}_t - \bar{y})^2\right] + E\left[(\bar{y} - \bar{Y})^2\right] + 2E\left[(\bar{y}_t - \bar{y})(\bar{y} - \bar{Y})\right],$$

where

$$\bar{y} = \frac{1}{n} \sum_{i=1}^{N} U_i Y_i.$$

Koch calls the three terms the *response variance* (RV), the *sampling variance* (SV), and the *interaction variance* (IV). These components are exactly analogous to the three components in the Hansen–Hurwitz–Bershad model.

In investigating these components further, Koch arrives at a general model in which the response or measurement variance is decomposed into three components with

$$\text{Var}(\bar{y}_t) = \text{SV} + \text{RV} + \text{IV}$$
$$= \text{SV} + \text{SRV} + \text{CRV} + \text{IRV} + \text{IV},$$

where the response variance now consists of three components, called the *simple response variance* (SRV), the *simple correlated response variance* (CRV) and the *interaction response variance* (IRV). The following demonstrates the nature of these components using notation similar to Koch's.

$$\text{RV} = E\left\{ \frac{1}{n^2} \left[\sum_{i=1}^{N} U_i (Y_{it} - Y_i)^2 + \sum_{i \neq i'}^{N} U_i U_{i'} (Y_{it} - Y_i)(Y_{i't} - Y_{i'}) \right] \right\}$$

$$= \frac{1}{n^2} \left\{ \sum_{i=1}^{N} E(U_i) E_t \left[(Y_{it} - Y_i)^2 | U_i = 1 \right] \right.$$

$$\left. + \sum_{i \neq i'}^{N} E(U_i U_{i'}) E_t \left[(Y_{it} - Y_i)(Y_{i't} - Y_{i'}) | U_i U_{i'} = 1 \right] \right\}$$

$$\text{SV} = \frac{1}{n^2} \left[\sum_{i=1}^{N} E\left(U_i - \frac{n}{N} \right)^2 Y_i^2 + \sum_{i \neq i'}^{N} E\left(U_i - \frac{n}{N} \right)\left(U_{i'} - \frac{n}{N} \right) Y_i Y_{i'} \right]$$

$$\text{IV} = \frac{1}{n^2} \left[\sum_{i=1}^{N} E(U_i)\left(U_i - \frac{n}{N} \right) \right] E_t \left[(Y_{it} - Y_i) Y_i | U_i = 1 \right]$$

$$+ \sum_{i \neq i'}^{N} E\left\{ U_i \left(U_{i'} - \frac{n}{N} \right) E_t \left[(Y_{it} - Y_i)(Y_{i'}) | U_i U_{i'} = 1 \right] \right\}.$$

Koch says that the interaction variance arises from the fact that

$$E_t [Y_{it} | U_i U_{i'} = 1] = Y_{ii'} \neq Y_i, \tag{11.3}$$

that is, "the expected response for the *i*th population element is different for those samples which contain both the *i*th and *i*'th element than for those that contain the ith but not the i'th element" (p. 908). This last statement can be thought of as meaning that the expected value of the measurements depends on the contents of the sample. Thus, for samples of size 2, say, we could imagine that the expected measurement for a particular element would depend on the other element that appeared in the sample. For example, suppose that we were trying to measure the physical fitness of people using as our measurement the judgements of interviewers. An average person i may have higher or lower expected measurements depending on whether i', the other member of the sample, is a superathlete or a semi-invalid. This situation occurs because the expected measurement assigned to a given person will vary depending on the contents of a particular sample.

Using this formulation, Koch decomposes the RV as follows:

$$
\text{RV} = \frac{1}{n}\left[\frac{1}{N}\sum_{i=1}^{N}E_t(Y_{it} - Y_i)^2 + (n-1)\frac{1}{N(N-1)}\right.
$$
$$
\left.\times \sum_{i \neq i'}^{N} E_t(Y_{it} - Y_{ii'})(Y_{i't} - Y_{i'i}) + \sum_{i \neq i'}^{N}(Y_{ii'} - Y_i)(Y_{i'i} - Y_{i'})\right],
$$

which are, respectively, the above-mentioned simple response variance, simple correlated response variance component, and interaction response variance.

The interaction variance component using (11.3) becomes

$$
\text{IV} = \frac{2(n-1)}{n}\left[\frac{1}{N(N-1)}\sum_{i \neq i'}^{N}(Y_{ii'} - Y_i)(Y_{i'} - \bar{Y})\right].
$$

This term is the one called covariance between response and sampling deviations in the Hansen–Hurwitz–Bershad model. The sampling variance is the same for both models.

We now consider the third variance decomposition model, which examines the interaction or covariance between response and sampling deviations. Based on an article by Koop (1974), the model uses the concept defined by Hansen, Hurwitz, and Bershad of the expected value of the response for a fixed sample. Let Y_{its} be the measurement for the ith element at the tth trial for the sth sample. Actually, the ideas become clear if we drop the trial terminology and use the idea that the general conditions induce the measurements to behave as a random variable (Raj, 1968; Tremblay et al., 1976). Thus we assume that for a fixed sample s for each element there is a distribution of measurements. However, to maintain the distinction between the expected value over samples and the expected value over the distribution of responses, the former will be denoted E_s, the latter E_t.

Now suppose that the sample influences the measurements. An example of how this situation could happen is in a survey to assess the market value of houses. If a house that has been recently sold is in the sample for a particular interviewer, the interviewer may adjust her estimates of all houses in the sample based on this knowledge (Hansen et al., 1961). Thus

$$E_t(Y_{its}|i \in s) = Y_{is}.$$

Then the expected value and variance of $\bar{y}_{ts}$ is given as follows:

$$\bar{y}_{ts} = \frac{1}{n} \sum_{i=1}^{n} Y_{its}$$

$$E(\bar{y}_{ts}) = E_s\left[E_t(\bar{y}_{ts|s})\right]$$

$$= E_s\left(\frac{1}{n} \sum_{i=1}^{n} Y_{is}\right)$$

$$= \sum_s p(s) \frac{1}{n} \sum_{i=1}^{n} Y_{is}$$

$$= \frac{1}{n} \sum_{i=1}^{N} \sum_{s \supset i} p(s) Y_{is}.$$

Here $\sum_s$ is the sum over all possible samples, and $p(s)$ is the probability of obtaining a particular sample. For a simple random sample of n from N, one notes that

$$p(s) = \frac{1}{\binom{N}{n}} = \frac{n/N}{\binom{N-1}{n-1}},$$

where $\binom{N}{n}$ denotes the number of combinations of N things taken n at a time. With this notation $E(\bar{y}_{ts})$ becomes

$$E(\bar{y}_{ts}) = \frac{1}{n} \sum_{i=1}^{N} \frac{n}{N} \sum_{s \supset i} \frac{Y_{is}}{\binom{N-1}{n-1}}.$$

Let

$$Y_i = \sum_{s \supset i} \frac{Y_{is}}{\binom{N-1}{n-1}}.$$

Then Y_i is average over all samples that contain i, or Y_i could be thought of as the expected value over all possible samples and responses given the ith element in that $p(s \supset i) = 1/\binom{N-1}{n-1}$. Hansen, Hurwitz, and Bershad use this terminology. Thus we have

$$E(\bar{y}_{ts}) = \frac{1}{N} \sum_{i=1}^{N} Y_i = \bar{Y}.$$

Using this approach, we show that the measurement variance has four components when the responses depend on the sample. Koop called these components simple and correlated response variance and the interaction variance and covariance. However, this interaction variance is not the same as that in Koch's formulation but is, rather, a subset of the response variance. Now using the above, we have

$$
\begin{aligned}
\text{RV} &= E\left[(\bar{y}_{ts} - \bar{y})^2\right] \\
&= E\left\{\left[\frac{1}{n} \sum_{i=1}^{n} (Y_{ist} - Y_i)\right]^2\right\} \\
&= \frac{1}{n^2} E\left[\sum_{i=1}^{n} (Y_{ist} - Y_i)^2 + \sum_{i \neq i'}^{n} (Y_{ist} - Y_i)(Y_{i'st} - Y_{i'})\right] \\
&= \frac{1}{n^2} E_s\left[\sum_{i=1}^{n} E_{t|s}(Y_{ist} - Y_i)^2 + \sum_{i \neq i'}^{n} E_{t|s}(Y_{ist} - Y_i)(Y_{i'st} - Y_{i'})\right] \\
&= \frac{1}{n^2} E_s\left[\sum_{i=1}^{n} E_{t|s}(Y_{ist} - Y_{is})^2 + \sum_{i \neq i'}^{n} E_{t|s}(Y_{ist} - Y_{is})(Y_{i'st} - Y_{i's}) \right. \\
&\qquad \left. + \sum_{i=1}^{n} (Y_{is} - Y_i)^2 + \sum_{i \neq i'}^{n} (Y_{is} - Y_i)(Y_{i's} - Y_{i'})\right].
\end{aligned}
$$

Letting

$$E_{t|s}(Y_{ist} - Y_{is})^2 = V_{is}^2$$

$$E_{t|s}(Y_{ist} - Y_{is})(Y_{i'st} - Y_{i's}) = C_{ii's}$$

$$I_{is}^2 = (Y_{is} - Y_i)^2$$

$$I_{ii's} = (Y_{is} - Y_i)(Y_{i's} - Y_{i'}),$$

we get

$$\text{RV} = \frac{1}{n^2}\left[\frac{n}{N} \sum_{i=1}^{N} \sum_{s \supset i} \frac{V_{is}^2 + I_{is}^2}{\binom{N-1}{n-1}} + \frac{n(n-1)}{N(N-1)} \sum_{i \neq i'}^{N} \sum_{s \supset i, i'} \frac{C_{ii's} + I_{ii's}}{\binom{N-2}{n-2}}\right].$$

Letting V_i^2 and I_i^2 be the averages over samples that contain i and $C_{ii'}$, and $I_{ii'}$ be averages over samples that contain i and i', we have

$$\mathrm{RV} = \frac{1}{n}\left[\frac{1}{N}\sum_{i=1}^{N}V_i^2 + \frac{n-1}{N(N-1)}\sum_{i\neq i'}^{N}C_{ii'}\right]$$
$$+ \frac{1}{n}\left[\frac{1}{N}\sum_{i=1}^{N}I_i^2 + \frac{n-1}{N(N-1)}\sum_{i\neq i'}^{N}I_{ii'}\right].$$

This expression shows that the response variance has two parts—a simple and correlated response variance due to the variance and covariance of the distribution of responses. Under our trial terminology, we would say that this situation is because of the trial-to-trial variability in response. The second part of RV is the variance and covariance across all samples of the expected response for a particular sample. This term will be zero when the expected measurement does not depend on the sample. Koop called these two terms the interaction variance and covariance.

Now the term that gives rise to the covariance between response and sampling deviations which Koch called interaction variance is

$$2E\left[(\bar{y}_{ts} - \bar{y})(\bar{y} - \bar{Y})\right]$$

$$= 2E\left[\frac{1}{n}\sum_{i=1}^{n}(Y_{ist} - Y_i)\frac{1}{n}\sum_{i'=1}^{n}(Y_{i'} - \bar{Y})\right]$$

$$= \frac{2}{n^2}E_s\left[\sum_{i=1}^{n}(Y_{is} - Y_i)\sum_{i'=1}^{n}(Y_{i'} - \bar{Y})\right]$$

$$= \frac{2}{n^2}\left\{\sum_{s}p(s)\left[\sum_{i=1}^{n}(Y_{is} - Y_i)(Y_i - \bar{Y}) + \sum_{i\neq i'}^{n}(Y_{is} - Y_i)(Y_{i'} - \bar{Y})\right]\right\}$$

$$= \frac{2}{n^2}\left[\sum_{i=1}^{N}\sum_{s\supset i}\frac{(Y_{is} - Y_i)(Y_i - \bar{Y})}{\binom{N}{n}} + \sum_{i\neq i'}^{N}\sum_{s\supset i,i'}\frac{(Y_{is} - Y_i)(Y_{i'} - \bar{Y})}{\binom{N}{n}}\right]$$

$$= \frac{2}{n^2}\left[\sum_{i=1}^{N}\frac{n}{N}(Y_i - \bar{Y})\sum_{s\supset i}\frac{Y_{is} - Y_i}{\binom{N-1}{n-1}}\right.$$

$$\left. + \frac{n(n-1)}{N(N-1)}\sum_{i\neq i'}^{N}(Y_{i'} - \bar{Y})\sum_{s\supset i,i'}\frac{Y_{is} - Y_i}{\binom{N-2}{n-2}}\right]. \qquad (11.4)$$

Now the first term in (11.4) is zero because of how Y_i is defined. However, the second term is not zero, and if one defined $\Sigma_{s \supset i, i'} Y_{is} / \binom{N-2}{n-2}$ as $Y_{ii'}$, we would have

$$2E\left[(\bar{y}_t - \bar{y})(\bar{y} - \bar{Y})\right] = \frac{2(n-1)}{n}\left[\frac{1}{N(N-1)}\sum_{i \neq i'}^{N}(Y_{ii'} - Y_i)(Y_{i'} - \bar{Y})\right]$$

$$= \frac{2n-1}{n}\sigma_{rs}.$$

Koop did not include this term in his model.

To summarize:

1. What Koop calls the *interactive variance* and *covariance* is a subset of the U.S. Census Bureau's *response variance*.
2. Koch's *interaction variance* is roughly equivalent to the *covariance between response* and *sampling deviations* in the U.S. Census formulation.
3. If the expected response for an individual depends on the sample, the response variance has two types of components, as illustrated above. This situation was not shown in the census model and only partially by Koch.
4. The Koop model contains no term equivalent to the covariance between response and sampling deviations. As shown in the previous derivation, this term in the variance expansion may be nonzero.

The discussion above illustrates well some of the uncertainties and confusions in terminology that have arisen over the years. Part of this confusion is because of the refinement of models and ideas over time. For example, we have the evolution of the U.S. Census model from Hansen et al. (1951) in which they define a discrete distribution of k possible responses occurring with probabilities p_{ik} to Hansen et al. (1961) in which the idea of a conceptual sequence of repeated trials is stressed to a publication on enumerator variance by the U.S. Bureau of the Census (1979) in which the following statement is made: "The mathematical model used in the Enumerator Variance Study is based on the model described by Hansen, Hurwitz and Bershad. The basic assumption of the model is that the response from a given unit [element] of the population is a random variable having some probability distribution" (p. 8).

Lack of standardization in terminology is also because different names have been given to components that have similar mathematical formulations. The lesson to be learned here is that the reader of an article on measurement error models cannot rely on understanding it from the names of the components and his familiarity with other models but must delve into the mathematics of their definitions.

11.2.2   Bias in Usual Sample Estimates of Variance in the Presence of Measurement Variability

When measurement variability is present, the usual estimators for the variance of the survey variables will be negatively biased. The following demonstrates this fact for a survey with simple random sampling in which the sampling error is estimated by the sample mean square and for a survey with unequal probability sampling and the Horvitz–Thompson estimators.

Assume simple random sampling of size n from a population N. The usual estimate of the variance of the sample mean is a function of the sample mean square

$$\frac{1}{n}\frac{N-n}{N}s_{yt}^2 = \frac{1}{n}\frac{N-n}{N}\sum_{i=1}^{n}\frac{(Y_{it}-\bar{y}_t)^2}{n-1}.$$

Assuming the Hansen–Hurwitz–Bershad model for measurement errors, we have

$$\bar{y}_t = \frac{1}{n}\sum_{i=1}^{n}Y_{it}$$

$$= \frac{1}{n}\sum_{i=1}^{n}(d_{it}+Y_i)$$

$$= \bar{d}_t + \bar{y}.$$

Thus

$$s_{\bar{y}t}^2 = \frac{1}{n-1}\sum_{i=1}^{n}\left[(d_{it}-\bar{d}_t)+(Y_i-\bar{y})\right]^2.$$

Now

$$E(s_{\bar{y}t}^2) = S_y^2 + \frac{1}{n-1}E\left[\sum_{i=1}^{n}(d_{it}-\bar{d}_t)\right]^2$$

$$+ \frac{1}{n-1}E\left[\sum_{i=1}^{n}2(d_{it}-\bar{d}_t)(Y_i-\bar{y})\right]. \tag{11.5}$$

The second term on the right-hand side of (11.5) gives the following:

$$\frac{1}{n-1}E\left[\sum_{i=1}^{n}(d_{it}-\bar{d}_t)^2\right]$$

$$=\frac{1}{n-1}E\left[\sum_{i=1}^{n}d_{it}^2-n(\bar{d}_t)^2\right]$$

$$=\frac{1}{n-1}\left[E_s\sum_{i=1}^{n}E_{t|s}d_{it}^2-nE_sE_{t|s}(\bar{d}_t^2)\right]$$

$$=\frac{1}{n-1}\left[\frac{n}{N}\sum_{i=1}^{N}\text{Var}_t(d_{it})\right]-\frac{n}{n-1}\left\{\frac{1}{n}\left[\sigma_d^2+(n-1)\rho\sigma_d^2\right]\right\}$$

$$=\sigma_d^2(1-\rho).$$

The third term is

$$\frac{2}{n-1}E\left[\sum_{i=1}^{n}(d_{it}-\bar{d}_t)(Y_i-\bar{y})\right]=\frac{2}{n-1}\left[E\left(\sum_{i=1}^{n}d_{it}Y_i\right)-nE(\bar{d}_t\bar{y})\right]$$

$$=\frac{-2n}{n-1}E(\bar{d}_t\bar{y})$$

since $E_{t|s}(d_{it})=0$. Now since $E(\bar{d}_t Y)=0$, the expectation above can be recast as

$$=\frac{-2n}{n-1}\left\{E\left[(\bar{d}_t)(\bar{y}-\bar{Y})\right]\right\}$$

$$=\frac{-2n}{n-1}\frac{n-1}{n}\sigma_{rs}$$

$$=-2\sigma_{rs}.$$

Thus

$$E(s_{\bar{y}t}^2)=S_y^2+\sigma_d^2(1-\rho)-2\sigma_{rs}.$$

Recall that the total variance under the Hansen–Hurwitz–Bershad model is

$$V(\bar{y}_t)=\frac{1}{n}\frac{N-1}{N}S_y^2+\frac{1}{n}\left[\sigma_d^2+(n-1)\rho\sigma_d^2\right]+\frac{2(n-1)}{n}\sigma_{rs}.$$

Thus the bias in the usual estimate of $\text{Var}(\bar{y}_t)$ is

$$B[\hat{v}(\bar{y}_t)] = E\left[\frac{1}{n}\left(\frac{N-n}{N}\right)s^2\right] - V(\bar{y}_t).$$

Ignoring the finite population correction factors, the bias equals

$$B[\hat{v}(\bar{y}_t)] = -\rho\sigma^2 - 2\sigma_{rs}.$$

This derivation illustrates that if there is no interaction between sampling and measurement errors and if the element measurement errors are uncorrelated, the usual estimate of the variance is unbiased in the case of negligible finite population correction factors and simple random sampling. This result was given in Hansen et al. (1961) for 0–1 variables, and by Cochran (1963) for the more general case. Note that the bias in the estimate is not dependent on the size of the sample and is therefore not reduced by increasing the sample size.

Koop (1974) illustrated the bias for the case of unequal probability sampling and the Horvitz–Thompson estimator.

Let p_i be the probability that the ith element is in the sample. Assume that there is no interaction variance, so that

$$E_t(Y_{it}|s) = Y_{is} = Y_i \quad \text{for all } s$$
$$\text{Var}_t(Y_{it}|s) = V_i^2$$
$$\text{Cov}_t(Y_{it}Y_{i't}) = C_{ii'}.$$

The Horvitz–Thompson estimator for a total is

$$\hat{T}_t = \sum_{i \in s} \frac{Y_{it}}{p_i}$$

$$E(\hat{T}_t) = \sum_{i=1}^{N} Y_i$$

$$\text{Var}(\hat{T}) = E(\hat{T} - Y)^2$$
$$= \text{Var}_s\left[E_{t|s}(\hat{T})\right] + E_s\left[\text{Var}_{t|s}(\hat{T})\right]. \tag{11.6}$$

Considering the second term on the right-hand side of (11.6) first, we have

$$\text{Var}_{t|s}(\hat{T}) = E_t\left\{\left[\sum_{i \in s}\frac{1}{p_i}(Y_{it} - Y_i)\right]^2\right\}$$

$$= \sum_{i \in s}\frac{V_i^2}{p_i^2} + \sum_{\substack{i \neq i' \\ i, i' \in s}}\frac{C_{ii'}}{p_i p_i'}$$

and

$$E_s\left[\mathrm{Var}_{t|s}(\hat{T})\right] = \sum_{i=1}^{N} \frac{V_i^2}{p_i} + \sum_{\substack{i \neq i'}}^{N} \frac{p_{ii'}C_{ii'}}{p_i p_{i'}}.$$

Now

$$\mathrm{Var}_s\left(E_{t|s}\sum_{i \in s} \frac{Y_{it}}{p_i}\right) = \sum_{i=1}^{N} \frac{Y_i^2(1-p_i)}{p_i} + \sum_{\substack{i \neq i'}}^{N} \frac{p_{ii'} - p_i p_{i'}}{p_i p_{i'}} Y_i Y_{i'}$$

is the usual variance of the Horvitz–Thompson estimator. Thus

$$\mathrm{Var}(\hat{T}) = \sum_{i=1}^{N} \frac{Y_i^2(1-p_i)}{p_i} + \sum_{\substack{i \neq i'}}^{N} \frac{p_{ii'} - p_i p_{i'}}{p_i p_{i'}} Y_i Y_{i'} + \sum_{i=1}^{N} \frac{V_i}{p_i} + \sum_{\substack{i \neq i'}}^{N} \frac{p_{ii'}C_{ii'}}{p_i p_{i'}}.$$

Now the Horvitz–Thompson estimator of the variance is

$$v(\hat{T}) = \sum_{i \in s} \frac{1-p_i}{p_i^2} Y_{it}^2 + \sum_{\substack{i \neq i' \\ i, i' \in s}} \frac{p_{ii'} - p_i p_{i'}}{p_i p_{i'} p_{ii'}} Y_{it} Y_{i',t}$$

$$E\{v(\hat{T})\} = E_s E_{t|s}\left[v(\hat{T})\right]$$

$$= E_s\left[\sum_{i \in s} \frac{1-p_i}{p_i^2}(V_i + Y_i^2) + \sum_{\substack{i \neq i' \\ i, i' \in s}} \frac{p_{ii'} - p_i p_{i'}}{p_i p_{i'} p_{ii'}}(C_{ii'} + Y_i Y_{i'})\right]$$

$$= \sum_{i=1}^{N} \frac{1-p_i}{p_i}(V_i + Y_i^2) + \sum_{\substack{i \neq i'}} \frac{p_{ii'} - p_i p_{i'}}{p_i p_{i'}}(C_{ii'} + Y_i Y_{i'}).$$

Using these expressions we find that the bias in the usual procedures is

$$\mathrm{Bias}\{v(\hat{T})\} = -\sum_{i=1}^{N} V_i - \sum_{\substack{i \neq i'}}^{N} C_{ii'}$$

$$= -\mathrm{Var}_t\left(\sum_{i=1}^{N} Y_{it}\right).$$

Again, we see that the bias does not depend on the sample design.

11.2.3   Estimating the Measurement Variance and Its Components

Estimation of the effect of measurement variability on survey statistics requires that some type of experimental procedure be introduced into the survey so that there is replication. Three types of approaches are used:

1. Repeat measurements on the same population elements.
2. Interpenetrated or replicated samples.
3. Methods that use a combination of the replicated samples and repeat measurements.

Under the assumptions of (1) independence between repeat measurements and (2) the same general conditions for the two sets of measurements, inducing the same distribution of measurements, repeat measurements will allow one to estimate simple response variance components. The technique of replicated or interpenetrated samples was developed by Mahalonobis and allows estimation of the total variance and correlated measurement variance.

A note on terminology is appropriate here. Hansen et al. (1961:367) called collecting repeat measurements the replication method. This variability in terminology has resulted in considerable misunderstanding. At several American Statistical Association meetings we have heard people vehemently deny that replication is required to measure the total variance, while others were confidently asserting that it was required. They are usually talking about different things—one group thinking that replication refers to repeat measurements on the same units; the other thinking that it refers to replicated or interpenetrated samples.

11.2.3.1 Use of Repeat Measurements
In this method more than one measurement is made on the same sample elements. Let

$$Y_{i1t} = \text{measurement made in the original survey}$$

$$Y_{i2t} = \text{measurement made in the resurvey.}$$

Repeat measurements can be made for the entire sample or for a subsample. Consider the case of a simple random sample of n units from N and a resurvey of the entire sample. There are now two statistics available for estimating the measurement variance and its components.

The first of these is the square of the element-by-element differences, termed the *gross difference rate* by some (Hansen et al., 1964; Bailar, 1968; Bailar and Dalenius, 1969). For a 0–1 variable it will be equivalent to the discrepancy rate, which is often calculated during the quality control activities of a survey (Lessler, 1976).

If one can assume that the repeat measurements are independent of the original measurements and that the Y_{it} have the same random distribution in the resurvey as they had in the original survey, this statistic is an unbiased estimate of the simple measurement variance under the Hansen–Hurwitz– and Bershad model. We will call this statistic the mean square within elements (MSWE).

The following shows that the MSWE is an unbiased estimator for the simple measurement variance. Let m denote the repeat measurements; $m = 1, 2$:

$$\text{MSWE} = \frac{1}{2n} \sum_{i=1}^{n} \sum_{m=1}^{2} (Y_{imt} - \bar{y}_{it})^2$$

$$= \frac{1}{2n} \sum_{i=1}^{n} (Y_{i1t} - Y_{i2t})^2.$$

Now

$$E(\text{MSWE}) = \frac{1}{2n} E \left(\sum_{i=1}^{n} \{ [Y_{i1t} - E_t(Y_{i1t})] - [Y_{i2t} - E_t(Y_{i2t})] \right.$$

$$\left. + [E_t(Y_{i1t}) - E_t(Y_{i2t})] \}^2 \right).$$

The assumption that the Y_{it} have the same distributions for the original and resurvey allows us to say $E_t(Y_{i1t}) = E_t(Y_{i2t}) = Y_i$. Thus

$$E(\text{MSWE}) = \frac{1}{2n} E_s \left\{ \sum_{i=1}^{n} E_t \{ [(Y_{i1t} - Y_i) - (Y_{i2t} - Y_i)]^2 \} \right\}$$

$$= \frac{1}{2n} E_s \left[\sum_{i=1}^{n} E_t(Y_{i1t} - Y_i)^2 + E_t(Y_{i2t} - Y_i)^2 \right.$$

$$\left. - 2E_t(Y_{i1t} - Y_i)(Y_{i2t} - Y_i) \right].$$

Assuming independence between the original and resurvey makes the cross-product term zero, giving

$$E(\text{MSWE}) = \frac{1}{N} \sum_{i=1}^{N} E_t(Y_{it} - Y_i)^2$$

$$= \sigma_d^2.$$

This estimate is also unbiased for the variance decomposition models presented by Koop (1974) and Koch (1973) under the same two assumptions.

Several authors discuss the implications of the assumptions of lack of independence between the two surveys and changes in the distributions of the Y_{it}. Hansen et al. (1964) discuss the effect of a positive correlation between the two measurements and the effect of the resurvey being an improved procedure relative to the original procedure. In each case, MSWE is an underestimate of the simple response variance. Bailar and Dalenius (1969) discuss situations in which one can make various assumptions about the presence or absence of positive correlations between the two sets of measurements.

A second statistic available for the repeat survey is the difference between the two means. Call this the *squared mean difference* (SMD), and we have

$$\text{SMD} = \tfrac{1}{2}(\bar{y}_{1t} - \bar{y}_{2t})^2.$$

Now

$$E(\text{SMD}) = \tfrac{1}{2}E\left\{\left[(\bar{y}_{1t} - \bar{y}_1) - (\bar{y}_{2t} - \bar{y}_2) + (\bar{y}_1 - \bar{y}_2)\right]^2\right\}.$$

Under the assumption that $E_t(Y_{i1t}) = E_t(Y_{i2t}) = Y_i$, $\bar{y}_1 = \bar{y}_2$,

$$E(\text{SMD}) = \tfrac{1}{2}E\left[(\bar{y}_{1t} - \bar{y})^2 + (\bar{y}_{2t} - \bar{y})^2 - 2(\bar{y}_{1t} - \bar{y})(\bar{y}_{2t} - \bar{y})\right].$$

Independence between the original and resurvey makes the expected value of the cross-product term zero, which gives

$$E(\text{SMD}) = \text{MV} = \frac{1}{n}\left[\sigma_d^2 + (n - 1)\rho\sigma_d^2\right].$$

This estimate is given by Hansen et al. (1961) and also by Bailar and Dalenius (1969), who allow for more than two repetition surveys each made by different agents and calculate the between-agent mean square.

11.2.3.2 *Use of Interpenetrated (or Replicated) Samples*

The method of replicated or interpenetrated samples entails selecting several independent samples and carrying out the survey measurement process for each sample. This method allows one to estimate the total variance, that is, the sum of the sampling variance and measurement variance. The between-sample mean square provides the estimate. Let s index samples and assume that k independent samples are selected. Assume that there is no correlation between sampling and measurement errors and that the covariance of measurement errors between samples is zero. Let

$$\bar{y}_{st} = \frac{1}{n} \sum_{i=1}^{N} Y_{it}$$

$$\bar{y}_{t} = \frac{1}{k} \sum_{s=1}^{k} \bar{y}_{st}$$

$$\text{BSMS} = \frac{1}{k-1} \sum_{s=1}^{k} (\bar{y}_{st} - \bar{y}_{t})^2$$

$$E(\text{BSMS}) = E\left\{ \frac{1}{k-1} \sum_{s=1}^{k} \left[(\bar{y}_{st} - \bar{Y}) - (\bar{y}_{t} - \bar{Y}) \right]^2 \right\}$$

$$= \frac{1}{k-1} \left[\sum_{s=1}^{k} \text{Var}(\bar{y}_{st}) - k\,\text{Var}(\bar{y}_{t}) \right].$$

Now

$$\text{Var}(\bar{y}_{t}) = \frac{1}{k^2} \sum_{s=1}^{k} \text{Var}(\bar{y}_{st}) = \frac{1}{k}\,\text{Var}(\bar{y}_{st}).$$

Therefore,

$$E(\text{BSMS}) = \frac{1}{k-1} \left[k\,\text{Var}(\bar{y}_{st}) - \text{Var}(\bar{y}_{st}) \right]$$

$$= \text{Var}(\bar{y}_{st}).$$

This method is used extensively in surveys with interviewers for which it was first developed by Mahalanobis in the 1940s (Mahalanobis, 1944, 1946) and is discussed in the next section. In that case the assumption is that between interviewer covariance is zero.

11.3 GENERAL MODEL FOR A SURVEY WITH INTERVIEWERS

A number of different models in the literature treat measurement errors for surveys that employ personal interviews. It is sometimes difficult to understand the distinction and similarities among these models because different approaches are used to derive the measurement variance components, some underlying assumptions are not explicitly stated, and components with the same names are based on different assumptions. To clarify some of the similarities and differences, we first develop a general model for surveys with interviewers and then use it as a framework for comparing several types of models in the literature.

In the most general finite population case we have a finite population of elements $i = 1, 2, \ldots, N$. Also consider a finite population of interviewers indexed by $j = 1, 2, \ldots, K$. We have a survey design in which we select a sample of elements, a sample of interviewers, and an assignment scheme for assigning elements to interviewers.

Let

Y_{ijt} = measurement made by the jth interviewer
on the ith element; the index t is used to denote that
Y_{ijt} is a random variable

$$U_i = \begin{cases} 1 & \text{if the } i\text{th element is selected for the sample} \\ 0 & \text{otherwise} \end{cases}$$

$$V_j = \begin{cases} 1 & \text{if the } j\text{th interviewer is selected for the survey} \\ 0 & \text{otherwise} \end{cases}$$

$$C_{ij} = \begin{cases} 1 & \text{if the } i\text{th element is assigned to the } j\text{th interviewer} \\ 0 & \text{otherwise.} \end{cases}$$

Given the sampling and interviewer assignment scheme, we can define the expected value of the measurement of the ith element by the jth interviewer. We let

$$E_t\left(Y_{ijt} | U_i V_j C_{ij} = 1\right) = Y_{ij}.$$

Thus the expected value of the measurements for a particular element does not depend on the contents of the sample but does depend on the interviewer.

An overall sample of $n = km$ elements is drawn with a subsample of m elements assigned to each of k selected interviewers. The sample mean is then

$$\bar{y}_t = \frac{1}{n} \sum_{j=1}^{K} \sum_{i=1}^{N} U_i V_j C_{ij} Y_{ijt}.$$

Following Hansen et al. (1961) we also define the sample mean of the expected measurements as

$$\bar{y} = \frac{1}{km} \sum_{j=1}^{K} \sum_{i=1}^{N} U_i V_j C_{ij} Y_{ij},$$

where m is the size of the interviewer sample.

The expected value of $\bar{y}_t$ is

$$E(\bar{y}_t) = E_s E_{t|s}(\bar{y}_t)$$

$$= \frac{1}{n} \sum_{j=1}^{K} \sum_{i=1}^{N} E(U_i V_j C_{ij}) E_t(Y_{ijt} | U_i V_j C_{ij} = 1)$$

$$= \frac{1}{n} \sum_{j=1}^{K} \sum_{i=1}^{N} E(U_i V_j C_{ij}) Y_{ij}$$

$$= \frac{1}{KN} \sum_{j=1}^{K} \sum_{i=1}^{N} Y_{ij}$$

$$= \bar{\bar{Y}}.$$

We have assumed that our sample selection and interviewer assignment is such that $E(U_i V_j C_{ij}) = n/KN$. Expectations for the separate indicators U_i, V_j, and C_{ij} are specified as needed in subsequent derivations.

Under the assumption of no interaction between sampling and measurement errors, we can write the variance of $\bar{y}_t$ as the sum of a measurement variance and a sampling variance component

$$\text{Var}(\bar{y}_t) = E\left\{ (\bar{y}_t - \bar{y})^2 \right\} + E\left\{ (\bar{y} - \bar{\bar{Y}})^2 \right\}$$

$$= MV + SV.$$

Expanding each of these terms, we get the following general expressions for the measurement variance and sampling variance:

$$\text{MV} = E\left\{ \left[\frac{1}{n} \sum_{j=1}^{K} \sum_{i=1}^{N} U_i V_j C_{ij} (Y_{ijt} - Y_{ij}) \right]^2 \right\}$$

$$= \frac{1}{n^2} \left\{ \sum_{j=1}^{K} \sum_{i=1}^{N} E(U_i V_j C_{ij}) E_t \left[(Y_{ijt} - Y_{ij})^2 | U_i V_j C_{ij} = 1 \right] \right.$$

$$+ \sum_{j=1}^{K} \sum_{i \neq i'}^{N} E(U_i U_{i'}, V_j C_{ij} C_{i'j})$$

$$\times E_t \left[(Y_{ijt} - Y_{ij})(Y_{i'jt} - Y_{i'j}) | U_i U_{i'} V_j C_{ij} C_{i'j} = 1 \right]$$

$$+ \sum_{j \neq j'}^{K} \sum_{i=1}^{N} E(U_i V_j V_{j'} C_{ij} C_{ij'})$$

$$\times E_t \left[(Y_{ijt} - Y_{ij})(Y_{ij't} - Y_{ij'}) | U_i V_j V_{j'} C_{ij} C_{ij'} = 1 \right]$$

$$+ \sum_{j \neq j'}^{K} \sum_{i \neq i'}^{N} E(U_i U_{i'} V_j V_{j'} C_{ij} C_{i'j'})$$

$$\left. \times E_t \left[(Y_{ijt} - Y_{ij})(Y_{i'j't} - Y_{i'j'}) | U_i U_{i'} V_j V_{j'} C_{ij} C_{i'j'} = 1 \right] \right\}.$$

Since the indicator variables take only the values 0 and 1, $E[(U_i V_j C_{ij})^2]$ is the same as $E(U_i V_j C_{ij})$, so the squares have been dropped. The $E_t[(Y_{ijt} - Y_{ij})^2 | U_i V_j C_{ij} = 1]$ is the variance of the measurements of the ith element by the jth interviewer given that both element i and interviewer j have been selected for the survey and the element i has been assigned to interviewer j. Similarly, $E_t[(Y_{ijt} - Y_{ij})(Y_{i'jt} - Y_{i'j}) | U_i U_{i'}, V_j C_{ij} C_{i'j} = 1]$ is the measurement covariance between the elements measured by the same interviewer. Also, $E_t[(Y_{ijt} - Y_{ij})(Y_{ij't} - Y_{ij'}) | U_i V_j V_{j'} C_{ij} C_{ij'} = 1]$ is the measurement covariance between interviewers within elements and $E_t[(Y_{ijt} - Y_{ij})(Y_{i'j't} - Y_{i'j'}) | U_i U_{i'} V_j V_{j'} C_{ij} C_{i'j'} = 1]$ is the between-element between-interviewer covariance. In many models these last two terms are assumed to be zero; that is, it is assumed that there is no between-interviewer correlation of measurement error.

For the sampling variance component we have

$$
SV = E\left\{\left[\frac{1}{n}\sum_{j=1}^{K}\sum_{i=1}^{N} U_i V_j C_{ij}\left(Y_{ij} - \overline{\overline{Y}}\right)\right]^2\right\}
$$

$$
= \frac{1}{n^2}\left[\sum_{j=1}^{K}\sum_{i=1}^{N} E\left(U_i V_j C_{ij}\right)\left(Y_{ij} - \overline{\overline{Y}}\right)^2\right.
$$

$$
+ \sum_{j=1}^{K}\sum_{i\neq i'}^{N} E\left(U_i U_{i'} V_j C_{ij} C_{i'j}\right)\left(Y_{ij} - Y\right)\left(Y_{i'j} - \overline{\overline{Y}}\right)
$$

$$
+ \sum_{j\neq j'}^{K}\sum_{i=1}^{N} E\left(U_i V_j V_{j'} C_{ij} C_{ij'}\right)\left(Y_{ij} - \overline{\overline{Y}}\right)\left(Y_{ij'} - \overline{\overline{Y}}\right)
$$

$$
+ \left.\sum_{j\neq j'}^{K}\sum_{i\neq i'}^{N} E\left(U_i U_{i'} V_j V_{j'} C_{ij} C_{i'j'}\right)\left(Y_{i'j} - \overline{\overline{Y}}\right)\left(Y_{i'j'} - \overline{\overline{Y}}\right)\right].
$$

This general approach can be used to examine several different interviewer assignment and selection schemes.

For simple random sampling without replacement of elements and interviewers, we have

$$
Pr(U_i = 1) = \frac{n}{N}; \qquad Pr(U_i = 1, U_{i'} = 1) = \frac{n}{N}\frac{n-1}{N-1}
$$

$$
Pr(V_j = 1) = \frac{k}{K} \qquad Pr(V_j = 1, V_{j'} = 1) = \frac{k}{K}\frac{k-1}{K-1}.
$$

The elements of the sample are assigned to the interviewers without replacement so that each element is interviewed by a single interviewer. Thus

$$
Pr\left(C_{ij} = 1 | U_i = 1, V_j = 1\right) = \frac{m}{n}
$$

$$
Pr\left(C_{ij} = 1, C_{i'j} = 1 | U_i U_{i'} = 1, V_j = 1\right) = \frac{m}{n}\frac{m-1}{n-1}
$$

$$
Pr\left(C_{ij} = 1, C_{i'j'} = 1 | U_i U_{i'} = 1, V_j V_{j'} = 1\right) = \frac{m}{n}\frac{m}{n-1}
$$

$$
Pr\left(C_{ij} = 1, C_{ij'} = 1 | U_i = 1, V_j V_{j'} = 1\right) = 0.
$$

Taking the expected values of the indicator random variables, we have

$$
\text{MV} = \frac{1}{n}\left[\frac{1}{NK} \sum_{j=1}^{K}\sum_{i=1}^{N} \text{Var}_t(Y_{ijt}) + \frac{m-1}{KN(N-1)} \sum_{j=1}^{K}\sum_{i \neq i'}^{N} \text{Cov}_t(Y_{ijt}Y_{i'jt}) \right.
$$

$$
\left. + \frac{m(k-1)}{KN(N-1)(K-1)} \sum_{j \neq j'}^{K}\sum_{i \neq i'}^{N} \text{Cov}(Y_{ijt}Y_{i'j't}) \right].
$$

Similarly, for the sampling variance, we have

$$
\text{SV} = \frac{1}{n}\left[\frac{1}{NK} \sum_{j=1}^{K}\sum_{i=1}^{N} \left(Y_{ij} - \overline{\overline{Y}}\right)^2 + \frac{m-1}{NK(N-1)} \sum_{j=1}^{K}\sum_{i \neq i'}^{N} \left(Y_{ij} - \overline{\overline{Y}}\right)\left(Y_{i'j} - \overline{\overline{Y}}\right) \right.
$$

$$
\left. + \frac{m(k-1)}{N(N-1)K(K-1)} \sum_{j \neq j'}^{K}\sum_{i \neq i'}^{N} \left(Y_{ij} - \overline{\overline{Y}}\right)\left(Y_{i'j'} - \overline{\overline{Y}}\right) \right].
$$

Now we can use this general model to look at some of the different models that appear in the literature.

11.3.1   Census Bureau Models / Cochran Model

The U.S. Census Bureau model appears in a number of places in the literature (Bailar, 1976; Bailey et al., 1978; U.S. Bureau of the Census, 1979) in connection with the description of experiments in which K interviewers are randomly assigned independent samples of elements—each sample of size m.

In these articles, the component caused by the interaction between sampling and measurement errors is included. We will not include it in our explication. The overall mean of the conditional expected value of the measurements for the ith element by the jth interviewer is defined as

$$
\overline{Y} = \frac{1}{N} \sum_{i=1}^{N} E_t(Y_{ij}|i).
$$

Thus the expected measurement does not depend on the interviewer, that is, $E_t(Y_{ijt}) = Y_i$. The variance for the mean for a single interviewer is given as

$$
\text{Var}(\bar{y}_t) = \frac{\sigma_S^2}{m} + \frac{\sigma_R^2}{m}[1 + (m-1)\rho]
$$

and for K interviewers is

$$\text{Var}(\bar{y}_t) = \frac{\sigma_S^2}{Km} + \frac{\sigma_R^2}{Km}[1 + (m - 1)\rho],$$

where the finite population correction factors are ignored.

A similar model is also presented in Cochran (1977), in which he writes

$$Y_{ijt} = Y_{ij} + d_{ijt},$$

and ignoring finite population correction factors,

$$\text{Var}(\bar{y}_{jt}) = \frac{1}{m}\{S_y^2 + \sigma_d^2[1 + (m - 1)\rho_w]\},$$

where ρ_w is the correlation between the d_{ijt} obtained by the same interviewer. It is assumed that there is no correlation of response deviations for different interviewers and that

$$\text{Var}(\bar{y}_t) = \frac{1}{K}\sum_{j=1}^{K}\text{Var}(\bar{y}_{jt})$$

$$= \frac{1}{n}\{S_y^2 + \sigma_d^2[1 + (m - 1)\rho_w]\}.$$

Previously, σ_d^2 had been defined as "the population average of the variance of the errors of measurement," so that $\sigma_d^2 = (1/N)\sum_{i=1}^{N}\sigma_i^2$; $\sigma_i^2 = \text{Var}_t(d_{it})$; $d_{it} = Y_{it} - E_t(Y_{it})$. Cochran gives no explicit expression for σ_d^2 in terms of his double subscript notation.

Turning to the general model above, we can obtain an expression similar to the ones by Cochran and the U.S. Census authors as follows:

Assume that there is a fixed set of interviewers K, and that we draw K independent samples of size m, one for each interviewer. Thus $n = Km$, and we have $V_j = 1$ for all K interviewers. Also, we do not need the indicator U_i since an independent sample is drawn for each interviewer:

$$\text{Pr}(C_{ij} = 1) = \frac{m}{N}$$

$$\text{Pr}(C_{ij}C_{i'j} = 1) = \frac{m}{N}\frac{m-1}{N-1}$$

$$\text{Pr}(C_{ij}C_{ij'} = 1) = \frac{m}{N}\frac{m}{N}$$

$$\text{Pr}(C_{ij}C_{i'j'} = 1) = \frac{m}{N}\frac{m}{N}.$$

Also assume that the between-interviewer measurement covariances are 0. For convenience we drop the conditional notation in $E_t[(\cdot)|U_iC_{ij} = 1]$ and simply write $E_t(\cdot)$.

Using these conventions, the measurement variance is

$$
\mathrm{MV} = \frac{1}{n^2}\left\{\frac{m}{N}\sum_{j=1}^{K}\sum_{i=1}^{N}E_t\left[(Y_{ijt} - Y_{ij})^2\right] + \frac{m(m-1)}{N(N-1)}\right.
$$

$$
\times \left.\sum_{j=1}^{K}\sum_{i\neq i'}^{N}E_t\left[(Y_{ijt} - Y_{ij})(Y_{i'jt} - Y_{ij})\right]\right\}
$$

$$
= \frac{1}{n}\left\{\frac{1}{KN}\sum_{j=1}^{K}\sum_{i=1}^{N}E_t\left[(Y_{ijt} - Y_{ij})^2\right] + \frac{m-1}{KN(N-1)}\right.
$$

$$
\times \left.\sum_{j=1}^{K}\sum_{i\neq i'}^{N}E_t\left[(Y_{ijt} - Y_{ij})(Y_{i'jt} - Y_{i'j})\right]\right\}.
$$

The first term in this expression is often called the simple measurement variance (SMV) or simple response variance; the second the correlated measurement variance (CMV) or correlated response variance. Similarly, the sampling variance is

$$
\mathrm{SV} = \frac{1}{n^2}\left\{\frac{m}{n}\sum_{j=1}^{K}\sum_{i=1}^{N}E_t\left[\left(Y_{ij} - \overline{\overline{Y}}\right)^2\right] + \frac{m}{n}\frac{m-1}{N-1}\sum_{j=1}^{K}\sum_{i\neq i'}^{N}\left(Y_{ij} - \overline{\overline{Y}}\right)\left(Y_{i'j} - \overline{\overline{Y}}\right)\right.
$$

$$
\left.+ \frac{m}{n}\frac{m}{n}\left[\sum_{j\neq j'}^{K}\sum_{i=1}^{N}\left(Y_{ij} - \overline{\overline{Y}}\right)\left(Y_{ij'} - \overline{\overline{Y}}\right) + \sum_{j\neq j'}^{K}\sum_{i\neq i'}^{N}\left(Y_{ij} - \overline{\overline{Y}}\right)\left(Y_{i'j'} - \overline{\overline{Y}}\right)\right]\right\}.
$$

If we let $Y_{ij} = Y_{i'}$, that is, we assume that the expected value over the distribution of measurements does not depend on the interviewers, we can write $\overline{Y} = (1/N)\sum_{i=1}^{N}Y_i = \overline{\overline{Y}}$. Then

$$
\mathrm{SV} = \frac{1}{n}\left\{\frac{1}{KN}\sum_{j=1}^{K}\sum_{i=1}^{N}\left(Y_i - \overline{Y}\right)^2 + \frac{m-1}{KN(N-1)}\sum_{j=1}^{K}\sum_{i\neq i'}^{N}\left(Y_i - \overline{Y}\right)\left(Y_{i'} - \overline{Y}\right)\right.
$$

$$
\left.+ \frac{m}{KN^2}\sum_{j\neq j'}^{K}\left[\sum_{i=1}^{N}\left(Y_i - \overline{Y}\right)\right]^2\right\}.
$$

The last term is zero, so we have

$$SV = \frac{1}{n}\left[\frac{1}{N}\sum_{i=1}^{N}(Y_i - \bar{Y})^2 - \frac{m-1}{N(N-1)}\sum_{i=1}^{N}(Y_i - \bar{Y})^2\right]$$

$$= \frac{1}{n}\left\{\frac{N-m}{N}\left[\frac{1}{N-1}\sum_{i=1}^{N}(Y_i - \bar{Y})^2\right]\right\}$$

$$\approx \frac{1}{n}S_y^2,$$

ignoring the impact of the finite population correction. Similarly, if we define

$$\sigma_d^2 = \frac{1}{KN}\sum_{j=1}^{K}\sum_{i=1}^{N}E_t(Y_{ijt} - Y_i)$$

and

$$\rho\sigma_d^2 = \frac{1}{KN(N-1)}\sum_{j=1}^{K}\sum_{i\neq i'}^{N}E_t\left[(Y_{ijt} - Y_i)(Y_{i'jt} - Y_{i'})\right],$$

then

$$\text{Var}(\bar{y}_t) = \frac{S_y^2}{n} + \frac{\sigma_d^2}{n}[1 + (m-1)\rho],$$

which is the expression obtained by Cochran (1977), Bailar (1976), Bailey et al. (1978), with the additional assumption that the measurement errors and sampling errors are uncorrelated.

11.3.2 Raj Model

Now consider the model that Raj (1968) presents in his textbook. In this case we select a simple random sample of k out of K available interviewers. Thus

$$\Pr(V_j = 1) = \frac{k}{K}$$

$$\Pr(V_jV_{j'} = 1) = \frac{k}{K}\frac{k-1}{K-1}.$$

Again we draw an independent sample of size m for each interviewer, so that we have

$$\Pr(V_j C_{ij} = 1) = \frac{mk}{NK}$$

$$\Pr(V_j C_{ij} C_{i'j} = 1) = \frac{m}{n} \frac{m-1}{N-1} \frac{k}{K}$$

$$\Pr(V_j V_{j'} C_{ij} C_{ij'} = 1) = \frac{m}{N} \frac{m}{N} \frac{k}{K} \frac{k-1}{K-1}$$

$$\Pr(V_j V_{j'} C_{ij} C_{i'j'} = 1) = \frac{m}{N} \frac{m}{N} \frac{k}{K} \frac{k-1}{K-1}.$$

Now $n = mk$,

$$\bar{\bar{y}}_t = \frac{1}{mk} \sum_{j=1}^{K} \sum_{i=1}^{N} U_i C_{ij} V_j Y_{ijt},$$

and

$$E(\bar{\bar{y}}_t) = \bar{\bar{Y}}.$$

If we substitute the expected value of our indicator variables for the sample selection and interviewer assignment, we get

$$\mathrm{MV} = \frac{1}{n} \left\{ \frac{1}{NK} \sum_{j=1}^{K} \sum_{i=1}^{N} \mathrm{Var}_t(Y_{ijt}) + \frac{m-1}{KN(N-1)} \sum_{j=1}^{K} \sum_{i=i'}^{N} \mathrm{Cov}_t(Y_{ijt}, Y_{i'jt}) \right\}$$

$$+ \frac{m(k-1)}{N^2 K(K-1)} \left[\sum_{j \neq j'}^{K} \sum_{i=1}^{N} \mathrm{Cov}_t(Y_{ijt} Y_{ij't}) + \sum_{j \neq j'}^{K} \sum_{i=i'}^{N} \mathrm{Cov}_t(Y_{ijt} Y_{ij't}) \right].$$

Raj also assumes that the between interviewer measurement covariance terms are zero, so that we have

$$\mathrm{MV} = \frac{1}{n} \left[\frac{1}{NK} \sum_{j=1}^{K} \sum_{i=1}^{N} \mathrm{Var}_t(Y_{ijt}) + \frac{m-1}{KN(N-1)} \sum_{j=1}^{K} \sum_{i \neq i'}^{N} \mathrm{Cov}_t(Y_{ijt}, Y_{i'jt}) \right].$$

Now, taking into account the sampling of elements and interviewers and the interviewer assignment for the sampling variance, we have

$$SV = \frac{1}{n} \left\{ \frac{1}{NK} \sum_{j=1}^{K} \sum_{i=1}^{N} \left(Y_{ij} - \bar{\bar{Y}}\right)^2 + \frac{m-1}{KN(n-1)} \sum_{j=1}^{K} \sum_{i \neq i'}^{N} \left(Y_{ij} - \bar{\bar{Y}}\right)\left(Y_{i'j} - \bar{\bar{Y}}\right) \right.$$

$$+ \frac{m(k-1)}{N^2 K(K-1)} \left[\sum_{j \neq j'}^{K} \sum_{i=1}^{N} \left(Y_{ij} - \bar{\bar{Y}}\right)\left(Y_{ij'} - \bar{\bar{Y}}\right) \right.$$

$$\left. \left. + \sum_{j \neq j'}^{K} \sum_{i \neq i'}^{N} \left(Y_{ij} - \bar{\bar{Y}}\right)\left(Y_{i'j'} - \bar{\bar{Y}}\right) \right] \right\}.$$

Raj derives variance components in terms of the mean across all elements of interviewer j's expected measurements, that is,

$$\bar{Y}_j = \frac{1}{N} \sum_{i=1}^{N} Y_{ij}.$$

Using this, we have

$$SV = \frac{1}{n} \left\{ \frac{1}{KN} \sum_{j=1}^{K} \sum_{i=1}^{N} \left(Y_{ijt} - \bar{Y}_j\right)^2 + \frac{m-1}{KN(N-1)} \sum_{j=1}^{K} \sum_{i \neq i'}^{N} \left(Y_{ij} - \bar{Y}_j\right)\left(Y_{i'j} - \bar{Y}_j\right) \right.$$

$$\left. + \frac{1}{k} \frac{K-k}{K} \left[\frac{1}{K-1} \sum_{j=1}^{K} \left(\bar{Y}_j - \bar{\bar{y}}\right)^2 \right] \right\}.$$

If we ignore the finite population correction factor for interviewers so that

$$\frac{1}{k} \frac{K-k}{K} \left[\frac{1}{K-1} \sum_{j=1}^{K} \left(\bar{Y}_j - \bar{\bar{Y}}\right)^2 \right] \approx \frac{1}{K} \sum_{j=1}^{K} \left(\bar{Y}_j - \bar{\bar{Y}}\right)^2,$$

we get exactly the same expression as Raj obtained.

Raj, however, rewrites the entire variance as

$$Var(\bar{\bar{y}}) = \frac{\sigma_y^2}{n} [1 + (m-1)\rho]$$

by defining

$$\sigma_y^2 = \frac{1}{KN} \sum_{j=1}^{K} \sum_{i=1}^{N} E\left(Y_{ijt} - \bar{\bar{y}}\right)^2,$$

which in turn equals

$$\sigma_y^2 = \frac{1}{KN} \sum_{j=1}^{K} \sum_{i=1}^{N} \left(Y_{ij} - \bar{Y}_j \right)^2 + \frac{1}{K} \sum_{j=1}^{K} \left(\bar{Y}_j - \bar{\bar{Y}} \right)^2 + \frac{1}{KN} \sum_{j=1}^{K} \sum_{i=1}^{N} \text{Var}_t(Y_{ijt}).$$

He also defined the covariance of the responses obtained by different interviewers as

$$C(Y, I) = \frac{1}{KN(N-1)} \sum_{j=1}^{K} \sum_{i \neq i'}^{N} E_t \left(Y_{ijt} - \bar{\bar{Y}} \right) \left(Y_{i'jt} - \bar{\bar{Y}} \right)$$

$$C(Y, I) = \frac{1}{KN(N-1)} \sum_{j=1}^{K} \sum_{i \neq i'}^{N} \text{Cov}_t(Y_{ijt} Y_{i'jt})$$

$$+ \frac{1}{K} \sum_{j=1}^{K} \left(\bar{Y}_j - \bar{\bar{Y}} \right)^2 + \frac{1}{KN(N-1)} \sum_{j=1}^{K} \sum_{i \neq i'}^{N} \left(Y_{ij} - \bar{Y}_j \right) \left(Y_{i'j} - \bar{Y}_j \right).$$

Thus

$$\text{Var}(\bar{y}) = \frac{1}{n}\sigma_y^2 + \left(\frac{1}{k} - \frac{1}{n} \right) C(Y, I)$$

$$= \frac{\sigma_y^2}{n} [1 + (m-1)\rho].$$

Here $\rho = C(Y, I)/\sigma_y^2$ and is called the *intra-interviewer correlation coefficient of response*. This model is similar to the Census Bureau model (CBM) in form and terminology; however, the decomposition that Raj uses combines terms, which in the CBM came from the measurement and sampling variance components.

11.3.3 Murthy Model

Now, consider a model in which the measurements are not random variables. Murthy (1967) presents such a model, and the expression that he gives for the sample mean has the same form as the CBM, namely

$$\text{Var}(\bar{\bar{y}}_t) = \frac{\sigma_s^2}{n} + \frac{\sigma_d^2}{n} [1 + (m-1)\rho_c],$$

where the terms are called, respectively, *sampling variance*, *simple* or *uncorrelated response variance*, and *correlated response variance*. Murthy's components are defined as

$$\sigma_s^2(\text{MUR}) = \frac{1}{N} \sum_{i=1}^{N} \left(\bar{Y}_i - \bar{\bar{Y}} \right)^2$$

$$\sigma_d^2(\text{MUR}) = \frac{1}{KN} \sum_{i=1}^{N} \sum_{j=1}^{K} \left(Y_{ij} - Y_i \right)^2$$

$$\rho_c \sigma_d^2(\text{MUR}) = \frac{1}{KN(N-1)} \sum_{j=1}^{K} \sum_{i \neq i'}^{N} (Y_{ij} - Y_i)(Y_{i'j} - Y_{i'}).$$

The response variance components however, do not arise from the random nature of measurements. Rather they arise from a decomposition of the sampling variance.

The following illustrates how Murthy arrives at his components. Under Murthy's design an independent random sample is not selected for each interviewer; rather, a random sample of size n is randomly assigned to a random sample of k interviewers. Under such a scheme, the same element is never assigned to two different interviewers and the sampling variance becomes

$$\text{SV} = \frac{1}{km} \left[\frac{1}{NK} \sum_{j=1}^{K} \sum_{i=1}^{N} \left(Y_{ij} - \bar{\bar{Y}} \right)^2 + \frac{m-1}{KN(N-1)} \sum_{j=1}^{K} \sum_{i \neq i'}^{N} \left(Y_{ij} - \bar{\bar{Y}} \right) \left(Y_{i'j} - \bar{\bar{Y}} \right) \right.$$

$$\left. + \frac{m(k-1)}{N(N-1)K(K-1)} \sum_{j \neq j'}^{K} \sum_{i \neq i'}^{N} \left(Y_{ij} - \bar{\bar{Y}} \right) \left(Y_{i'j'} - \bar{\bar{Y}} \right) \right].$$

There is no measurement variance term because Murthy does not let the measurements be random variables.

Using $\bar{Y}_i = (1/K) \sum_{j=1}^{K} Y_{ij}$, it can be shown that the SV component is

$$\text{SV} = \frac{1}{n} \left[\frac{1}{NK} \sum_{j=1}^{K} \sum_{i=1}^{N} \left(Y_{ij} - \bar{Y}_i \right)^2 + \frac{m-1}{KN(N-1)} \sum_{j=1}^{K} \sum_{i \neq i'}^{N} \left(Y_{ij} - \bar{Y}_i \right) \left(Y_{i'j} - \bar{Y}_{i'} \right) \right]$$

$$+ \frac{1}{n} \frac{N-m}{N-1} \left[\frac{1}{N} \sum_{i=1}^{N} \left(\bar{Y}_i - \bar{\bar{Y}} \right)^2 \right] - \frac{1}{k} \frac{k-1}{K-1} \left[\frac{1}{K} \sum_{j=1}^{K} \left(\bar{Y}_j - \bar{\bar{Y}} \right)^2 \right].$$

The last term equals

$$-\frac{1}{k} \frac{k-1}{K-1} \left(\frac{\sigma_d^2}{N} + \frac{N-1}{N} \rho_c \sigma_d^2 \right),$$

so that

$$SV \approx \frac{\sigma_s^2(MUR)}{n} + \frac{\sigma_d^2(MUR)}{n}[1 + (m-1)\rho_c].$$

Thus, we see that we get different expressions for the variance depending on how one chooses to define the measurement scheme and how one chooses to define the differing components.

Some authors (Lessler, 1976; Koch, 1973) have defined element, interviewer, and element interviewer interaction components when decomposing the sampling variance. If we let

$$I_i = \bar{Y}_i - \bar{\bar{Y}}$$

$$J_j = \bar{Y}_j - \bar{\bar{Y}}$$

$$(IJ)_{ij} = Y_{ij} - \bar{Y}_i - \bar{Y}_j + \bar{\bar{Y}},$$

then

$$Y_{ij} - \bar{\bar{Y}} = I_i + J_j + (IJ)_{ij}.$$

11.3.4 Linear Models Approach

All of the previous models can be thought of as variance decomposition models, in that the various components are derived by adding and subtracting different means in the overall expression $E(\bar{y}_t - \bar{\bar{Y}})^2$. Another approach to the study of measurement error is the linear model approach. In this approach the measurement for a survey element is defined as the sum of fixed and random effects. Sukhatme and Sukhatme (1970), Kish (1962), Sukhatme and Seth (1952), and others discuss models of this type. We will examine the linear model approach in the context of our general approach above.

Let

$$Y_{ijt} = X_i + \alpha_j + \varepsilon_{ij}.$$

We assume that X_i, α_j, and ε_{ij} are mutually independent and that the α_j and ε_{ij} arise, respectively, from an infinite population of interviewer effects and an infinite population of random effects. The α_j are independently distributed; $E_t(\alpha_j) = 0$ and $Var(\alpha_j) = \sigma_\alpha^2$; similarly, ε_{ij} are independently distributed with $E_t(\varepsilon_{ij}) = 0$, $\text{Var}(\varepsilon_{ij}) = \sigma_e^2$. In this simple case there is no overall bias in the measurement process and

$$E_t(Y_{ijt}) = X_i.$$

Note that a bias could be defined by letting either the expected value of α_j or ε_{ij} be nonzero.

We assume that we have a simple random sample of k interviewers and that we have selected an independent sample of m elements from a finite population of N for each interviewer. Using notation similar to the above, we let

$$C_{ij} = \begin{cases} 1 & \text{if element } i \text{ is selected} \\ & \text{for interviewers } j\text{'s sample} \\ 0 & \text{otherwise.} \end{cases}$$

Therefore,

$$E(C_{ij}) = \frac{m}{N}$$

$$E(C_{ij}C_{i'j}) = \frac{m}{N}\frac{m-1}{N-1}$$

$$E(C_{ij}C_{ij'})E(C_{ij}C_{i'j'}) = \frac{m}{N}\frac{m}{N}$$

$$\bar{y}_t = \frac{1}{km}\sum_{j=1}^{k}\sum_{i=1}^{N}C_{ij}Y_{ijt}$$

$$E(\bar{y}_t) = E_s\big[E_{t|s}(\bar{y}_t)\big]$$

$$= \frac{1}{m}\sum_{i=1}^{N}E(C_{ij})X_i$$

$$= \bar{X}.$$

Now using our prior method of decomposing the variance, we have

$$\text{Var}(\bar{y}_t) = \text{MV} + \text{SV}$$

$$\text{SV} = E_s\left\{\left[\frac{1}{mk}\sum_{j=1}^{k}\sum_{i=1}^{N}C_{ij}(X_i - \bar{X})\right]^2\right\}$$

$$\text{SV} = E_s\frac{1}{(mk)^2}\left\{\left[\sum_{j=1}^{k}\sum_{i=1}^{N}C_{ij}(X_i - \bar{X})^2\right.\right.$$

$$+ \sum_{j=1}^{k}\sum_{i\neq i'}^{N}C_{ij}C_{i'j}(X_i - \bar{X})(X_{i'} - \bar{X})$$

$$+ \sum_{j\neq j'}^{k}\sum_{i=1}^{N}C_{ij}C_{ij'}(X_i - \bar{X})^2$$

$$\left.\left.+ \sum_{j\neq j'}^{k}\sum_{i\neq i'}^{N}C_{ij}C_{i'j'}(X_i - \bar{X})(X_{i'} - \bar{X})\right]\right\}$$

$$= \frac{1}{(mk)^2}\left\{\frac{m}{N}\sum_{j=1}^{k}\sum_{i=1}^{N}(X_i - \bar{X})^2 + \frac{m}{n}\frac{m-1}{N-1}\right.$$

$$\times \sum_{j=1}^{k}\sum_{i\neq i'}^{N}(X_i - \bar{X})(X_{i'} - \bar{X})$$

$$\left.+ \frac{m}{N}\frac{m}{N}\sum_{j\neq j'}^{k}\left[\sum_{i=1}^{N}(X_i - \bar{X})\right]^2\right\}$$

$$= \frac{1}{mk}\left[\frac{1}{N}\sum_{i=1}^{N}(X_i - \bar{X})^2 + \frac{m-1}{N(N-1)}\sum_{i\neq i'}^{N}(X_i - \bar{X})(X_{i'} - \bar{X})\right]$$

$$= \frac{1}{k}\frac{1}{m}\frac{N-m}{N}\left[\frac{1}{N-1}\sum_{i=1}^{N}(X_i - \bar{X})^2\right].$$

Using the same procedure, we have the following for the measurement variance component:

$$\text{MV} = \frac{1}{(mk)^2}\left\{\frac{m}{N}\sum_{j=1}^{k}\sum_{i=1}^{N}E\left[(\alpha_j - \varepsilon_{ij})^2\right]\right.$$

$$+ \frac{n(m-1)}{N(N-1)}\sum_{j=1}^{k}\sum_{i\neq i'}^{N}E\left[(\alpha_j + \varepsilon_{ij})(\alpha_j + \varepsilon_{i'j})\right]$$

$$+ \frac{m}{N}\frac{m}{N}\sum_{j\neq j'}^{k}\sum_{i=1}^{N}E\left[(\alpha_j + \varepsilon_{ij})(\alpha_{j'} + \varepsilon_{ij'})\right]$$

$$+ \frac{m}{N} \frac{m}{N} \sum_{j \neq j'}^{k} \sum_{i=i}^{N} E\big[(\alpha_j + \varepsilon_{ij})(\alpha_{j'} + \varepsilon_{i'j'})\big]\bigg\}$$

$$= \frac{1}{(mk)^2} \left[\frac{m}{N} \sum_{j=1}^{k} \sum_{i=1}^{N} \sigma_\alpha^2 + \sigma_\varepsilon^2 + \frac{m(m-1)}{N(N-1)} \sum_{j=1}^{k} \sum_{i \neq i'}^{N} \sigma_\alpha^2 \right]$$

$$= \frac{1}{mk} \big[\sigma_\alpha^2 + \sigma_\varepsilon^2 + (m-1)\sigma_\alpha^2\big]$$

$$= \frac{\sigma_\varepsilon^2}{n} + \frac{\sigma_\alpha^2}{k}.$$

The term that corresponds to the simple response variance in the Hansen et al. (1961) model is $\sigma_\alpha^2 + \sigma_\varepsilon^2$. That which corresponds to the correlate response variance is σ_α^2. Thus defining ρ as

$$\rho = \frac{\sigma_\alpha^2}{\sigma_\alpha^2 + \sigma_\varepsilon^2},$$

we can write the measurement variance as

$$\mathrm{MV} = \frac{\sigma_\alpha^2 + \sigma_\varepsilon^2}{n} \big[1 + (m-1)\rho\big],$$

a form similar to that in the Hansen–Hurwitz–Bershad model. Using a slightly different approach, O'Muircheartaigh (1977a) also shows the linkage of the linear models approach to the Hansen–Hurwitz–Bershad model.

As was the case for the variance decomposition models, there are various versions of this model in the literature. Sukhatme and Sukhatme (1970) have a model of this general type in which they consider a fixed population of K interviewers and a finite population of fixed effects α_j. In this case the α_j have a mean

$$\bar{\alpha} = \frac{1}{K} \sum_{j=1}^{K} \alpha_j \quad \text{and} \quad \text{variance } S_\alpha^2 = \frac{1}{K-1} \sum_{j=1}^{K} (\alpha_j - \bar{\alpha})^2.$$

Assuming a sample and measurement design with

1. A simple random sample of n elements from N
2. A simple random sample of k interviewers from K
3. Subsamples randomly assigned to interviewers

Sukhatme and Sukhatme give

$$\text{Var}(\bar{y}) = S_X^2 \left(\frac{1}{n} - \frac{1}{N} \right) + S_\alpha^2 \left(\frac{1}{k} - \frac{1}{n} \right) + \frac{S_\varepsilon^2}{n}.$$

The authors then derive the correlation between responses obtained by the same interviewer. The variance for a single observation is approximately $S_y^2 = S_X^2 + S_\alpha^2 + S_e^2$. The correlation ρ is defined by

$$
\begin{aligned}
\rho S_y^2 \frac{N-1}{N} &= E\{[y_{ij} - E(y_{ij})][y_{i'j} - E(y_{i'j})]\} \\
&= E\Big(E\{[y_{ij} - E(y_{ij})][y_{i'j} - E(y_{i'j})]|j\}\Big) \\
&= -\frac{S_X^2}{N} + \left(1 - \frac{1}{K}\right) S_\alpha^2.
\end{aligned}
$$

Thus, $\rho S_y^2 \simeq S_\alpha^2$, implying that

$$\rho = \frac{S_\alpha^2}{S_x^2 + S_\alpha^2 + S_\varepsilon^2}.$$

This gives $V(\bar{y}) = (S_y^2/n)[1 + \rho(n/k - 1)]$.

Kish (1962) uses a linear model similar to that of Sukhatme and Sukhatme for the effect of interviewers; however, he pools the S_x^2 and S_e^2 components arguing that "these errors cannot be distinguished from sampling error unless replicate measurements are taken on the respondents" (see Section 11.3.5). Thus Kish uses

$$Y_{ij} = Y_{ij}' + A_i$$

and ρ is defined as

$$\rho = \frac{S_a^2}{S_{y'}^2 + S_a^2}.$$

The linear model approach has given rise to the widespread statement in the literature that the correlated response variance can be attributed to the biases of the interviewers and the impression that the correlated measurement (response) variance is always positive. The various linear model approaches do require the ρ be nonnegative and that there be a variability between interviewers in interviewer bias for α to be greater than zero. If one accepts the variance decomposition approach, however, there is no constraint that ρ be positive.

What are the limits on ρ for the variance decomposition approach? As noted above, if one accepted Murthy's intracluster correlation model, the lower limit on ρ would be $-1/(N-1)$.

In the Census Bureau model the lower limit on ρ is $-1/(n-1)$. This results comes from

$$\text{RV} = \frac{1}{n}S_d^2[1 + (n-1)\rho] \geq 0$$

$$\rho \geq \frac{-1}{n-1}.$$

In the case where the correlation is confined to being within an interviewer's assignment, the low limit on the correlation is found to be $-1/(m-1)$ from

$$\text{RV} = \frac{1}{n}S_d^2[1 + (m-1)\rho],$$

where n is the total sample size and m is the interviewer work load for each of k interviewers, $n = mk$.

Several authors have reported negative values of ρ (Kish, 1962; Groves and Magilavy, 1980). In most cases, however, negative estimates are suppressed because it is assumed they are caused by the instability of the estimates (Bailey et al., 1978) which are arrived at by subtraction (see below). The negative correlation could arise, however, when interviewers expect greater differences between population elements than actually exist, thus resulting in a within-interviewer variance that is greater than the between-interviewer variance.

11.3.5 Estimating the Measurement Variance and Its Components for Models with Interviewers

All of the models require use of procedures similar to those described for the general case in order to estimate the total variance and the different components of variance. The same three methods are used: repeat measurements on the same population element, interpenetrating samples, and methods that are combinations of these two methods.

The following passage illustrates how these methods are used to make estimates of the overall variance and its components. We will first consider the method of interpenetrated samples. The between-interviewer mean square provides an estimate of the total variance under certain assumptions.

First consider the Census Bureau model, in which we have a fixed population of K interviewers. An independent sample of size m is drawn for each interviewer, and the indicator variable C_{ij} is used to denote presence in the jth interviewer's sample. Then

$$\bar{y}_{jt} = \frac{1}{m} \sum_{i=1}^{N} C_{ij} Y_{ijt}$$

$$E(\bar{y}_{jt}) = \frac{1}{N} \sum_{i=1}^{N} Y_{ij}$$

$$= \bar{Y}.$$

Under the CBM,

$$\text{Var}(\bar{y}_{jt}) = \frac{1}{m} \left[\sigma_d^2 + (m-1)\rho\sigma_d^2 \right] + \frac{1}{m} \frac{N-m}{N} S_y^2.$$

Now

$$\bar{y}_t = \frac{1}{K} \frac{1}{m} \sum_{j=1}^{K} \sum_{i=1}^{N} C_{ij} Y_{ijt}$$

$$= \frac{1}{K} \sum_{j=1}^{K} \bar{y}_{jt} \qquad \text{the mean of } K \text{ independent samples.}$$

Therefore,

$$\text{Var}(\bar{y}_t) = \frac{1}{K^2} \sum_{k=1}^{K} \text{Var}(\bar{y}_{jt})$$

$$= \frac{1}{Km} \left[\sigma_d^2 + (m-1)\rho\sigma_d^2 \right] + \frac{1}{Km} \frac{N-m}{N} S_y^2.$$

Consider the between-interviewer mean square:

$$\text{BIMS} = \frac{1}{K-1} \sum_{j=1}^{K} \left(\bar{y}_{jt} - \bar{y}_t \right)^2$$

$$E(\text{BIMS}) = \frac{1}{K-1} E\left\{ \sum_{j=1}^{K} \left[\left(\bar{y}_{jt} - \bar{Y} \right) - \left(\bar{y}_t - \bar{Y} \right) \right]^2 \right\}$$

$$= \frac{1}{K-1} E\left[\sum_{j=1}^{K} \left(\bar{y}_{jt} - \bar{Y} \right) - K\left(\bar{y}_t - \bar{Y} \right) \right]$$

$$= \frac{1}{K-1} \left[\sum_{j=1}^{K} \text{Var}(\bar{y}_{jt}) - K \text{Var}(\bar{y}_t) \right]$$

$$= \frac{1}{K-1} \left[K^2 \text{Var}(\bar{y}_t) - K \text{Var}(\bar{y}_t) \right]$$

$$= \frac{1}{K-1} \left[K(K-1)\text{Var}(\bar{y}_t) \right]$$

$$= K\left[\text{Var}(\bar{y}_t) \right].$$

Thus for the Census Bureau model, $(1/K)$ (BIMS) is an estimator for the total variance.

Now consider the case in which the interviewers are viewed as a sample of size k from a fixed population of K interviewers. In this case we have

$$
\begin{aligned}
\text{BIMS} &= \frac{1}{k-1} \sum_{j=1}^{K} \left(V_j \bar{y}_{jt} - \frac{1}{k} \sum_{j=1}^{K} V_j \bar{y}_{jt} \right)^2 \\
&= \frac{1}{k-1} \left[\sum_{j=1}^{K} V_j \bar{y}_{jt}^2 - \frac{1}{k} \left(\sum_{j=1}^{K} V_j \bar{y}_{jt} \right)^2 \right] \\
&= \frac{1}{k-1} \left(\frac{k-1}{k} \sum_{j=1}^{K} V_j \bar{y}_{jt}^2 - \frac{1}{k} \sum_{j \neq j'}^{K} V_j V_{j'} \bar{y}_{jt} \bar{y}_{j't} \right) \\
&= \frac{1}{km^2} \sum_{j=1}^{K} V_j \sum_{i=1}^{N} C_{ij} Y_{ijt}^2 + \sum_{i \neq i'}^{N} C_{ij} C_{i'j} Y_{ijt} Y_{i'jt} \\
&\quad - \frac{1}{k(k-1)m^2} \sum_{j \neq j'}^{K} V_j V_{j'} \left(\sum_{i=1}^{N} C_{ij} Y_{ijt} \right) \left(\sum_{i=1}^{N} C_{ij'} Y_{ij't} \right) \\
&= \frac{1}{km^2} \left(\sum_{j=1}^{K} \sum_{i=1}^{N} V_j C_{ij} Y_{ijt}^2 + \sum_{j=1}^{K} \sum_{i \neq i'}^{N} V_j C_{ij} C_{i'j} Y_{ijt} Y_{i'jt} \right) \\
&\quad - \frac{1}{m^2 k(k-1)} \left(\sum_{j \neq j'}^{K} \sum_{i=1}^{N} V_j V_{j'} C_{ij} C_{ij'} Y_{ijt} Y_{ij't} \right. \\
&\quad \left. + \sum_{j \neq j'}^{K} \sum_{i \neq i'}^{N} V_j V_{j'} C_{ij} C_{i'j'} Y_{ijt} Y_{i'j't} \right). \quad (11.7)
\end{aligned}
$$

Taking the expected value, we have

$$
\begin{aligned}
E(\text{BIMS}) &= \frac{1}{km^2} \left\{ \frac{km}{KN} \sum_{j=1}^{K} \sum_{i=1}^{N} \left[\text{Var}_t(Y_{ijt}) + Y_{ij}^2 \right] \right. \\
&\quad \left. + \frac{k}{K} \frac{m}{N} \frac{m-1}{N-1} \sum_{j=1}^{K} \sum_{i \neq i'}^{N} \left[\text{Cov}_t(Y_{ijt}, Y_{i'jt}) + Y_{ij} Y_{i'j} \right] \right\} \\
&\quad - \frac{1}{m^2 k(k-1)} \frac{k}{K} \frac{k-1}{K-1} \frac{m}{N} \frac{m}{N} \left(\sum_{j \neq j'}^{K} \sum_{i=1}^{N} Y_{ij} Y_{ij'} + \sum_{j \neq j'}^{K} \sum_{i \neq i'}^{N} Y_{ij} Y_{i'j'} \right) \\
&= k(\text{MV}) + \frac{1}{m} \left[\frac{1}{KN} \sum_{j=1}^{K} \sum_{i=1}^{N} Y_{ij}^2 + \frac{m-1}{KN(N-1)} \sum_{j=1}^{K} \sum_{i \neq i'}^{N} Y_{ij} Y_{i'j} \right. \\
&\quad \left. - \frac{1}{N^2 K(K-1)} \left(\sum_{j \neq j'}^{K} \sum_{i=1}^{N} Y_{ij} Y_{ij'} + \sum_{j \neq j'}^{K} \sum_{i \neq i'}^{N} Y_{ij} Y_{i'j'} \right) \right].
\end{aligned}
$$

If we use the decomposition that Raj uses, we find that

$$E\left(\frac{\text{BIMS}}{k}\right) = \text{Var}(\bar{y}_t) - \frac{1}{K}\left[\frac{1}{K-1}\sum_{j=1}^{K}\left(\bar{Y}_j - \bar{\bar{Y}}\right)^2\right].$$

The extra term is attributable to the without-replacement sampling of interviewers. It is analogous to the results achieved in without-replacement sampling of population elements.

That the BIMS/k also estimates the overall variance in the linear models case can be seen by substituting $X_i + \alpha_j + \varepsilon_{ij}$ for Y_{ijt} in expression (11.7) and then taking the expected value.

Another sample statistic is the within-interviewer mean square:

$$\text{WIMS} = \frac{1}{k(m-1)}\sum_{j=1}^{k}V_j\sum_{i=1}^{N}C_{ij}\left(Y_{ijt} - \bar{y}_{jt}\right)^2$$

$$= \frac{1}{k(m-1)}\sum_{j=1}^{K}V_j\left[\sum_{i=1}^{N}C_{ij}Y_{ijt}^2 - \frac{1}{m}\left(\sum_{i=1}^{N}C_{ij}Y_{ijt}\right)^2\right]$$

$$= \frac{1}{k(m-1)}\sum_{j=1}^{K}V_j\left(\frac{m-1}{m}\sum_{i=1}^{N}C_{ij}Y_{ijt}^2 - \frac{1}{m}\sum_{i\neq i'}^{N}C_{ij}C_{i'j}Y_{ijt}Y_{i'jt}\right)$$

$$= \frac{1}{k(m-1)}\left(\frac{m-1}{m}\sum_{j=1}^{K}\sum_{i=1}^{N}V_jC_{ij}Y_{ijt}^2 - \frac{1}{m}\sum_{j=1}^{K}\sum_{i\neq i'}^{N}V_jC_{ij}C_{i'j}Y_{ijt}Y_{i'jt}\right).$$

$$(11.8)$$

Taking the expected value, we have

$$E(\text{WIMS}) = \frac{1}{KN}\sum_{j=1}^{K}\sum_{i=1}^{N}\text{Var}_t(Y_{ijt}) - \frac{1}{KN(N-1)}\sum_{j=1}^{K}\sum_{i\neq i'}^{N}\text{Cov}_t(Y_{ijt}Y_{i'jt})$$

$$+ \frac{1}{KN}\sum_{j=1}^{K}\sum_{i=1}^{N}Y_{ij}^2 - \frac{1}{KN(N-1)}\sum_{j=1}^{K}\sum_{i\neq i'}^{N}Y_{ij}Y_{i'j}.$$

For a fixed population of interviewers and the assumption that $Y_{ij} = Y_i$, the expect value of the within-interviewer mean square is

$$E(\text{WIMS}) = \sigma_d^2(1-\rho) + S_Y^2.$$

Under the linear models approach, we substitute $X_i + \alpha_j + \varepsilon_{ij}$ for Y_{ijt} in the expression (11.8). This gives

$$E(\text{WIMS}) = \sigma_\alpha^2 + \sigma_e^2 - \sigma_\alpha^2 + s_x^2$$

$$= \sigma_e^2 + s_x^2.$$

When interviewers are sampled without replacement, the E(WIMS) is only approximately equal to the above-variance components sums. However, if we use the approximate expressions Raj uses, then

$$E(\text{WIMS}) = V(y) - C(y, I).$$

Now consider the case in which repeat measurements are made. Assume that each interviewer reinterviews each element in his sample. Then using the square of the difference between measurements on the same element (BMWE),

$$\text{BMWE} = \frac{2}{n} \sum_{j=1}^{K} \sum_{i=1}^{N} C_{ij} (Y_{ij1t} - Y_{ij2t})^2.$$

Under the assumption that $E(Y_{ij1t}) = E(Y_{ij2t}) = Y_{ij}$, we have

$$E(\text{BMWE}) = \frac{1}{2} \frac{1}{NK} \sum_{j=1}^{K} \sum_{i=1}^{N} \left[\text{Var}(Y_{ij1t}) + \text{Var}(Y_{ij2t}) - 2\,\text{Cov}(Y_{ij1t}, Y_{ij2t}) \right].$$

This equals SMV if we assume that the repeat measurements are independent of the original measurements and $\text{Var}(Y_{ijlt}) = \text{Var}(Y_{ijzt}) = \text{Var}(Y_{ijt})$. This holds for all of the variance decomposition models we have discussed.

For the linear models approach,

$$E\{\text{BMWE}\} = \sigma_\alpha^2 + \sigma_e^2.$$

Several more complex extensions of these models and more complex experimental designs make use of these basic procedures. Fellegi (1964) presents a design with the following features:

1. A simple random sample of mk elements is selected without replacement.
2. The sample is divided into k subsamples of m elements each.
3. Each subsample is randomly paired with another subsample.
4. The pairs of samples are assigned at random to k interviewers.
5. The interviewers interview the first member of the pair of an original survey.
6. The interviewers interview the second member of the pair for the repeat survey.

Fellegi derives the expected value of the mean under this design. He does not assume that the between-interviewer measurement covariance is zero and he allows for an interaction between sampling and measurement errors.

Although Fellegi uses a different approach to derive the overall variance, the indicator variable approach above could be used. We can assume that the interviewers are fixed and define indicator variables associated with the selection of the original sample, the assignment to subsamples, the pairing of subsamples, and the assignment of interviewers to pairs.

This design then yields several sample statistics that can be used to estimate the overall variance and its components. These include sums of squares:

Between interviewers, within survey

Within interviewer, within survey

Within subsample, between survey

Between interviewers, within subsample, between survey, and

within interviewer, between subsamples, between survey sum of squares.

Similar models have also been developed for more complex sample designs and statistics other than the mean. Koop's (1974) article is formulated for unequal probability sampling. Koch (1973) formulates his model in the multivariate case and considers subclass means that are complex ratios. Nathan (1973) considers estimators based on different samples and ratio estimates. Chai (1971) examines the effect of correlated measurement error on ordinary least squares estimators of the regression coefficient. Publications by the U.S. Bureau of the Census (1979) also consider more complex situations.

Folsom (1980) considers a complete crossover design in which a sample of all combinations of interviewer pairs are considered. Under this scheme repeat observations by the same interviewer on the same element are allowed.

The linear models approach has been extended by several researchers. An early paper by Sukhatme and Seth (1952) presents several different designs that consider stratification, repeat measurements, and interpenetration. Hartley and Rao (1978) and Biemer (1978) present a linear model that includes interviewer and coder effects. Complex multistage sample designs in which only the last stage of selection uses equal probability selection are considered. Biemer discusses the estimability conditions. These methods are extended in Biemer and Stokes (1985), who examine the optimum design of interviewer variance experiments.

11.4   MODELS FOR CATEGORICAL DATA

11.4.1   Model for Dichotomous Variables

In the previous sections we have explicated the measurement error models in the context of numeric measurements. In this section we focus on categorical data. The Hansen et al. (1961) model was originally formulated for a dichotomous variable. Let

$$
Y_{it} = \begin{cases} 1 & \text{if the } i\text{th element is assigned to a particular} \\ & \text{class or } i \text{ is said to have a certain characteristic} \\ & \text{when the survey measurements are made} \\ \\ 0 & \text{if otherwise} \end{cases}
$$

X_i = true measurement for the ith element.

The goal of the survey is to measure

$$
\pi = \frac{1}{N} \sum_{i=1}^{N} X_i.
$$

the proportion of the population that is in the particular class or has the particular characteristic. The expected measurement for a particular element over all possible samples and trials is

$$
E_{s \supset i} \left[E_t(Y_{it}|i) \right] = P_i.
$$

Note that if there is any variability in the measurements for the ith element, P_i will not equal X_i and there is necessarily an individual response bias $B_i = P_i - X_i$. In the presence of measurement variability for element i, $0 < P_i < 1$, so that B_i is negative for $X_i = 1$ and positive for $X_i = 0$. The net bias across individuals could sill be zero; however, a net bias of zero is not likely since this would require an improbably fortuitous collusion among the individual biases, B_i.

Assuming a simple random sample of n from a population of N, Hansen et al. (1961) derive the variance of the sample mean

$$
p_t = \frac{1}{n} \sum_{i=1}^{n} Y_{it}
$$

using $E_s[E_t(p_t)] = P$ with $P = (1/N)\sum_{i=1}^{N} P_i$. They also define the response deviation

$$
d_{it} = y_{it} - P_i
$$

and the expected measurement for the ith element for a fixed sample,

$$
E_t(y_{it}|s) = P_{is}.
$$

The variance of p_t is

$$
\begin{aligned}
\text{Var}(p_t) &= E\left[(p_t - P)^2\right] \\
&= E\left[(p_t - p)^2 + 2(p_t - p)(p - P) + (p - P)^2\right],
\end{aligned}
$$

where

$$p = \frac{1}{n} \sum_{i=i}^{n} P_i.$$

These terms are equivalent to the three terms in (11.2) and are the measurement or response variance, the covariance between response and sampling deviations, and the sampling variance.

Assume that the covariance between response and sampling deviations is zero. For the measurement variance we have

$$E\left[(p_t - p)^2\right] = E_s\left\{ E_t\left[\frac{1}{n^2}\left(\sum_{i=1}^{n} Y_{it} - P_i \right)^2 \right] \right\}$$

$$= E_s\left[\frac{1}{n^2} \sum_{i=1}^{n} E_t(Y_{it} - P_i)^2 + \sum_{i \neq i'}^{n} E_t(Y_{it} - P_i)(Y_{i't} - P_{i'}) \right]$$

$$= \frac{1}{n}\left[\frac{1}{N} \sum_{i=1}^{N} \mathrm{Var}_t(Y_{it}) + \frac{n-1}{N(N-1)} \sum_{i \neq i'}^{N} \mathrm{Cov}_t(Y_{it}, Y_{i't}) \right].$$

To write this term in the form that Hansen, Hurwitz, and Bershad use, we note that the variance over trials of Y_{it} is $P_i(1 - P_i)$ and define $\sigma_d^2 = (1/N)\sum_{i=1}^{N}P_i(1 - P_i)$; then

$$\rho = \frac{\mathrm{Cov}(d_{it}, d_{i't})}{\sigma_d^2}$$

$$= \frac{[1/N(N-1)]\sum_{i=1}^{N}\mathrm{Cov}_t(Y_{it}, Y_{i't})}{(1/N)\sum_{i=1}^{N}P_i(1 - P_i)}.$$

This allows us to write

$$E\left[(p_t - p)^2\right] = \frac{\sigma_d^2}{n}[1 + \rho(n - 1)].$$

In terms of the P_i the sampling variance component is

$$SV = \frac{N-n}{N} \frac{1}{N-1} \sum_{i=1}^{N} (P_i - P)^2.$$

The model is also expanded to reflect correlations among response deviations associated with processors of the survey data, crew leaders, and interviewers. The idea behind such decompositions is that correlations among response deviations are introduced through the actions of the various personnel who collect and process the survey data. If one can isolate and estimate the contributions of the various sources of correlation, one can identify ways to reduce these contributions to measurement errors.

Although the usual estimates of the sampling variance include the contribution of the simple response variance (assuming negligible finite population correction factors), there has been considerable interest in estimating the simple response variance because it is indicative of the reliability of the survey measurement process. Hansen et al. (1964) define an *index of inconsistency* which is the ratio of the simple response variance to the total variance under the assumption that the correlated measurement variance (called *within-trial covariance of response deviations* in their article) is zero. In the notation of this section, the index of inconsistency, I, is

$$ I = \frac{\sigma_{dG}^2}{P_G(1 - P_G)}, $$

where G denotes the general conditions of the survey.

Hansen, Hurwitz, and Pritzker discuss the mean square within elements (MSWE) estimator given in Section 11.3.2 under various assumptions about the repeat measurements. They define the *gross difference rate*, g, as

$$ g = \frac{1}{n} \sum_{i=1}^{n} (Y_{itG} - Y_{it'G})^2. $$

Thus $g/2$ (as is shown in Section 11.3.2) is an unbiased estimator for the simple response variance when "G and G' represent the same survey conditions" (i.e., the measurements have the same random distribution in the resurvey as they had in the original survey) and "t and t' denote independent repetitions on a single element" (Hansen et al., 1964:119) (i.e., the repeat measurements are independent of the original measurements).

Bailar (1968) reviews U.S. Census Bureau research into the use of reinterview procedures to measure the quality of census results. The general conditions G are important in Bailar's article because she considers five sets of reinterview data in which the reinterview protocol differs from the original survey. The goal of the investigation was to determine how reinterview data could be used to evaluate a census or one time survey. The effect of the time lag between the original and reinterviews and the effect of allowing the reinterviewers to have access to the original responses was examined. Three samples of people enumerated in the 1960 Census were selected. In sample 1 reinterviews were conducted in July 1960 and the reinterviewers had no knowledge of the original results. Differences between the original census data and the reinterview were noted, and as far as logistically possible, a reconciliation interview as conducted in October 1960. In sample 2, the reinterviews were conducted in October 1960. The interviewer was first to conduct the reinterview and to then conduct an "on-the-spot" reconciliation using the original responses. Sample 3 was also interviewed in October 1960; no reconciliation was carried out. The five data sets could then be used to estimate the effects that reconciliation had on estimates of the gross difference rate, where it is assumed that reconciliation introduces a positive correlation between the original and resurvey results. Bailar notes that the design does not permit estimating the effect of "conditioning" of the interviewed household, which is another source of positive correlation between the original and reinterview measurements.

Bailer (1976) estimates the simple response variances for a number of variables measured in the 1970 Census. These variances were estimated using data from the Current Population Survey, March 1970. The results in the paper illustrate a finding that has been replicated in many studies, namely, that response variance differs from characteristic to characteristic. Some results from Bailar's article are shown in Table 11.1. Variables more likely to be subject to interpretation by interviewers and respondents generally have larger measurement variability, as also illustrated by the following example from the U.S. National Crime Survey.

The United States National Crime Survey is a large survey directed at studying the extent to which the U.S. population is vicitimized by major crimes, such as assault, burglary, larceny, auto theft, and robbery. The survey began in 1972, and by the mid-1970s encompassed a sample of some 72,000 housing units. In 1975, the U.S. Census Bureau, which collects data for the survey, conducted an interviewer variance study in eight major metropolitan areas. Results of this study as reported by Bailey et al. (1978) were reviewed below.

The method of interpenetrated samples was the experimental method used for making the estimates. The following quote from the Bailey et al. (1978:16) describes the experimental design:

> In each of the eight impact cities, the field staff delineated 144 interviewer assignment areas and 18 crew leader districts, each containing eight geographically contiguous interview assignment areas. Then, within each crew leader district, pairs of interviewer assignment areas were formed. Interviewers were assigned to crew leader districts and interviewer assignment areas based on the geographical proximity of their homes to each of the areas. A random selection of 36 of the possible 72 assignment pairs from each impact city was then made in Washington. Finally, within each pair of assignment areas, housing units were assigned by a systematic method, so that each interviewer was assigned approximately half of the units. For each city, this procedure permitted the comparison of the work of pairs of interviewers and the estimation of an average correlated response variance for an NCS interviewer assignment areas.

The squared mean difference between interviewers within an assignment area was used to estimate the total variance. The sampling variance was

Table 11.1   Ratio of Simple Response Variance to Sampling Variance for Selected Characteristics from the 1980 Census

Population Characteristics	Ratio of Simple Response to Sampling Variance	Population Characteristics	Ratio of Simple Response to Sampling Variance
Age, all persons		*Educational attainment,*	
Under 15 years	0.01	*Negro persons 14 years*	
15 to 24 years	0.04	*and over*	
25 to 34 years	0.06	Less than elementary 8	0.27
35 to 44 years	0.07	Elementary 8	0.51
45 to 54 years	0.06	High school 1 to 3	0.40
55 to 64 years	0.07	High school 4	0.34
65 years and over	0.05	College 1 to 3	0.41
		College 4 or more	0.19
Marital status, all persons		*Employment status, persons*	
14 years and over		*14 years and over*	
Married	0.02	In labor force	0.16
Widowed	0.07	Employed	0.16
Divorced or separated	0.19	Agriculture	0.28
Never married	0.02	Nonagriculture	0.15
		Unemployed	0.61
		Not in labor force	0.16
Educational attainment,			
persons 25 years		*Industry, employed persons*	
and over		*16 years and over*	
Elementary 0 to 4	0.34	Agriculture, forestry,	
Elementary 5 to 7	0.47	fisheries	0.14
Elementary 8	0.42	Mining	0.30
High school 1 to 3	0.41	Construction	0.21
High school 4	0.28	Manufacturing—	
College 1 to 3	0.32	durables	0.15
College 4	0.26	Manufacturing—	
College 5 or more	0.22	nondurables	0.18
Educational attainment,		Transportation,	
white persons 14 years		communication	0.13
and over		Wholesale trade	0.48
Less than elementary 8	0.30	Retail trade	0.19
Elementary 8	0.36	Finance, insurance,	
High school 1 to 3	0.29	real estate	0.11
High school 4	0.25	Business and repair	0.37
College 1 to 3	0.29	Personal services	0.14
College 4 or more	0.13	Entertainment,	
		recreation	0.41
		Professional services	0.10
		Public administration	0.15

Table 11.2 Estimated Sampling Relvariances, Correlated Response Relvariances, and Total Relvariances for Major Personal Victimization Rates for Baltimore[a] (Base for Rates: 4744 Persons)[b]

Kind of Victimizations	Number[b] of Victimizations (1)	Victimization rates (2)	Total (3)	Sampling (4)	Correlated Response (5)	Ratios of Correlated Response to Sampling (6)
Total	517	0.1087	0.1589	0.0988	0.0601	0.61
Assaultive violence total	287	0.0604	0.2624	0.1609	0.1015	0.63
Assaultive violence with theft	55	0.0115	0.7531	0.6454	0.1077	0.17
Assaultive violence without theft	232	0.0489	0.2940	0.1912	0.1028	0.54
Personal theft without assault	230	0.0484	0.2216	0.1711	0.0505	0.30

Source: Table 1 in Bailey et al. (1978).

[a]These estimates are applicable to an area approximately equal in size to an average NCS interviewer assignment area in Baltimore.
[b]These numbers are weighted counts.

Table 11.3 Ratio of Sum Correlated Response Variance and Covariance of Response and Sampling Deviations to Sampling Variance for the 1975 Impact Cities

Kinds of Personal Victimizations	Atlanta (1)	Baltimore (2)	Cleveland (3)	Dallas (4)	Denver (5)	Newark (6)	Portland (7)	St. Louis (8)
a. Personal Victimizations								
Total[a]	0.73	0.61	0.18	0.51	0.34	1.40	0.59	0.96
Assaultive violence	0.72	0.63	0.31	0.45	0.09	0.85	0.68	1.16
Assaultive violence with theft	0.00	0.17	0.33	0.00	0.00	0.10	0.24	0.00
Assaultive violence without theft	0.80	0.54	0.37	0.50	0.12	1.21	0.70	1.16
Personal theft without assault	0.28	0.30	0.00	0.22	0.47	0.79	0.27	0.06
b. Household Victimizations								
Total[a]	1.24	0.75	0.40	0.37	0.43	1.00	0.56	0.40
Burglary, forcible entry something taken	0.01	0.00	0.09	0.44	0.10	0.00	0.00	0.28
Burglary, unlawful entry without force	0.00	0.70	0.04	0.56	0.02	0.08	0.35	0.00
Burglary, attempted forcible entry	0.64	0.07	0.00	0.21	0.00	0.00	0.17	0.17
Larceny under $50	1.03	0.54	0.31	0.64	0.37	0.60	0.64	0.41
Larceny $50 or more	0.54	0.31	0.28	0.28	0.28	0.10	0.55	0.56
Larceny NA amount	0.45	0.00	0.00	0.48	0.00	0.39	0.03	0.00
Attempted Larceny	0.41	0.48	0.36	0.00	0.14	0.42	0.31	0.38
Auto theft, theft of car	0.00	0.13	0.13	0.00	0.00	0.00	0.03	0.40
Auto theft, attempted theft of car	0.39	0.17	0.00	0.07	0.20	0.00	0.42	0.00

Source: Tables 2 and 3, Bailey et al. (1978).

[a]The total variance equals 1 plus the ratio multiplied by the sampling variance.

estimated as the average of the within-interviewer sampling deviations under the assumption of no ordering in the population. The sum of the correlated measurement variance and the covariance between response and sampling deviations was obtained by subtraction as described above. Estimates were made for each pair of assignment areas and averaged over all pairs of assignment areas in the city. The results obtained are summarized in Tables 11.2 and 11.3.

The results are not encouraging. In six of the areas the ratio of the correlated response variance to the sampling variance was greater than 0.5 for total personal victimizations. Ratios are also high for the overall household victimization rate. The importance of these high ratios was noted by the authors when they compare the rates of two cities. The household victimization rate for Newark was approximately 0.26 and that for Baltimore 0.52. The authors noted that under normality assumptions and using only the sampling variance to measure reliability, these two differences would be statistically different at the 0.01 level. However, this significant difference disappears when the total variance is used. Thus it appears that measurement variability has a detrimental effect on our ability to determine if real differences exist between groups. This detrimental effect on the ability to detect differences is very important because the examination of subgroup differences is one of the major ways by which one may try to find the determinants of variability in social conditions since true experimental designs are not possible.

One of the results evident in Tables 11.2 and 11.3 is that the measurement variance is larger for items that are likely to be more susceptible to interviewer interpretations. An example is the difference seen for assault with and without theft. The authors speculate that this difference may be because of variability among the interviewers' interpretations of violent assaults by the victims acquaintances, relatives, or friends. This result—that measurement variability is higher for questions requiring interpretation by the interview—is often seen.

11.4.2 Model for Categorical Data with L Categories

Some of the characteristics examined under this categorical data model are not strictly 0–1 variables. For example, Bailar (1968) presents a table in which characteristics such as age, education, and income are expressed as categorical variables with multiple categories (e.g., income for females is given for 15 categories of income). Her approach is to treat the data as if 15 separate 0–1 variables had been measured, one corresponding to each category of income. A multinomial analog of the model is illustrated below.

Suppose that we have a population of N people, and we wish to estimate the proportion of the population that is in the various classes of an L-fold categorical variables. Thus we wish to estimate a vector

$$\underset{L \times 1}{\boldsymbol{\pi}} = (\pi_1, \pi_2, \ldots, \pi_L)',$$

where $\pi_l = N_l/N$ and N_l is the number elements in the population in the lth class.

The true measure for the ith element is $\mathbf{X}_{i(L\times 1)}$, a vector of 0's and a single 1 denoting the true category classification of the ith element, and

$$\frac{1}{N} \sum_{i=1}^{N} \mathbf{X}_i = \boldsymbol{\pi}$$

is the population parameter to be estimated. We need an $(L \times 1)$ column vector $\mathbf{X}_i$ to describe the measurement for the ith individual and a corresponding $L \times 1$ vector $\boldsymbol{\pi}$ to denote the population values because the scale of measurement does not contain enough information for a difference to be informative for the characteristic in question.

For a simple random sample of size n, $\bar{\mathbf{x}}$ will be an unbiased estimate of $\boldsymbol{\pi}$:

$$\bar{\mathbf{x}} = \frac{1}{n} \sum_{i=1}^{n} \mathbf{X}_i.$$

The variance over samples of $\bar{\mathbf{x}}$ is given by

$$\mathrm{Var}_s(\bar{\mathbf{x}}) = E_s\left[(\bar{\mathbf{x}} - \boldsymbol{\pi})(\bar{\mathbf{x}} - \boldsymbol{\pi})'\right]$$
$$= \frac{1}{n}\frac{N-n}{N-1}\mathbf{V}_\pi,$$

where $\mathbf{V}_\pi$, the variance covariance matrix, is an $L \times L$ symmetric matrix with $\pi_l(1 - \pi_l)$ on the diagonal and $-\pi_l\pi_l'$ as off-diagonal elements, that is,

$$\mathrm{Var}_s(\bar{\mathbf{x}}) = \frac{1}{n}\frac{N-n}{N-1}\begin{bmatrix} \pi_1(1-\pi_1) & -\pi_1\pi_2 & \cdots & -\pi_1\pi_L \\ & & -\pi_l\pi_{l'} & \\ \text{Symmetric} & & & \\ & & & \pi_L(1-\pi_L) \end{bmatrix}.$$

The vector analog of the usual estimate of the sampling variance can be used to estimate $\mathbf{V}_\pi$. This is seen as follows:

$$s_{\bar{\mathbf{x}}}^2 = \frac{1}{n-1} \sum_{i=1}^{n} (\mathbf{X}_i - \bar{\mathbf{x}})(\mathbf{X}_i - \bar{\mathbf{x}})'$$

is an $L \times L$ matrix in which the diagonal terms are

$$s_{\bar{x}}^2 = \frac{1}{n-1} \sum_{i=1}^{n} (X_{il} - \bar{x}_l)^2$$

and the cross-product terms are

$$s^2_{\bar{x}_{ll'}} = \frac{1}{n-1} \sum_{i=1}^{n} (x_{il} - \bar{x}_l)(x_{il'} - \bar{x}_{l'}).$$

Using the indicator random variables U_i to denote presence in the sample, we may write

$$s^2_{\bar{x}_l} = \frac{1}{n-1} \sum_{i=1}^{N} U_i (X_{il} - \bar{x}_l)^2$$

$$= \frac{1}{n-1} \left[\sum_{i=1}^{N} U_i X_{il}^2 - \frac{1}{n} \left(\sum_{i=1}^{N} U_i X_{il} \right)^2 \right].$$

Since X_{il} equals either 0 or 1, $X_{il}^2 = X_{il}$, and

$$E_s(s^2_{\bar{x}_l}) = E_s \left\{ \frac{1}{n-1} \left[\sum_{i=1}^{N} U_i X_{il} - \frac{1}{n} \left(\sum_{i=1}^{N} U_i X_{il} + \sum_{i \neq i'}^{N} U_i U_{i'} X_{il} X_{i'l} \right) \right] \right\}$$

$$= \frac{1}{n-1} \left\{ \frac{n}{N} \sum_{i=1}^{N} X_{il} - \frac{1}{n} \left[\frac{n}{N} \sum_{i=1}^{N} X_{il} + \frac{n(n-1)}{N(N-1)} \sum_{i \neq i'}^{N} X_{il} X_{i'l} \right] \right\}$$

$$= \frac{N_1}{N} - \frac{N_1(N_1 - 1)}{N(N-1)}$$

$$= \frac{N_1}{N} \frac{N - N_1}{N - 1}$$

$$= \frac{N}{N-1} [\pi_1(1 - \pi_1)].$$

For the cross-product terms we have

$$E_s(s^2_{\bar{x}_{ll'}}) = E_s \left[\frac{1}{n-1} \sum_{i=1}^{N} U_i (X_{il} - \bar{x}_l)(X_{il'} - \bar{x}_{l'}) \right]$$

$$= E_s \left[\frac{1}{n-1} \sum_{i=1}^{N} U_i \left(X_{il} X_{il'} - X_{il} \bar{x}_{l'} - \bar{x}_l X_{il'} + \bar{x}_l \bar{x}_{l'} \right) \right].$$

The ith element can only be in one category, so $X_{il}X_{il'} = 0$, which gives

$$
E_s\left(s_{\bar{x}_{ll'}}^2\right) = \left[\frac{1}{n-1}\sum_{i=1}^N(-n\bar{x}_l\bar{x}_{l'})\right]
$$

$$
= -\frac{1}{n(n-1)}\left(\sum_{i=1}^N U_iX_{il}X_{il'} + \sum_{i\neq i'}^N U_iU_{i'}X_{il}X_{il'}\right)
$$

$$
= -\frac{1}{N(N-1)}\sum_{i\neq i'}^N X_{il}X_{il'}
$$

$$
= -\frac{N_lN_{l'}}{N(N-1)}
$$

$$
= -\frac{N}{N-1}\pi_l\pi_{l'}.
$$

This illustration of the measurement of "true" values for an L-fold categorical variable is helpful in understanding the following explication of a measurement error model for L-fold categorical variables analogous to the Hansen–Hurwitz–Bershad model for dichotomous variables.

Let $\mathbf{Y}_{it(L\times 1)}$ = measurement for the ith element. This is a vector of 0's and 1's in which one component equals 1. $\mathbf{Y}_{it}$ is a random variable, and

$$
E\left(\underset{L\times 1}{\mathbf{Y}_{it}}\right) = \underset{L\times 1}{\mathbf{P}_i},
$$

where P_{il} is the probability that any particular trial of the measurement process results in outcome l for the ith element. Recall that $\mathbf{X}_i$ is a vector of 0's and a single 1, so that, as in the dichotomous case, the presence of measurement variability results in a measurement process in which the expected measurements for the ith element are biased. Thus

$$
\underset{L\times 1}{\mathbf{b}_i} = \underset{L\times 1}{\mathbf{X}_i} - \underset{L\times 1}{\mathbf{p}_i}
$$

for a simple random sample

$$
\underset{L\times 1}{\bar{\mathbf{y}}_t} = \frac{1}{n}\sum_{i=1}^N U_i\underset{L\times 1}{\mathbf{y}_{it}}
$$

is an estimate of the vector $\boldsymbol{\pi}$. The expected value is

$$
E(\bar{\mathbf{y}}_t) = E_S[E_t(\bar{\mathbf{y}}_t)] = \frac{1}{N}\sum_{i=1}^N \mathbf{P}_i.
$$

We will assume that there is no covariance between response and sampling deviations, so that

$$\mathrm{Var}(\bar{\mathbf{y}}_t) = \mathbf{MV} + \mathbf{SV}.$$

The measurement variance **MV** is given by

$$\mathbf{MV} = E_s\big\{E_{t|s}\big[(\bar{\mathbf{y}}_t - \bar{\mathbf{p}})(\bar{\mathbf{y}}_t - \bar{\mathbf{p}}')\big]\big\},$$

where $\bar{\mathbf{p}} = (1/n)\sum_{i=1}^{N} U_i \mathbf{P}_i$. Now

$$E_{t|s}\big[(\bar{\mathbf{y}}_t - \mathbf{p})(\bar{\mathbf{y}}_t - \bar{\mathbf{p}}')\big] = \frac{1}{n^2} E_{t|s}\left\{\left[\sum_{i=1}^{N} U_i\left(\underset{L\times 1}{\mathbf{Y}_{it}} - \mathbf{p}_i\right)\right]\left[\sum_{i=1}^{N} U_i(\mathbf{Y}_{it} - \mathbf{P}_i)'\right]\right\}$$

$$= \frac{1}{n^2}\left\{\sum_{i=1}^{N} U_i E_{t|s}\big[(\mathbf{Y}_{it} - \mathbf{P}_i)(\mathbf{Y}_{it} - \mathbf{P}_i)'\big]\right.$$

$$\left. + \sum_{i\neq i'}^{N} U_i U_{i'} E_t\big[(\mathbf{Y}_{it} - \mathbf{P}_i)(\mathbf{Y}_{i't} - \mathbf{P}_{i'})'\big]\right\}$$

$$= \frac{1}{n^2}\sum_{i=1}^{N} U_i\underset{L\times L}{\boldsymbol{\sigma}_{id}^2} + \sum_{i\neq i'}^{N} U_i U_{i'}\underset{L\times L}{\boldsymbol{\rho}_{ii'}}.$$

The component matrices are of the following forms:

$$\boldsymbol{\sigma}_{id}^2 = E_{t|s}(\mathbf{d}_{it}\mathbf{d}_{it}')$$

$$= \begin{bmatrix} P_{il}(1 - P_{il}) & \\ & -P_{il}P_{il'} \\ & \end{bmatrix}$$

an $L \times L$ symmetric matrix with $P_{il}(1 - P_{il})$ on the diagonal and $-P_{il}P_{il}'$ as the off-diagonal elements.

$\boldsymbol{\rho}_{ii'}$ is an $L \times L$ matrix with $\rho_{ii'l}$ on the diagonal, where

$$\rho_{ii'l} = \frac{E_{t|s}\big[(Y_{ilt} - P_{il})(Y_{i'lt} - P_{i'l})\big]}{\big[P_{il}(1 - P_{il})P_{i'l}(1 - P_{i'l})\big]^{1/2}}$$

and off-diagonal elements $\rho_{ii'll'}$:

$$\rho_{ii'll'} = \frac{E_{t|s}\big[(Y_{ilt} - P_{il})(Y_{i'l't} - P_{i'l'})\big]}{\big[P_{il}(1 - P_{il})P_{i'l'}(1 - P_{i'l'})\big]^{1/2}}.$$

Now taking the expected value over samples allows us to write the vector analog of the Hansen–Hurwitz–Bershad model as

$$E_s\left[E_{t|s}(\mathbf{y}_t - \mathbf{p})(\mathbf{y}_t - \mathbf{p})'\right] = \frac{1}{n}\left[\frac{1}{N}\sum_{i=1}^{N}\sigma_{id}^2 + \frac{n-1}{N(N-1)}\sum_{i \neq i'}^{N}\mathbf{\rho}_{ii'}\right]$$

$$= \frac{1}{n}\left[\underset{L \times L}{SMV} + (n-1)\underset{L \times L}{CMV}\right].$$

Bershad (1969) defines an index of inconsistency for an L-fold classification system as the sum of the diagonal elements of SMV divided by the sum of the $P_l(1 - P_l)$, where $P_l = (1/N)\sum_{i=1}^{N}P_{il}$. Thus

$$I_{(L)} = \frac{\sum_{l=1}^{L}(1/N)\sum_{i=1}^{N}P_{il}(1 - P_{il})}{\sum_{l=1}^{L}P_l(1 - P_l)}$$

$$= \frac{\sum_{l=1}^{L}\sigma_{dl}^2}{\sum_{l=1}^{L}P_l(1 - P_l)}.$$

Note that the dichotomous index of inconsistency is

$$I_{(l)} = \frac{\sigma_{dl}^2}{P_l(1 - P_l)}.$$

$I_{(L)}$ becomes a weighted average of the indices calculated by considering each category of the L-fold variable a single dichotomous measure, with the weights being

$$w_l = \frac{P_l(1 - P_l)}{\sum_l P_l(1 - P_l)},$$

that is,

$$I_{(L)} = \sum_{l=1}^{L} w_l I_{(l)}.$$

Methods analogous to those given in previous sections can be used to estimate the components of the L-fold model. We will show only one of these for illustrative purposes. Assume that repeat measurements are made, so that we have

$$\bar{y}_{1t} = \frac{1}{n}\sum_{i=1}^{N} U_i Y_{i1t}$$

$$\bar{y}_{2t} = \frac{1}{n}\sum_{i=1}^{N} U_i Y_{i2t}.$$

The matrix analog of the mean square within elements is

$$\mathbf{MSWE} = \frac{1}{2n} \sum_{i=1}^{N} U_i(\mathbf{Y}_{i1t} - \mathbf{Y}_{i2t})(\mathbf{Y}_{i1t} - \mathbf{Y}_{i2t})'.$$

Now $(\mathbf{Y}_{i1t} - \mathbf{Y}_{i2t})(\mathbf{Y}_{i1t} - \mathbf{Y}_{i2t})'$ is an $L \times L$ matrix with diagonal elements $(Y_{i1lt} - Y_{i2lt})^2$. Previous results have demonstrated that $E[(1/n)\sum_{i=1}^{N} U_i(Y_{i1lt} - Y_{i2lt})^2]$ equals $P_{il}(1 - P_{il})$. The cross-products terms of this matrix are $(Y_{i1lt} - Y_{i2lt})(Y_{il't} - Y_{i2l't})$. Adding and subtracting P_{il} and under the assumption that $P_{il1} = P_{il2} = P_{il}$ and under the assumption of independence between the original and resurvey, the expected value of these cross product terms is $-p_{il}p_{il'}$. Thus the matrix analog of the mean square within elements is an estimator of SMV.

11.5   METHODS TO ADJUST FOR RESPONSE ERRORS

Whereas earlier we dealt with procedures for adjusting for bias under fixed-bias models, these procedures can also be used in the context of measurement variability. Double sampling schemes are employed in which a subsample of the original sample is selected and more accurate measurements obtained for members of the subsample either by means of record check studies or more expensive interviews in which the accurate values can be obtained. Hansen et al. (1953) present a ratio estimator using the double sampling scheme; Madow (1965) discusses a regression estimator; and Madow (1965), Lessler (1974, 1976), and Brackstone et al. (1975) discuss difference estimators.

Consider the difference estimator. Let

$\bar{y}_t$ = mean of the inaccurate values for the original sample of the size n

$\bar{y}_s$ = mean of the inaccurate values for a subsample of size n_1

$\bar{x}_s$ = mean of the accurate values for the subsample.

Using these quantities, an unbiased estimate of the population mean is

$$\bar{w}_t = \bar{y}_t - (\bar{y}_{st} - \bar{x}_s).$$

Now the variance of $\bar{w}_t$, assuming the Hansen–Hurwitz–Bershad model and no interaction variance, is

$$\text{Var}(\bar{w}_t) = \frac{n - n_1}{nn_1}\left[\sigma_d^2(1 - \rho)\right] + \frac{n - n_1}{nn_1}\sigma_B^2 + \frac{1}{n}\frac{N - n}{n}\sigma^2,$$

where

$$\sigma_B^2 = \frac{1}{N-1} \sum_{i=1}^{N} \left[(Y_i - X_i) - (\bar{Y} - \bar{X}) \right]^2$$

$$= \frac{1}{N-1} \sum_{i=1}^{N} (B_i - \bar{B})^2.$$

This last component is the population variability of the element bias terms. The $\text{Var}(\bar{w}_t)$ is arrived at by noting that

$$\bar{w}_t - \bar{X} = \bar{y}_t - \bar{y}_{st} + \bar{x}_s - \bar{X}$$

$$= (\bar{y}_t - \bar{y}) - (\bar{y}_{st} - \bar{y}_s) + (\bar{y} - \bar{x}) - (\bar{y}_s - \bar{x}_s) + \bar{x} - \bar{X}$$

$$= (\bar{d}_t - \bar{d}_{st}) + (\bar{b} - \bar{b}_s) + (\bar{x} - \bar{X}).$$

Also, $E_{ss}(\bar{d}_t \bar{d}_{st}) = \bar{d}_t^2$; similarly for $E_{ss}[(\bar{b} - \bar{B})(\bar{b}_s - \bar{B})]$. This gives

$$\text{Var}(\bar{w}_t) = \frac{1}{n_1} \left[\sigma_d^2 + (n-1)\sigma_d^2 \rho \right] - \frac{1}{n} \left[\sigma_d^2 + (n-1)\rho\sigma_d^2 \right]$$

$$+ \frac{1}{n_1}\sigma_B^2 - \frac{1}{n}\sigma_B^2 + \frac{1}{n}\frac{N-n}{N}\sigma^2.$$

This method may be adapted for use with interviews and in multistage designs (Lessler, 1974, 1977). Estimators for components are also given.

Hansen et al. (1953b) formulated a ratio estimator. The estimator is

$$\bar{x} = \frac{\bar{y}_t}{\bar{y}_{st}} \bar{x}_s.$$

Using the usual approximations for a ratio estimator, we have that

$$E(\bar{z}_t) \simeq \frac{\bar{Y}\bar{X}}{\bar{Y}} = \bar{X}$$

and

$$\text{Var}(\bar{z}_t) = \frac{\sigma_{\bar{y}_{st}}^2 - \sigma_{\bar{y}_t}^2}{\bar{y}^2} + \frac{\sigma_{\bar{x}_s}^2}{\bar{x}^2} - \frac{2\sigma_{\bar{x}_s \bar{y}_{st}} - \sigma_{\bar{y}_t \bar{x}}}{\bar{y}\bar{x}}.$$

The material above is an extensive discussion of the various methods that have been used for the study of measurement errors in surveys. The usefulness of any particular method depends on the resources, goals, and design of the survey.

One area of research on measurement variability that has not been discussed is those investigations into the quality of measurements that focus on the degree to which a particular variable actually measures the characteristic it aims to measure. References to this work were given in Chapter 9; in addition, the articles by Andrews (1984), Heise and Bohrnstedt (1970), and Alwin (1977) explicate this approach.

These three chapters on measurement error have dealt with its definition and quantification. Despite the considerable work in the area, assessing and controlling measurement errors remain a difficult problem. Experiments to measure these effects are expensive and complex both operationally and conceptually. Despite numerous studies that document the extent of measurement error, most surveys do not involve a formal assessment of the extent and impact of these errors.

CHAPTER 12

Total Survey Design: More General Error Models

The development of total error models for survey estimates is important, given the evidence of interrelationships among sources of error. Attempts to reduce or control errors of one type may have adverse effects on some other component of the total error. For example, very vigorous attempts to increase the response rate may result in including a group of very poorly motivated respondents. The answers this group of respondents gives may be of such poor quality that any reduction in nonresponse bias is more than offset by increases in measurement bias to the overall detriment of the survey.

For example, Weeks et al. (1983) compare results of the telephone and personal interview modes in a health interview survey. The same population was sampled and interviewed by the two modes using identical questionnaires. Respondents were interviewed about a wide variety of health-related topics. All respondents who reported one or more hospital stays were included in a validation study, and the respondent's permission to obtain verifying information from the hospital was sought. Results were reported for those from whom permission forms were obtained. Nearly 25 percent of the telephone respondents did not specify the condition associated with the reported hospital stay; whereas fewer than 8 percent of the personal interview respondents failed to specify a condition. Thus it appeared that the adverse effect of item nonresponse was much greater for the telephone respondents than for the personal interview respondents. However, the authors also noted that the personal interview respondents had a much higher error rate in the conditions they reported. They were twice as likely as the telephone respondents (15 percent versus 7.8 percent) to report a condition that clearly disagreed with the hospital-reported condition. Thus the data indicate that the well-recognized advantage of the personal interview mode in terms of greater completeness of data may carry with it a disadvantage of increased measurement errors.

12.1 AN EXPANDED ERROR MODEL

In this chapter we examine the nature of total error models, that is, those that accommodate several sources of error. We first look at a model that considers errors arising from sampling, measurement, and nonresponse. We examine this model in the context of two methods of adjusting for nonresponse—the weighting class method and the random hot-deck method. Following this exploration we examine the general nature of such a model that would include these three types of error plus frame error. The models presented here are in the early stages of development. Much additional work needs to be done, and we hope that by including them we will stimulate additional work on total error models.

The error model presented in this section is based on work by Platek et al. (1977) and Lessler (1983) using the concept of response probabilities discussed in Chapter 7. The survey consists of four processes assumed to be stochastic: (1) sample selection, (2) efforts to obtain a response from sample members, (3) measurement of those agreeing to participate, and (4) imputation for those selected but unwilling to participate. The symbols s, r, t, and z are used to denote these processes, respectively. Thus, in deriving the expected value of a survey statistic, the following expectations are needed:

E_s denotes expectation over all possible samples.

$E_{r|s}$ denotes expectation over outcomes of the response mechanism given the outcome of the sampling process.

$E_{t|rs}$ denotes expectation over measurements given selection in the sample and participation in the survey.

$E_{z|rs}$ denotes expectation over outcomes of the imputation process given sample selection and survey nonparticipation.

We let X_i denote the "true" value for the ith member of the population.

We assume that the actual measurements are subject to measurement variability and that some type of adjustment is made for nonresponse in the sample. Moreover, suppose that this nonresponse adjustment can be viewed as equivalent to imputing a more or less appropriate value for the missing measurements. Each member of the sample has a probability of response. If we let

$$R_i = \begin{cases} 1 & \text{if the } i\text{th population element responds} \\ 0 & \text{otherwise,} \end{cases}$$

then the measurement for a particular member of the sample can be written as

$$Z_{istrz} = R_i(X_i + \varepsilon_{1it}) + (1 - R_i)(X_i + \varepsilon_{0it})$$
$$= R_i Y_{it} + (1 - R_i) Y'_{iz}. \tag{12.1}$$

The quantities ε_{1it} and ε_{0it} are, respectively, the elementary measurement or response error and the elementary imputation error. The Z_{istrz} is the measurement for the ith respondent in a particular sample. The subscripts t, r, and z indicate that the responses are subject to the other three stochastic sources of error as well.

For example, suppose that a weighting class adjustment is used to adjust the overall estimates for nonresponse. For estimates of means and totals these adjustments are equivalent to imputing the average of the respondents within the weighting class to members of the weighting class who do not respond. In the simplest case, the entire sample is considered a weighting class, and the mean for the respondents is substituted for the nonrespondents. This procedure may be done by actual substitution, or, most frequently, by means of an adjustment to the weights of the respondents. The following demonstrates the derivation of a total error model under this situation.

Assume an unequal probability sample design with

π_i = probability that the ith element is in the sample

$\pi_{ii'}$ = probability that the i and i'th elements are in the sample.

Consider estimation of the total of the true values

$$X = \sum_{i=1}^{N} X_i.$$

Let Y_{it} denote the measurement obtained for the ith responding element. Allowing for the possibility of a net bias in the measurement process for an individual so that $E_t(Y_{it}) \neq X_i$, then letting $Y_{it} = X_i + B_i + e_{it}$, we will assume that

$$E_t(Y_{it}) = B_i$$

$$E_t(e_{it}) = 0$$

$$\text{Var}_t(e_{it}) = \sigma_{ei}^2$$

$$\text{Cov}_t(e_{it}, e'_{it}) = \sigma_{eii'}^2.$$

The σ_{ei}^2 represent the contribution that the ith element makes to the simple measurement variance in a Hansen–Hurwitz–Bershad model for measurement errors; the $\sigma_{eii'}$ are the contributions to the correlated component of measurement variance.

In the overall weight adjustment procedure, weights of each responding element are inflated by the ratio of the total weights to weights of the nonresponding individuals. In the context of Eq. (12.1) this procedure is equivalent to substituting the following value for Y'_{iz}:

$$Y'_{iz} = \frac{\Sigma_{i \in s} R_i Y_{it}/\pi_i}{\Sigma_{i \in s} R_i/\pi_i}.$$

Under this scenario, there is a single value for Y'_{iz} for a joint outcome of t, r, and s. Thus we can drop the subscript z and write

$$Z_{istr} = R_i Y_{it} + (1 - R_i) \frac{\Sigma_{i \in s} R_i Y_{it}/\pi_i}{\Sigma_{i \in s} R_i/\pi_i}.$$

The Horvitz–Thompson estimator for the total is then

$$\hat{X}_{\text{WC}} = \sum_{i \in s} \frac{Z_{istr}}{\pi_i}.$$

The estimator $\hat{X}_{\text{WC}}$ is not unbiased, and there are two components to the bias, one from the nonresponse and the other from bias in the individual measurements. We assume that the response process is mutually independent among population elements, so that

$$\Pr(R_i = 1) = p_i, \quad \Pr(R_i = 0) = 1 - p_i, \quad \text{and} \quad \Pr(R_i = 1, R_{i'} = 1) = p_i p_{i'}.$$

Using the conditional expected values defined above, we have the following for the expected value of $\hat{X}_{\text{WC}}$:

$$
\begin{aligned}
E(\hat{X}_{\text{WC}}) &= E_s E_{r|s} E_{t|rs}(\hat{X}_{\text{WC}}) \\
&= E_s E_{r|s} E_{t|rs}\left(\sum_{i \in s} \frac{R_i Y_{it}}{\pi_i} \right) \\
&\quad + E_s E_{r|s} E_{t|rs}\left[\sum_{i \in s} (1 - R_i) \frac{1}{\pi_i} \frac{\Sigma_{i \in s} R_i Y_{it}/\pi_i}{\Sigma_{i \in s} R_i/\pi_i} \right] \\
&= E_s E_{r|s}\left[\sum_{i \in s} \frac{R_i(X_i + B_i)}{\pi_i} \right] \\
&\quad + E_s E_{r|s}\left[\sum_{i \in s} (1 - R_i) \frac{1}{\pi_i} \frac{\Sigma_{i \in s} R_i(X_i + B_i)/\pi_i}{\Sigma_{i \in s} R_i/\pi_i} \right] \\
&= E_s\left[\sum_{i \in s} \frac{p_i(X_i + B_i)}{\pi_i} \right] \\
&\quad + E_s\left[\sum_{i \in s} (1 - p_i) \frac{1}{\pi_i} \frac{\Sigma_{i \in s} p_i(X_i + B_i)/\pi_i}{\Sigma_{i \in s} p_i/\pi_i} \right].
\end{aligned}
$$

Ignoring the bias of the product ratio estimator in the second term gives

$$
\begin{aligned}
E(\hat{X}_{\text{WC}}) &\doteq \sum_{i=1}^{N} p_i(X_i + B_i) + \sum_{i=1}^{N} (1 - p_i) \left[\frac{\sum_{i=1}^{N} p_i(X_i + B_i)}{\sum_{i=1}^{N} p_i} \right] \\
&= \sum_{i=1}^{N} \frac{p_i}{\bar{p}} (X_i + B_i),
\end{aligned}
$$

where $\bar{p} = (1/N)\sum_{i=1}^{N} p_i$, the average response probability. Now the bias in $\hat{X}_{\text{WC}}$ is therefore approximately

$$
\begin{aligned}
\text{Bias}(\hat{X}_{\text{WC}}) &\doteq \sum_{i=1}^{N} \left[\frac{p_i}{\bar{p}} (X_i + B_i) - X_i \right] \\
&= \sum_{i=1}^{N} X_i \frac{p_i - \bar{p}}{\bar{p}} + \sum_{i=1}^{N} \frac{p_i}{\bar{p}} B_i.
\end{aligned}
$$

The first component of the bias is from nonresponse and the second from errors in measurement. If all response probabilities are the same, $p_i = \bar{p}$, and the only bias is because of inherent bias in the measurements and equals $\sum_{i=1}^{N} B_i$. Under the deterministic view of nonresponse in which the population of N elements is viewed as consisting of N_1 respondents and N_0 nonrespondents, there are N_1 elements with $p_i = 1$ and N_0 elements with $p_i = 0$. The bias in $\hat{X}_{\text{WC}}$ can be written as

$$
\text{Bias}(\hat{X}_{\text{WC}}) \doteq N(\bar{X}_1 - \bar{X}) + N(\bar{B}_1)
$$

because $\bar{p} = N_1/N$ and $\bar{B}_1 = (1/N_1)\sum_{i=1}^{N_1} B_i$, the average bias for the responding stratum. An alternative form is

$$
\text{Bias}(\hat{X}_{\text{WC}}) \doteq N_0(\bar{X}_1 - \bar{X}_0) + N\bar{B}_1,
$$

which illustrates that there is no bias in the weighting class estimator from nonresponse under the deterministic view if the means of the responding and nonresponding strata are equal.

The variance of the weighting class estimator will have three components. This can be seen by writing the variance expression in the following form

$$
\begin{aligned}
\text{Var}(\hat{X}_{\text{WC}}) = \text{Var}_s\, E_{r|s} E_{t|rs}(\hat{X}_{\text{WC}}) &+ E_s\, \text{Var}_{r|s}\, E_{t|rs}(\hat{X}_{\text{WC}}) \\
&+ E_s E_{r|s}\, \text{Var}_{t|rs}(\hat{X}_{\text{WC}}).
\end{aligned} \tag{12.2}
$$

The three components in Eq. (12.2) correspond to the sampling variance, the variance across outcomes of the response mechanism, and the measurement variance.

To derive the variance we make use of the Taylor series linearization approach. To do this we rewrite $\hat{X}_{\text{WC}}$ in the following equivalent form:

$$\hat{X}_{\text{WC}} = \sum_{i \in s} \frac{R_i Y_{it}}{\pi_i} \frac{\sum_{i \in s} 1/\pi_i}{\sum_{i \in s} R_i/\pi_i}.$$

Thus $\hat{X}_{\text{WC}}$ is a product ratio estimator of the form uv/w, where

$$u = \sum_{i \in s} \frac{1}{\pi_i}$$

$$v = \sum_{i \in s} \frac{R_i Y_{it}}{\pi_i}$$

$$w = \sum_{i \in s} \frac{R_i}{\pi_i}.$$

The approximate variance of $\hat{X}_{\text{WC}}$ may be found by using the linearized value l_i. The estimator $\hat{X}_{\text{WC}}$ is of the form $\sum_{i \in s} l_{it}/\pi_i$, where

$$\frac{l_{it}}{\pi_i} = \frac{[E(u)E(v)/E(w)][u_i/E(u) + v_i/E(v) - w_i/E(w)]}{\pi_i}$$

with $u_i = 1$, $v_i = R_i Y_{it}$, and $w_i = R_i$.

The variance is found by substituting the linearized value for the estimator $\hat{X}_{\text{WC}}$ in Eq. (12.2) and then taking the appropriate conditional expected values and variances.

To evaluate the right-hand side of Eq. (12.2), we will first consider the *measurement variance* term. We first substitute for $E(u)$, $E(v)$, and $E(w)$ in l_{it}/π_i and obtain

$$\frac{l_{it}}{\pi_i} = \frac{N\sum_{i=1}^N p_i(X_i + B_i)}{\sum_{i=1}^N p_i} \left[\frac{1/N + R_i Y_{it}/\sum_{i=1}^N p_i(X_i + B_i) - R_i/\sum_{i=1}^N p_i}{\pi_i} \right].$$

This result is then substituted into the expression for the conditional variance across measurements given the sample and the response pattern so that

$$\text{Var}_{t|rs}[\hat{X}_{\text{WC}}] = E_{t|rs}\left[\left(\sum_{i \in s} \frac{l_{it}}{\pi_i} - E_{t|rs} \sum_{i \in s} \frac{l_{it}}{\pi_i} \right)^2 \right]$$

$$= E_{t|rs}\left[\left(\frac{N}{\sum_{i=1}^N p_i} \sum_{i \in s} \frac{R_i \varepsilon_{1it}}{\pi_i} \right)^2 \right]$$

$$= \left(\frac{1}{\bar{p}} \right)^2 \left(\sum_{i \in s} \frac{R_i \sigma_{ei}^2}{\pi_i^2} + \sum_{\substack{i \neq i' \\ i,i' \in s}} \frac{R_i R_{i'} \sigma_{eii'}}{\pi_i \pi_{i'}} \right).$$

This expression for variance assumes that the measurements do not depend on either the sample (no covariance between response and sampling deviations) and also that they do not depend on the response pattern.

Now taking the expected value over outcomes of the response process given the sample and then over samples, we have

$$\text{measurement variance} = \left(\frac{1}{\bar{p}}\right)\left(\sum_{i=1}^{N} \frac{p_i \sigma_{ei}^2}{\pi_i} + \sum_{i \neq i'}^{N} \frac{p_i p_{i'} \pi_{ii'} \sigma_{eii'}}{\pi_i \pi_{i'}}\right).$$

The first term in this expression corresponds to the simple response or measurement variance in the model suggested by Hansen et al. (1961), while the second term is the correlated measurement (response) variance.

Next consider the *nonresponse variance* term. We again use the linearized value and first take its expected value over measurements. Next we note that terms not involving R_i will not contribute to the nonresponse variance, so that if we let $Y_i = X_i + B_i$, we get

$$E_s \, \text{Var}_{r|s} \, E_{t|rs}(\hat{X}_{\text{WC}})$$

$$= E_s \, \text{Var}_{r|s}\left\{\sum_{i \in s} \frac{1}{\pi_i} \frac{N\sum_{i=1}^{N} p_i Y_i}{\sum_{i=1}^{N} p_i}\left[R_i \frac{Y_i \sum_{i=1}^{N} p_i - \sum_{i=1}^{N} p_i Y_i}{\Sigma p_i \Sigma p_i Y_i}\right]\right\}$$

$$= E_s\left[\sum_{i \in s} \frac{1}{\pi_i^2}\left(\frac{N\sum_{i=1}^{N} p_i Y_i}{\sum_{i=1}^{N} p_i}\right)^2 p_i(1 - p_i)\left(\frac{Y_i \sum_{i=1}^{N} p_i - \sum_{i=1}^{N} p_i Y_i}{\sum_{i=1}^{N} p_i \sum_{i=1}^{N} p_i Y_i}\right)^2\right]$$

$$= \frac{1}{\bar{p}^2}\sum_{i=1}^{N} \frac{p_i(1 - p_i)}{\pi_i}\left(Y_i - \frac{\sum_{i=1}^{N} p_i Y_i}{\sum_{i=1}^{N} p_i}\right)^2.$$

The term involving $\sum_{i \neq i'}^{N}$ is missing because of the assumption $\Pr(R_i = 1, R_{i'} = 1) = p_i p_{i'}$, which causes $E_{r|s}[(R_i - p_i)(R_{i'} - p_{i'})] = 0$.

For arriving at the *sampling variance*, we again make use of the linearized value and substitute it directly into the usual variance formula, giving us

$$\text{Var}_s \, E_{r|s} E_{t|rs}(\hat{X}_{\text{WC}})$$

$$= \sum_{i=1}^{N} \frac{1 - \pi_i}{\pi_i}(E_{r|s} E_{t|rs} l_{it})^2 + \sum_{i \neq i'}^{N} \frac{\pi_{ii'} - \pi_i \pi_{i'}}{\pi_{ii'}}(E_{r|s} E_{t|rs} l_{it})(E_{r|s} E_{t|rs} l_{i't})$$

$$= \text{Var}_s\left[\sum_{i \in s} \frac{1}{\pi_i} \frac{N\sum_{i=1}^{N} p_i Y_i}{\Sigma p_i}\left(\frac{1}{N} + \frac{p_i Y_i}{\sum_{i=1}^{N} p_i Y_i} - \frac{p_i}{\Sigma p_i}\right)\right]$$

$$= \left(\frac{\sum_{i=1}^{N} p_i Y_i}{\bar{p}}\right)^2\left(\sum_{i=1}^{N} \frac{1 - \pi_i}{\pi_i}\right)\left(\frac{1}{N} + \frac{p_i Y_i}{\Sigma p_i Y_i} - \frac{p_i}{\Sigma p_i}\right)^2$$

$$+ \sum_{i \neq i'}^{N} \frac{\pi_{ii'} - \pi_i \pi_{i'}}{\pi_i \pi_{i'}}\left(\frac{1}{N} + \frac{p_i Y_i}{\Sigma p_i Y_i} - \frac{p_i}{\Sigma p_i}\right)\left(\frac{1}{N} + \frac{p_{i'} Y_{i'}}{\Sigma p_i Y_i} - \frac{p_{i'}}{\Sigma p_i}\right).$$

The weighting class adjustment is most often used to adjust for missing elements those from which no data were obtained. Often, only certain data items are missing for a particular element. In this case a common procedure in large-scale surveys is to substitute the data of some other element for the missing items. When the substitute data are drawn from the responding elements in the same survey, it is often called *hot-deck imputation*. If previous data from the respondent are used, as may occur in a panel survey, the term *historical data substitution* is sometimes used. Sometimes the substitute data are selected from a previous survey. This is called *cold-deck imputation*.

When the data to be substituted are selected randomly, it is possible to obtain the expected value and variance of the estimates. The following describes a total error model under a random hot-deck imputation procedure.

Imputation is to be done for missing data items. For example, income may be missing for some respondents. The type of analysis that is anticipated may be facilitated by having a value for income available. In addition, it is often considered desirable for the imputation adjusted estimates to be equal in expectation to the estimates from the respondents. Cox (1980) describes a weighted sequential hot-deck imputation procedure that has this property. The procedure is usually carried out within imputation cells; however, to simplify the presentation, we will assume a single imputation cell.

In addition to the three steps described above we now have a fourth, the random imputation of missing data items. For a particular data item such as income, the entire sample can be classified into responding and nonresponding elements. Responding elements will be randomly selected and their item data randomly assigned to a nonresponding element.

We will define an additional indicator random variable

$$\lambda_j(i) = \begin{cases} 1 & \text{if the } i\text{th responding element is} \\ & \text{substituted for the } j\text{th nonresponding} \\ & \text{element; that is, } R_i = 1, R_j = 0 \\ \\ 0 & \text{otherwise.} \end{cases}$$

Here we have let i index the responding elements and j index the nonresponding elements.

Let Y'_{iz} denote the value substituted for the jth nonresponding sample member at the zth imputation. Thus the measurements are

$$Z_{istrz} = R_i Y_{it} + (1 - R_i) Y'_{iz}$$
$$= R_i Y_{it} + (1 - R_i) \sum_{i \in s} R_i \lambda_j(i) Y_{it}. \qquad (12.3)$$

Assume that there are $m = \sum_{j \in s}(1 - R_j)$ nonresponding elements. Responding elements will be selected with probabilities proportional to size using a with-replacement sampling scheme. Under this scheme, a single responding unit may be selected more than once, and the expected number of selections for the ith responding unit is

$$E[r(i)] = \frac{m(1/\pi_i)}{\sum_{i \in s} R_i/\pi_i},$$

where $r(i)$ is the number of times the ith responding element is selected for substitution. Now given $r(i)$,

$$\Pr[\lambda_j(i) = 1] = \frac{r(i)}{m}$$

$$\Pr[\lambda_j(i)\lambda_{j'}(i) = 1 | r(i), m] = \frac{r(i)}{m} \frac{r(i) - 1}{m - 1}$$

$$\Pr[\lambda_j(i)\lambda_j(i') = 1 | r(i), r(i'), m] = \frac{r(i)}{m} \frac{r(i')}{m - 1}.$$

This formulation is equivalent to the single and double draw formulation of a sample design in which elements are randomly selected from a frame and randomly assigned a sampling unit label (Folsom, 1980). Thus the expected values over imputations of the $\lambda_j(i)$ and $\lambda_j(i)\lambda_{j'}(i')$ are

$$E[\lambda_j(i)] = \frac{E[r(i)]}{m} = \phi(i)$$

$$E[\lambda_j(i)\lambda_{j'}(i')] = \begin{cases} \dfrac{E\{r(i)[r(i) - 1]\}}{m(m - 1)} & \text{for } i = i', j \neq j' \\[2mm] \dfrac{E[r(i)r(i')]}{m(m - 1)} & \text{for } i \neq i', j \neq j' \\[2mm] \dfrac{E[r(i)]}{m} & \text{for } i = i', j = j' \\[2mm] 0 & \text{for } i \neq i', j = j' \end{cases}$$

$$= \phi(ii').$$

We can now write the random hot-deck estimator as

$$\hat{X}_{\text{RHD}} = \sum_{i \in s} \frac{Z_{istrz}}{\pi_i}$$

$$\hat{X}_{\text{RHD}} = \sum_{i \in s} \frac{R_i Y_{it}}{\pi_i} + \sum_{j \in s} \frac{(1 - R_j)}{\pi_j} \sum_{i \in s} R_i \lambda_j(i) Y_{it}.$$

Substituting for the expected value of $r(i)$, we get

$$E(\hat{X}_{\mathrm{RHD}}) = E_s E_{r|s} E_{t|rs} E_{z|trs}(\hat{X}_{\mathrm{RHD}})$$

$$= E_s E_{r|s} E_{t|rs} \left(\sum_{i \in s} \frac{R_i Y_{it}}{\pi_i} + \sum_{i \in s} \frac{1 - R_j}{\pi_i} \frac{\sum_{i \in s} R_i Y_{it}/\pi_i}{\sum_{i \in s} R_i/\pi_i} \right)$$

$$= E(\hat{X}_{\mathrm{WC}}).$$

Thus we see that the two procedures have the same expected values. The random hot-deck method does, however, have an additional component of variance.

Now

$$\mathrm{Var}(\hat{X}_{\mathrm{RHD}}) = \mathrm{Var}_{srt} E_{z|srt}(\hat{X}_{\mathrm{RHD}}) + E_s E_{r|s} E_{t|sr} \mathrm{Var}_{z|tsr}(\hat{X}_{\mathrm{RHD}}).$$

Now the first term is equivalent to the variance of the weighting-class-adjusted procedure and was given in the preceding section. The additional component of variance in the random hot-deck procedure is derived as follows:

$$\mathrm{Var}_{z|srt}(\hat{X}_{\mathrm{RHD}}) = E_{z|srt}\left[\sum_{j \in s} \frac{1 - R_j}{\pi_j} \sum_{i \in s} R_i \lambda_j(i) Y_{it} \right.$$

$$\left. - \sum_{j \in s} \frac{1 - R_j}{\pi_j} \sum_{i \in s} \frac{R_i Y_{it}/\pi_i}{\sum_{i \in s} R_i/\pi_i} \right]^2.$$

Recalling that $E_{z|srt}[\lambda_j(i)] = \phi(i)$, this expression may be rewritten as

$$E_{z|srt}\left\{ \sum_{i \in s} [\lambda_j(i) - \phi(i)] R_j Y_{it} \sum_{j \in s} \frac{1 - R_j}{\pi_j} \right\}^2$$

$$= E_{z|srt}\left\{ \sum_{i \in s} \sum_{j \in s} [\lambda_j(i) - \phi(i)]^2 R_j Y_{it}^2 \frac{1 - R_j}{\pi_j^2} \right.$$

$$+ \sum_{i \neq i' \in s} \sum_{j \in s} [\lambda_j(i) - \phi(i)][\lambda_j(i') - \phi(i')] R_i R_{i'} Y_{it} Y_{i't} \frac{1 - R_j}{\pi_j^2}$$

$$+ \sum_{i \in s} \sum_{j \neq j' \in s} [\lambda_j(i) - \phi(i)]^2 R_i Y_{it}^2 \frac{1 - R_j}{\pi_j} \frac{1 - R_{j'}}{\pi_{j'}}$$

$$+ \sum_{i \neq i' \in s} \sum_{j \neq j' \in s} [\lambda_j(i) - \phi(i)][\lambda_{j'}(i') - \phi(i')]$$

$$\left. \times R_i R_{i'} Y_{it} Y_{i't} \frac{1 - R_j}{\pi_j} \frac{1 - R_{j'}}{\pi_j} \right\}.$$

For sampling with replacement $\phi(ii') = \phi(i)\phi(i')$, which gives

$$
\begin{aligned}
\mathrm{Var}_{z|srt}\big(\hat{X}_{\mathrm{RHD}}\big) = {}& \sum_{i\in s}\sum_{j\in s}\phi(i)[1-\phi(i)]R_iY_{it}^2\frac{1-R_j}{\pi_j^2} \\
&+ \sum_{i\in s}\sum_{j\neq j\in s}\phi(i)[1-\phi(i)]R_iY_{it}^2\frac{1-R_j}{\pi_j}\frac{1-R_{j'}}{\pi_{j'}} \\
&- \sum_{i\neq i'\in s}\sum_{j\in s}\phi(i)\phi(i')R_iY_{it}R_{i'}Y_{i't}\frac{1-R_j}{\pi_j^2}.
\end{aligned}
$$

Now taking the expected value over measurements, we have

$$
\begin{aligned}
E_{t|sr}&\,\mathrm{Var}_{z|srt}\big(\hat{X}_{\mathrm{RHD}}\big) \\
={}& \sum_{i\in s}\phi(i)[1-\phi(i)]R_i\big[(X_i+B_i)^2+\sigma_{ei}^2\big]\Bigg(\sum_{j\in s}\frac{1-R_j}{\pi_j}\Bigg)^2 \\
&- \sum_{i\neq i'\in s}\phi(i)\phi(i')R_iR_{i'}\big[(X_i+B_i)(X_{i'}+B_{i'})+\sigma_{eii'}\big]\sum_{j\in s}\frac{1-R_j}{\pi_j^2}.
\end{aligned}
$$

Substituting for $\phi(i)$ yields

$$
\begin{aligned}
E_{t|sr}&\,\mathrm{Var}_{z|srt}\big\{\hat{X}_{\mathrm{RHD}}\big\} \\
={}& \frac{\left[\displaystyle\sum_{i\in s}\frac{R_i(X_i+B_i)^2+\sigma_{ei}^2}{\pi_i}\right]\left(\displaystyle\sum_{j\in s}\frac{1-R_j}{\pi_i}\right)^2}{\Sigma_{i\in s}R_i/\pi_i} \\
&- \frac{\left[\displaystyle\sum_{i\in s}\frac{R_i(X_i+B_i)^2+\sigma_{ei}^2}{\pi_i^2}\right]\left(\displaystyle\sum_{j\in s}\frac{1-R_j}{\pi_j}\right)^2}{\big(\Sigma_{i\in s}R_i/\pi_i\big)^2} \\
&- \frac{\left\{\displaystyle\sum_{i\neq i'\in s}\frac{R_i}{\pi_i}\frac{R_{i'}}{\pi_{i'}}\big[(X_i+B_i)(X_{i'}+B_{i'})+\sigma_{eii'}\big]\right\}\left(\displaystyle\sum_{j\in s}\frac{1-R_j}{\pi_j^2}\right)}{\big(\Sigma_{i\in s}R_i/\pi\big)^2}.
\end{aligned}
$$

$$(12.4)$$

The three terms in Eq. (12.4) are complex products of the form uv^2/w, uv^2/w^2, and uv/w^2, respectively. One method of computing this variance would be to substitute the expectation over outcomes of the sampling and response processes of the linearized values. However, we will use another approximation that simply substitutes the expected values over the sampling and response processes for the various terms in Eq. (12.4). This substitution gives the following for the imputation variance:

$$\text{imputation variance} = \frac{\left(\sum_{i=1}^{N} p_i Y_i^2 + \sum_{i=1}^{N} p_i \sigma_{ei}^2\right)\left[\sum_{i=1}^{N}(1-p_i)\right]^2}{\sum_{i=1}^{N} p_i}$$

$$- \frac{\left(\sum_{i=1}^{N} p_i Y_i^2/\pi_i + \sum_{i=1}^{N} p_i \sigma_{ei}^2/\pi_i\right)\left[\sum_{i=1}^{N}(1-p_i)\right]^2}{\left(\sum_{i=1}^{N} p_i\right)^2}$$

$$- \frac{\left[\sum_{i \neq 1}^{N} \pi_{ii'}/\pi_i \pi_{i'} p_i p_{i'}(Y_i Y_{i'} + \sigma_{eii'})\right]\left[\sum_{i=1}^{N}(1-p_i)/\pi_i\right]}{\left(\sum_{i=1}^{N} p_i\right)^2}.$$

　　Despite the apparent complexity of the formulations above, they represent a considerable simplification of what may occur in actual practice. In many surveys, a single weighting class is not used; rather, several weighting classes are formed and the adjustments are done within each weighting class. In addition, the derivation above assumes that the elements in the sample respond independently. This situation may not be true when interviewers are used because the approach of a particular interviewer may result in a covariance between the events of responding and not responding for the elements assigned to the same interviewer. In an area household sample, the number of callbacks the interviewer makes to a particular segment may cause the elements residing in that segment to have a positive covariance. An especially friendly or unfriendly reception in a certain neighborhood may influence the interviewer's approach to other elements in the neighborhood. Platek and Gray (1983) present a similar weighting class model in which more than one weighting class is used and in which $\text{Cov}(R_i, R_{i'})$ is not assumed to be zero. Platek and Gray also present a model for imputed values (they call it *duplication*); however, it is somewhat different from the current model.

12.2　A MORE GENERAL MODEL

In the content of the preceding section, we illustrate the nature of survey error models that consider sampling and measurement errors in conjunction with the assumption that participation in the survey has a random component and that some method is used for adjusting for missing data. Each model presented was produced by creating a set of specific assumptions concerning the nature of assumed survey errors and of efforts made to reduce or remove the effects of the assumed errors. Our objective in this section is to present a more general error model for estimating totals of measurements, subject to the four types of survey errors presented in this book: frame, sampling, nonresponse, and measurement errors.

The model we develop is of sufficient generality that the components of existing models can be extracted as special cases. The model also illustrates the basic components of a total error model. Any total error model developed must consider components of this general nature.

Suppose that the object of a survey is to estimate the total of some measure in a target population consisting of N elements. We use the symbol U_i to denote the ith member of this population. The true measure for population element U_i is X_i $(i = 1, 2, \ldots, N)$ so that we wish to estimate

$$X = \sum_{i=1}^{N} X_i.$$

Realities of the survey operation prevent us from measuring X_i without error. Instead, we must use

$$M_i = R_i(X_i + \varepsilon_{1i}) + (1 - R_i)(X_i + \varepsilon_{0i}), \tag{12.5}$$

where

$$R_i = \begin{cases} 1 & \text{if population element } U_i \\ & \text{responds, when selected} \\ 0 & \text{if otherwise,} \end{cases}$$

ε_{1i} is the error in measuring X_i if the population element U_i responds, and ε_{0i} is the error in imputing a value for X_i if population element U_i fails to respond. We refer to the two types of measurement error, ε_{1i} and ε_{0i}, as *elementary response error* and *elementary imputation error*, respectively. Thus we have specified two stochastic sources of error, the first because of our inability to obtain a perfect measure of X_i and the second because of our inability to obtain a response (leading to direct survey measurement) from selected members of the target population. Measurement, imputation, and nonresponse are stochastic in the sense that the outcomes of ε_{1i}, ε_{0i}, and R_i are viewed as random though not randomized by a synthetic device.

A form of synthetic randomization produces the third general stochastic source of error, that from sampling. It is assumed that some type of probability sampling is used to select a set of elements from which statistical inference to the target population is made. The sample is obtained from a frame of M elements whose kth member is denoted by U_k. Randomization, in the form of random number tables or computer-generated numbers, is used so that marginal and joint selection probabilities for all members of the sampling frame can be determined. No other restrictions on the manner of sample selection are made, however. The symbol λ_k is used to refer to the randomized outcome measuring the number of times frame element U_k is selected in the sample. Thus λ_k can assume any integral value between 0 and M.

The fourth source of survey error occurs when the sampling frame, the list of frame elements used for sample selection, is imperfect. Imperfection in the frame may be thought of as the presence of certain kinds of matching problems between the set of M frame elements and N population elements, where "elements" in both instances refers to the elementary units of observation in the survey. The match-up of frame and population is established by the indicator variable,

$$\theta_{ik} = \begin{cases} 1 & \text{if population element } U_i \text{ is linked to} \\ & \text{frame element } U_k \\ \\ 0 & \text{otherwise,} \end{cases}$$

which is viewed as a deterministic rather than stochastic outcome, since sampling is often conditioned on an existing sampling frame thus making the linkage defined by θ_{ik} predetermined and fixed in practice.

Three types of frame problems can be identified by noting different combinations of the θ_{ik}. One is undercoverage, when the population element U_i is absent from the frame, in which case $\theta_{ik} = 0$; $k = 1, 2, \ldots, M$. A second is multiplicity when the population element U_i appears more than once on the frame,

$$\theta_i = \sum_{k=1}^{M} \theta_{ik} > 1.$$

Finally, overcoverage occurs when frame element U_k is not a member of the target population,

$$\theta_k = \sum_{i=1}^{N} \theta_{ik} = 0.$$

A number of expected values are needed over the three stochastic sources of error described above.

$E_{t|rs}(\cdot)$ denotes expectation over repeated applications of the measurement process following a fixed protocol and conditioned on an outcome of the response and sampling processes.

$E_{r|s}(\cdot)$ denotes expectation over repeated applications of a fixed protocol for obtaining survey response, conditioned on an outcome of the sampling process.

$E_s(\cdot)$ denotes expectation over all possible applications of the sampling protocol following a specified sampling design.

Using these definitions, a number of important expectations are listed below:

$E_{t|rs}(\varepsilon_{1i}) = B_{1i}$ elementary measurement bias

$E_{t|rs}(\varepsilon_{1i} - B_{1i})^2 = \sigma_{e1i}^2$ elementary measurement variance

$E_{t|rs}(\varepsilon_{0i}) = B_{0i}$ elementary imputation bias

$E_{t|rs}(\varepsilon_{0i} - B_{0i})^2 = \sigma_{e0i}^2$ elementary imputation variance

$E_{t|rs}[(\varepsilon_{1i} - B_{1i})(\varepsilon_{1i'} - B_{1i'})] = \sigma_{e1ii'}$ elementary measurement covariance

$E_{t|rs}[(\varepsilon_{0i} - B_{0i})(\varepsilon_{0i'} - B_{0i'})] = \sigma_{e0ii'}$ elementary imputation covariance

$E_{t|rs}[(\varepsilon_{1i} - B_{1i})(\varepsilon_{0i'} - B_{0i'})] = \sigma_{e1i'0i'}$ elementary covariance between measurement and imputation errors

$E_{r|s}(R_i) = p_i$ elementary response probability

$E_{r|s}(R_i R_{i'}) = p_{ii'}$ elementary joint response probability

$E_s(\lambda_k) = \eta_k$ expected number of times the kth frame element will be selected in the sample

$E_s[(\lambda_k - \eta_k)^2] = \sigma_k^2$ elementary selection variance

$E_s[(\lambda_k - \eta_k)(\lambda_1 - \eta_1)] = \sigma_{k1}$ elementary selection covariance

We now consider the various error components of the estimator of X,

$$\hat{X} = \sum_{i=1}^{N} \sum_{k=1}^{M} \frac{M_i \lambda_k \theta_{ik}}{\eta_k}, \tag{12.6}$$

which is a generalized form of the estimator Horvitz and Thompson (1952) propose with no attempt to accommodate frame, nonresponse, or measurement error problems. When we have no frame problems, random sampling without replacement, nonresponse, or measurement error, Eq. (12.6) reduces to the more familiar form of the Horvitz–Thompson estimator,

$$\hat{X} = \sum_{i=1}^{n} \frac{X_i}{\pi_i},$$

where n is the sample size and π_i is the selection probability for the ith element selected in the sample. Assuming that the outcome of the stochastic processes during sampling, response solicitation, and measurement are unrelated, we can partition the mean square error of $\hat{X}$ as

$$
\begin{aligned}
\mathrm{MSE}(\hat{Y}) &= \xi_3[\hat{X} - X]^2 \\
&= \xi_3[\hat{X} - \xi_1(\hat{X})]^2 + \xi_3[\xi_1(\hat{X}) - \xi_2(\hat{X})]^2 \\
&\quad + \xi_3[\xi_2(\hat{X}) - \xi_3(\hat{X})]^2 + [\xi_3(\hat{X}) - X]^2, \quad (12.7)
\end{aligned}
$$

where, in general,

$$
\begin{aligned}
\xi_1(\cdot) &= E_{t|rs}(\cdot) \\
\xi_2(\cdot) &= E_{r|s} E_{t|rs}(\cdot) \\
\xi_3(\cdot) &= E_s E_{r|s} E_{t|rs}(\cdot).
\end{aligned}
$$

The first three terms of Eq. (12.7) correspond to variable error components from measurement, nonresponse, and sampling, respectively. The fourth component is the square of the total bias of $\hat{X}$. Thus we have one variable error component corresponding to each assumed stochastic source of error and a fourth component reflecting the total bias resulting from each stochastic source. This broad partitioning of total survey error is similar to the one Kish (1965, Sect. 13.2) suggests.

Let us consider first the variable error associated with measurement. Noting that

$$
\xi_1(\hat{X}) = \sum_{i=1}^{N} \sum_{k=1}^{M} [X_i + R_i B_{1i} + (1 - R_i)B_{0i}]\lambda_k \frac{\theta_{ik}}{\eta_k},
$$

we have

$$
\begin{aligned}
&\xi_3\big[\hat{X} - \xi_1(\hat{X})\big]^2 \\
&= \xi_3\left\{ \sum_{i=1}^{N} \sum_{k=1}^{M} [R_i(\varepsilon_{1i} - B_{1i}) + (1 - R_i)(\varepsilon_{0i} - B_{0i})]\lambda_k \frac{\theta_{ik}}{\eta_k} \right\}^2 \\
&= \sum_{i=1}^{N} \theta_i^2(1 + \mu_{ii})\big[p_i \sigma_{e1i}^2 + (1 - p_i)\sigma_{e0i}^2 \big] + \sum_{i \neq i'}^{N} \theta_i \theta_{i'}(1 + \mu_{ii'}) \\
&\quad \times \big\{ [p_{ii'}\sigma_{e1ii'} + (1 - p_i - p_{i'} + p_{ii'})\sigma_{e0ii'}] + 2(p_i - p_{ii'})\sigma_{e1i0i'} \big\}, \\
&\hspace{10cm} (12.8)
\end{aligned}
$$

where

$$\mu_{ii'} = E_s \left[\sum_{k=1}^{M} (\lambda_k - \eta_k) \frac{\theta_{ik}}{\theta_i \eta_k} \right] \left[\sum_{k=1}^{M} (\lambda_k - \eta_k) \frac{\theta_{i'k}}{\theta_{i'} \eta_k} \right].$$

Components of the error model proposed by Platek et al. (1977) can be produced as a special case of the results of Eq. (12.8). This particular model assumes a complete enumeration (no sampling error), no correlation among elementary measurement errors, and no frame problems, in which case $\mu_{ii'} = 0$ and $\sigma_{e1ii'} = 0$ for all population elements U_i and $U_{i'}$ ($i \neq i'$) and $\theta_i = 1$ for all population elements U_i. Thus we have

$$\xi_3 \left[\hat{X} - \xi_1(\hat{X}) \right]^2$$

$$= \sum_{i=1}^{N} p_i \sigma_{e1i}^2 + \left[\sum_{i=1}^{N} (1 - p_i) \sigma_{e0i}^2 + \sum_{i \neq 1}^{N} (1 - p_i - p_{i'} - p_{ii'}) \sigma_{e0ii'} \right]$$

$$+ 2 \sum_{i \neq i'}^{N} (p_i - p_{ii'}) \sigma_{e1i0i'}$$

$$= (\text{measurement variance}) + (\text{imputation variance})$$

$$+ 2(\text{covariance between measurement and imputation errors}).$$

$$(12.9)$$

A term to yield a correlated measurement variance,

$$\sum_{i \neq 1'}^{N} p_{ii'} \sigma_{e1ii'},$$

appears in Eq. (12.9) when the assumption about $\sigma_{e1ii'} = 0$ is removed.

Moving on to the second basic component of MSE($\hat{X}$) from variable error associated with nonresponse, we note that

$$\xi_2(\hat{X}) = \sum_{i=1}^{N} \sum_{k=1}^{M} [X_i + p_i B_{1i} + (1 - p_i) B_{0i}] \lambda_k \frac{\theta_{ik}}{\eta_k}.$$

Therefore,

$$\xi_3\left[\xi_1(\hat{X}) - \xi_2(\hat{X})\right]^2 = \xi_3\left[\sum_{i=1}^{N}\sum_{k=1}^{M}(B_{1i} - B_{0i})(R_i - p_i)\lambda_k\frac{\theta_{ik}}{\eta_k}\right]^2$$

$$= \sum_{i=1}^{N}\theta_i^2(B_{1i} - B_{0i})^2(1 + \mu_{ii})p_i(1 - p_i)$$

$$+ \sum_{i \neq i'}^{N}\theta_i\theta_{i'}(B_{1i} - B_{0i})(B_{1i'} - B_{0i'})(p_{ii'} - p_ip_{i'}).$$

$$(12.10)$$

Referring again to the assumptions of the Platek et al. model (1977), we note that Eq. (12.10), subject to the additional assumption that $p_{ii'} = p_ip_{i'}$ and the other assumptions mentioned above, reduces to

$$\xi_3\left[\xi_1(\hat{X}) - \xi_2(\hat{X})\right]^2 = \sum_{i=1}^{N}(B_{1i} - B_{0i})^2p_i(1 - p_i)$$

$$= \text{nonresponse variance.} \qquad (12.11)$$

The third variable error component of the general error model for $\hat{X}$ is from the sampling process. Noting this time that

$$\xi_3(\hat{X}) = \sum_{i=1}^{N}\theta_iZ_i$$

where $Z_i = X_i + p_iB_{1i} + (1 - p_i)B_{0i}$, we have

$$\xi_3\left[\xi_2(\hat{X}) - \xi_3(\hat{X})\right]^2 = \left[\sum_{i=1}^{N}\sum_{k=1}^{M}Z_i(\lambda_k - \eta_k)\frac{\theta_{ik}}{\eta_k}\right]^2$$

$$= \sum_{i=1}^{N}\theta_i^2Z_i^2\mu_{ii'} + \sum_{i \neq i'}^{N}\theta_i\theta_{i'}Z_iZ_{i'}\mu_{ii'}. \qquad (12.12)$$

Assuming no frame problems, random sampling without replacement, no response, and no response measurement error, Eq. (12.12) becomes the well-known result attributable to Horvitz and Thompson (1952),

$$\xi_3\left[\xi_2(\hat{X}) - \xi_3(\hat{X})\right]^2 = \sum_{i=1}^{N}\pi_i(1 - \pi_i)\frac{X_i^2}{\pi_i^2} + \sum_{i \neq i'}^{N}(\pi_{ii'} - \pi_i\pi_{i'})\frac{X_iX_{i'}}{\pi_i\pi_{i'}},$$

$$(12.13)$$

where π_i is the probability of selecting population element U_i and $\pi_{ii'}$ is the joint probability of selecting population element U_i and $U_{i'}$.

The absence of frame problems implies that $\lambda_i = 1$ and, without loss of generality $\lambda_{ii} = 1$ for each population element U_i corresponding to frame element U_k. Random sampling without replacement implies that λ_i can equal either 0 or 1, so that $E_s(\lambda_i) = \pi_i$ and $E_s(\lambda_i \lambda_{i'}) = \pi_{ii'}$. Thus, under these conditions, $\mu_{ii'} = (\pi_{ii'} - \pi_i \pi_{i'})/\pi_i \pi_{i'}$. Moreover, complete response and no measurement error implies that $Z_i = X_i$.

The final component of the $\text{MSE}(\hat{X})$ is the square of the total bias of $\hat{X}$. This total bias can, in turn, be expressed as the sum of three bias subcomponents,

$$\xi_3(\hat{X}) - X = \sum_{i=1}^{N} (\theta_i Z_i - X_i)$$

$$= \sum_{i=1}^{N} \theta_i p_i B_{1i} + \sum_{i=1}^{N} \theta_i (1 - p_i) B_{0i} + \sum_{i=1}^{N} (\theta_i - 1) X_i \quad (12.14)$$

$$= \text{response bias} + \text{nonresponse bias} + \text{frame bias}.$$

Although the second term of Eq. (12.14) is usually called the *nonresponse bias*, a term such as *imputation bias* might be more appropriate, since it is a weighted sum of elementary imputation biases, B_{0i}. It is also of interest to note that when one assumes a deterministic view of nonresponse in which elementary response probabilities are either zero or 1, the nonresponse bias term in Eq. (12.14) becomes the well-known result

$$- \sum_{i=1}^{N} (1 - p_i) X_i = -X_0,$$

where X_0 is the total of X_i for members of the nonrespondent stratum since $p_i = 0$ for members of the nonrespondent stratum and $p_i = 1$ for members of the respondent stratum under the deterministic view. Also, we note for each population element U_i that $\theta_i = 1$ because frame problems do not exist and that $B_{0i} = -X_i$ because no imputation is used. When frame problems are absent we also note that the response bias term becomes $\sum_{i=1}^{N} p_i B_{1i'}$, the same term that Platek et al. (1977) suggest.

Before proceeding with some extensions of the model presented above, we must take two important observations concerning the components of the general error model of Eq. (12.7). First, we note that the variable error component associated with any given stochastic source of error is affected by the other sources of error. For example, we see that the variable error from nonresponse in Eq. (12.11) is affected by terms involving frame problems (i.e., θ_i, $\theta_{i'}$, θ_{ii}, and $\theta_{ii'}$), sampling (i.e., $\mu_{ii'}$ and μ_{ii}), and measurement (i.e., B_{1i}, B_{0i}, $B_{1i'}$, and $B_{0i'}$), as well as the principal terms involving elementary variance $[p_i(1 - p_i)]$ and covariance $(p_{ii'} - p_i p_{i'})$ of the stochastic random variables R_i because of nonresponse. Second, we note that the total bias is a sum of individual biases because of frame problems, nonresponse, and measurement but not sampling per se. Also, absent is any estimation bias attributable to the nature of the estimator one chooses. Many survey estimators do exhibit a degree of estimation bias. The size of estimation bias, especially for a large sample, is usually small relative to the size of other error components in these cases, however.

Several extensions to the general model presented in this section are possible, although none consider sources of error beyond those already presented: namely, frame, sampling, nonresponse, and measurement error. One extension would be to isolate frame, sampling, and nonresponse errors associated with each of several stages of sampling. Each stage clearly contributes to errors arising from frame development, sample selection, and nonresponse, the later in instances where the sampling unit is an individual, an institution, or an organization from which an agreement to participate must be sought before survey measurements can be made. Another extension would be to expand the elementary measurement error terms, ε_{1i} and ε_{0i}, and include both random and fixed effects attributable to specific failures in the survey process. For example, we might view ε_{1i} as consisting of a sum of contributions from specific error sources, say due to problems with questionnaire wording, respondent recall, interviewer differences, recording responses on the questionnaire, coding, and keypunching. Similarly, we might consider the contributions to ε_{0i} from a combination of several nonresponse accommodation procedures used in a survey, such as nonresponse adjustments, machine imputation of missing data items, and Politz–Simmons estimators (see Chapter 8).

A final extension of the model is to consider population parameters other than totals. It can be demonstrated easily that the same four components of Eq. (12.7), as well as corresponding subcomponents presented in Eqs. (12.8), (12.9), (12.10), and (12.12), emerge for any linear estimator or Taylorized linear approximation to a nonlinear estimator. This result is especially encouraging since we seek error models that have comparable structure regardless of the type of estimator.

To illustrate the impact of a Taylor linearized approximation on a nonlinear estimator, consider the parameter $\phi(\mathbf{X})$, a nonlinear function $\phi(\cdot)$ of

$$\mathbf{X} = (X_1, X_2, \ldots, X_t, \ldots, X_T)',$$

a T-dimensional vector of totals. If our estimator of $\phi(\mathbf{X})$ takes the form $\phi(\hat{\mathbf{X}})$, where

$$\hat{\mathbf{X}} = \left(\hat{X}_1, \hat{X}_2, \ldots, \hat{X}_t, \ldots, \hat{X}_T\right)'$$

is the corresponding vector of estimators of the members of $\mathbf{X}$, then for J

stochastic sources of error,

$$\mathrm{MSE}\big[\phi(\hat{\mathbf{X}})\big] = \xi_J\big[\phi(\hat{\mathbf{X}}) - \phi(\mathbf{X})\big]^2 \simeq \xi_J(\mathbf{d}'\mathbf{k})^2,$$

where $\mathbf{d} = \hat{\mathbf{X}} - \mathbf{X}$ and

$$\mathbf{k} = (k_1, k_2, \ldots, k_t, \ldots, k_T)'$$

is a T-dimensional vector of constants whose tth member is the first partial derivative of $\phi(\mathbf{X})$ with respect to X_t. We can partition $\mathrm{MSE}[\phi(\hat{\mathbf{X}})]$ into J variable error components (labeled in the reverse order of their occurrence in the survey) plus a total squared bias component by noting that

$$\mathbf{d} = \hat{\mathbf{X}} \pm \sum_{j=1}^{J} \hat{\mathbf{X}}^{(j)} - \mathbf{X},$$

where for $j = 1, 2, \ldots, J$,

$$\hat{\mathbf{X}}^{(j)} = \big[\xi_j(\hat{X}_1), \xi_j(\hat{X}_2), \ldots, \xi_j(\hat{X}_t), \ldots, \xi_j(\hat{X}_T)\big]'.$$

The operator "$\pm$" refers to adding *and* subtracting the reference term, and in general,

$$\xi_j(\cdot) = E_{j|j+1,\ldots,J} E_{j-1|j,j+1,\ldots,J} \ldots E_{1|2,3,\ldots,J}(\cdot),$$

consisting of a sequence of conditional expectations over certain variable error components. What we observe in the case of three stochastic sources of error from sampling, nonresponse, and measurement plus a fourth source from frame problems is a four-component model similar to Eq. (12.7) with similar subcomponents (e.g., response variance, nonresponse variance, frame bias), only here expressed as functions of $\mathbf{k}$ and four $T \times T$ matrices: Λ_1, Λ_2, Λ_3, and Λ_4 whose stth members are

$$\xi_3\big[\hat{X}_s - \xi_1(\hat{X}_s)\big]\big[\hat{X}_t - \xi_1(\hat{X}_t)\big],$$

$$\xi_3\big[\xi_1(\hat{X}_s) - \xi_2(\hat{X}_s)\big]\big[\xi_1(\hat{X}_t) - \xi_2(\hat{X}_t)\big],$$

$$\xi_3\big[\xi_2(\hat{X}_s) - \xi_3(\hat{X}_s)\big]\big[\xi_2(\hat{X}_t) - \xi_3(\hat{X}_t)\big],$$

and

$$\xi_3\big[\xi_3(\hat{X}_s) - X_s\big]\big[\xi_3(\hat{X}_t) - X_t\big],$$

respectively. We note that an error model constructed from a linearized approximation of a nonlinear function of estimated totals will not contain a term for estimation bias, since sources of stochastic error aside,

$$\xi_J\left[\phi(\hat{\mathbf{X}})\right] \neq \phi\left[\xi_J(\hat{\mathbf{X}})\right].$$

Besides the effect of the linear approximation, this result follows since no member of $\hat{\mathbf{X}}$ has estimation bias.

The discussion above has illustrated the nature of total error models that would accommodate each of the four major sources of error we have identified. As indicated, these models are preliminary steps that must be more highly specified in an actual survey situation.

Given the complexity of total error models, we can see that there is a great need for accumulating a systematic body of knowledge as to the nature of errors in survey data. Developing a total error model for a survey can be an expensive and complex process. Often these models—or partial models—are developed during the design phase of a survey for use with an appropriate cost model in a formal survey optimization study. One can easily spend several hundred thousand dollars developing and quantifying a total error model and carrying out a design optimization. Such expense is justified for a large-scale survey that will continue for a number of years and cost millions of dollars.

However, the results from complex total error models must be synthesized into simpler models that can be applied to surveys of smaller magnitude. Simplified models summarizing the results from more complex investigations are needed for more modest surveys. Some summary measures that reflect the relationship among alternative design are needed—overall measures that can be used in a manner similar to the way we use the design effect to summarize the relative efficiency of alternative sample designs.

Compendium of Nonsampling
Error Terminology

Definitions of some of the terms discussed in the text are presented in this compendium. Our purpose is to indicate how the terminology linked to various sources of survey error are defined by individuals and groups that have contributed to the research literature. The set of definitions given, while far from complete, is intended to indicate the diversity of interpretations one finds for some key concepts. The major categories covered below correspond, as indicated in parentheses, to earlier sections where these terms were discussed.

FRAMES AND FRAME ERRORS

Definitions of Frames

By the term "frame" we understand the list of units. These lists are used in sample surveys for the selection of samples. (Zarkovich, 1966:97)

Before selecting the sample, the population must be divided into parts which are called sampling units, or units. These units must cover the whole of the population and they must not overlap, in the sense that every element of the population belongs to one and only one unit. ... The construction of this list of units, called a *frame*, is often one of the major practical problems. (Cochran, 1963:7)

... we are using the term list in a broad sense, to cover sources such as card files, reels of magnetic tape or microfilm, punch cards, forms, and other data storage devices.

We should also like to distinguish between units of analyses, reporting units, and sampling units. The *units of analysis* are those which are treated as units in the survey tabulations, ... The *reporting unit* is often the same as the unit of analysis, but may be different, e.g., a company may report for all or some of its stores, one members of a household may report for all members or all over a certain age, etc. The *sampling units* are those units to which a probability selection procedure is directly applied. ...

Broadly speaking, all samples we use in our [U.S. Census Bureau] survey programs are list samples, since even when land areas are the ultimate sampling units, we normally start with some kind of list of areas. ...

Consider a list with units

$$l_1, l_2, \ldots l_m$$

and a target population with reporting units

$$t_1, t_2, \ldots t_n$$

Establish *rules of association* between the listed units and those in the target population in such a way that the selection of listed units, l_i, with known probabilities leads directly to the selection of reporting units t_j, also with known probabilities. (Hansen et al., 1963:498–501)

... all material which describes the components of the target population (or an adequate part of that population) in such a way that it is possible to determine in the course of the survey individual components and to delimit them from other components. (Szameitat and Schäffer, 1963:518)

A sampling frame is a convenient listing or arrangement of the elements in the universe so that one can assure that probabilities of selection are as they should be for the sampling type devised. (Jessen, 1978:43)

Preferred Frame

Of particular use for a sample survey is a frame which not only provides a description of the individual components but also supplies additional information, such as size of the components or their inclusion into specific parts of the target population. If there are several frames of the same level available for a target population, that one will generally be used which contains the most exact information and can, at the same time, be easily applied. (Szameitat and Schäffer, 1963:519)

Serial List Frame

One of the simplest frames is a listing of each element in the universe of interest, numbering the entries serially from 1 to N. (Jessen, 1978:161)

Mixable Objects

In this case the frame units (FUs) are physical elements that are more or less alike and may be physically mixed to simulate a randomization of position or order. (Jessen, 1978:161)

Count Frame

Neither the list nor the numbering is necessary, however, for assuring adequate control of the sample section. If the limits are ordered in some identifiable manner, than all we need to do is count the units to obtain particular "numbered" units. (Jessen, 1978:162)

Area Sampling

When the entire area containing the population under study is subdivided into smaller area segments and each element in the population is associated with one and only one such area segment, the procedure is called area sampling. (Sukhatme and Sukhatme, 1970:222)

Area Sample

In an area sample the sampling units are land areas and the reporting units in the sample are identified through geographic rules in the field. (Raj, 1968:219)

Area Frames: Grid and Plot Types

... two types of area sampling units can be distinguished: (i) that in which the area contains a cluster of elements, which we shall call the grid type, and (ii) that in which the area itself is measured, the plot type. (Jessen, 1978:171)

Multiplicity

The multiplicity rule distributes population elements among enumeration units such that every element is linked to at least one enumeration unit and possibly some elements are linked to more than one enumeration unit. ... The multiplicity estimator does, however, require the multiplicity (i.e., the number of links to enumeration units) of each element in the sample. (Sirken and Levy, 1974:68)

Undercoverage

The error in estimates that results from failure to include in the frame all units belonging to the defined population. (Deighton et al., 1978:11)

A type of nonsampling error that results form either failure to include all appropriate sampling units in the frame or failure to include some of the units already on the frame. (Deighton et al., 1978:11)

Noncoverage

Noncoverage denotes failure to include some units, or entire sections, of the defined survey population in the actual operational sampling frame. (Kish, 1965:528)

Missing Elements, Inadequate Frame, Incomplete Frame

First is the problem of missing elements, which means that elements that should be included in the population are not on the sampling frame. ... In the first place the frame may be inadequate in the sense that it does not cover the whole of the population to be surveyed. ... The second way in which the missing elements problem occurs is when the frame is incomplete. By this is meant that some of the population members who are suppose to be on it are in fact not on it. (Moser and Kalton, 1972:154, 155)

[These definitions sound very similar—the authors are using *inadequate* to mean missing entirely some large subclass of the target population; *incomplete* refers to individually missing elements.]

Incomplete Coverage

The accurary of the estimate is also affected by errors arising from other causes such as incomplete coverage. (Sukhatme and Sukhatme, 1970:381)

Coverage Errors

Ascertainment errors may be further sub-divided into (i) coverage errors owing to over or under-enumeration of the population or sample, resulting from duplication or omission of units, and from nonresponse, ... (Murthy, 1967:452)

... errors of coverage, which may involve either the failure to identify and count units which should have been included or the erroneous inclusion of units not properly part of a census or survey. (Eckler and Pritzker, 1951:7)

Noncoverage refers to the negative errors of failure to include elements that would properly belong in the sample. There occur also positive errors of *overcoverage*, due to the inclusion in the sample of elements that do not belong there. The term *gross coverage error* refers to the sum of the absolute values of noncoverage and overcoverage error rates. The *net noncoverage* refers to the excess of noncoverage over overcoverage, and it is their algebraic sum. (Kish, 1965:259)

[Kish is careful to distinguish between *nonresponse* and *noncoverage*. Other authors appear to allow the term *coverage errors* to include both terms.]

Target Population

Here it is useful to distinguish between the population for which the results are required, the target population, and the population actually covered, the survey population. (Moser and Kalton, 1972:53)

The objectives of the survey should define the population the survey is intended to cover. But practical difficulties in handling certain segments of the population may point to their elimination from the scope of the survey . . . . Thus the target population would generally be different from the population actually sampled. (Raj, 1968:23)

The population to be sampled (the sampled population) should coincide with the population about which the information is wanted (the target population). (Cochran, 1963:6)

Sampled Population

The aggregate of all sampling units delimited by the frame defines a population which, in order to distinguish it from the target population, will be referred to as the "sampled population." (Szameitat and Schäffer, 1963:519)

Target Universe

Normally one wishes to have the universe of interest or target universe completely represented in the frame universe. . . . The truncated universe may be of sufficient interest to justify the investigation. (Jessen, 1978:162)

Out-of-Scope Units

There is probably no list of any size used for sampling in our program which does not contain some superfluous units, i.e., listed units for which the rules of association do not lead to any reporting units in the target population. Such units are sometimes referred to as "out-of-scope" units. (Hansen et al., 1963:503)

Foreign Elements

Another problem with sampling frames arises from blanks or foreign elements: an element is given a listing but is not a member of the survey [target] population. (Moser and Kalton, 1972:156)

Empty Listings, Blanks

. . . problems occur when the target population is a subclass M of the entire listed frame population N. Listings are empty if they contain no elements of the target population: either they are blanks containing no elements of any kind, or they contain only foreign elements. (Kish, 1965:386)

Boundary Problems

Although area frames can be made to be very rigorous conceptually, in practice estimates based on area samples sometimes have been found to have serious biases. These biases are usually ascribable to faulty work by investigators accounting for elements in the sample area because of indefiniteness about where the boundaries actually are. (Jessen, 1978:173)

Net Coverage Error

The net coverage error equals the number of counted units minus the number that should have been counted. [This definition is expressed as a formula in the article. Components of the definition implicitly defined in the formula.]

Coverage error in an estimate results from the failure to include in the frame all units belonging to the defined population; for example to include specific units in the conduct of a survey. This is referred to as undercoverage. Coverage error can also take the form of the inclusion of some units erroneously either because of a defective frame or because of inclusion of unspecified units or inclusion of specified units more than once, in the actual survey. This is referred to as overcoverage. (U.S. Bureau of the Census, 1979:2)

Duplicate Listings

Lists containing duplicates: Duplication occurs when more than one unit on a list leads to the same reporting unit. (Hansen et al., 1963:505)

Erroneous Inclusions (Deviation in Coverage)

Deviations in coverage:

(a) Reporting units belonging to the target population are not included in the sampled population

(b) Reporting units belonging to the large population are several times contained in the sampled population

(c) Reporting units contained in the sampled population do not belong to the target population. (Szameitat and Schäffer, 1963:519)

Element Multiplicity

The multiplicity of a population element with respect to a particular counting rule is the total number of sampling elements to which it is linked by the rule. (Casady and Sirken, 1980:601)

Deviations in Content

Deviations in content:

(a) The frame provides incorrect auxiliary information on reporting units

(b) Auxiliary information for reporting units is lacking in the frame. (Szameitat and Schäffer, 1963:519)

Out-of-Date or Inaccurate Frame

The composition of a frame may be accurate or inaccurate. The frame is called accurate if all units, as resulting from the adopted definition of the population, are listed once and once only. If some unit is listed twice or several times or is omitted from listings, the frame is called inaccurate. The frame is also inaccurate if it contains units which do not exist in the population. (Zarkovich, 1966:98)

The frame should not be merely a list of units, but should also contain other information: for one thing, the frame must contain sufficient details to ensure that each unit is identified with certainty; and for another it must contain the information required to enable the united to be located—in the case of individuals this might well be their home addresses. And clearly this information should be accurate. The requirement of accuracy usually means that the frame must be up-to-date for information on location in particular is likely to change with time. . . . an up-to-date frame is also of course more likely to be complete and contain fewer blanks. (Moser and Kalton, 1972:157)

THE NOTION OF NONRESPONSE

General Nonresponse

. . . failure to elicit response for units of analysis in a population or sample because of various reasons such as absence from home, failure to return questionnaires, refusals, omission of one or more entries in a form, vacant houses, etc. (U.S. Bureau of the Census, 1975:50)

Unit Nonresponse

In sample surveys, the failure to obtain information from a designated individual for any reason (death, absence, refusal to reply). . . . (Kendall and Buckland, 1960:200)

. . . many sources of failure to obtain observations (responses, measurements) on some elements selected and designated for the sample. (Kish, 1965:532)

... data are not collected for all units. (Zarkovich, 1966:145)

Many respondents do not reply and the available sample of returns is incomplete. (Sukhatme and Sukhatme, 1970:417)

... the failure to measure some of the units of the selected sample. (Cochran, 1977:359)

... occurs when no information is collected from a sample unit. (Kalton, 1983:4)

... units in the selected sample and eligible for the survey do not provide the requested information, or the provided information is unusable. (Madow et al., 1983a:3)

Item Nonresponse

... the type of nonresponse in which some questions, but not all, are answered for a particular unit. (U.S. Bureau of the Census, 1976:914)

... occurs when the unit cooperates in the survey but fails to provide answers to some of the questions. (Kalton, 1983:4)

... eligible units in the selected sample provide some, but not all, of the requested information, or the information provided for some items is unusable. (Madow et al., 1983a:3)

RELATIVE MEASURES OF UNIT NONRESPONSE

Response Rate

... the percentage of times an interviewer obtains interviews at sample addresses where contacts are made, i.e., $\dfrac{\text{number of interviews}}{\text{number of contacts}}$. (Hauck and Steinkamp, 1964:13)

The percentage of an eligible sample for whom information is obtained. For an interview survey the numerator of the formula is the number of interviews. The denominator is the total sample size minus noneligible respondents; that is, minus those not meeting the criteria for a potential respondent as defined for that particular study. (U.S. Department of Health, Education and Welfare, 1977:47)

... the proportion of the eligible respondents in the sample who were successfully interviewed. (Warwick and Lininger, 1975:294).

... should reflect the degree to which a researcher succeeds in obtaining the cooperation of all potential respondents included in the sample, [i.e.,] ... as the proportion of all sample members who are eligible to participate in the survey from whom a complete and usable set of data is collected, ...

$$\text{Response Rate} = \frac{\text{Number of completed interviews/questionnaires}}{\text{Number of eligible sample members}}.$$

(Kviz, 1977:265)

...the number of eligible sample units responding divided by the total number of eligible sampled units. (Bailar and Lanphier, 1978:51)

For Telephone Survey

The number of completed interviews divided by the number of completions, plus partials, plus callbacks, plus refusals, plus undetermined status, plus numbers never reached. (O'Neil et al., 1979:254)

...a summary measure (which should be used to designate the ratio of the number of interviews to the number of eligible units in the sample). The response rate is a measure of the result of all efforts, properly carried out, to execute the study. In determining a response rate, completion rates are used to evaluate the component steps. These component steps are then combined to form the response rate, ...

$$\text{Response rate} = \frac{\text{Number of completed interviews with reporting units}}{\text{Number of eligible reporting units in sample}}.$$

(CASRO Task Force on Completion Rates, 1982:3)

From a stochastic view of nonresponse:

$$\lambda_1 = \frac{\int_0^1 f(p)(1 - (1 - p))\, dp}{\int_0^1 f(p)\, dp},$$

where response probabilities (p) follow a beta distribution with the density, $f(p)$, and c is the number of allowable call attempts made to each population member. (Frankel and Dutka, 1983:75)

Nonresponse Rate

...the proportion of [nonresponding] individuals of the sample aimed at. ...(Kendall and Buckland, 1960:200)

The complement of [the] response rate. The numerator is those eligible respondents selected in a sample for whom information is not available because of refusals, not found at home, unavailable by reason of illness,

368 COMPENDIUM OF NONSAMPLING ERROR TERMINOLOGY

incompetence, language difficulty, etc. The denominator is the total number of eligible respondents initially selected for the sample. (U.S. Department of Health, Education, and Welfare, 1977:46)

... the ratio (l_0) of the number of nonrespondent eligible units $(n - n_1)$ to the number of eligible units (n) in the sample: $l_0 = 1 - (n_1/n)$.... Although the nonresponse rate (l_0) ... is used very widely, it is not appropriate for all survey designs. (Madow et al., 1983a:30)

Refusal Rate

In the sampling of human populations, the proportion of individuals who, though successfully contacted, refuse to give the information sought. The proportion is usually (and preferably) calculated by dividing the number of refusals by the total number of the sample which it was originally desired to achieve. (Kendall and Buckland, 1960:244)

For recording expected study results:

$$\text{Refusal rate} = \frac{\text{Refusals}}{\text{Refusals plus interviews}}.$$

(Hauck, 1974:2-151)

Operational definitions for household surveys where the first person encountered may refuse (household refusal) or where the person designated to complete the interview refuses (respondent refusal):

$$\text{Refusal rate}_1 = \frac{\text{Respondent refusals}}{\text{Respondent refusals} + \text{terminations} + \text{completed interviews}}$$

$$\text{Refusal rate}_2 = \frac{\text{Respondent refusals} + \text{household refusals}}{\text{All potential respondents and/or households that were contacted}}$$

$$\text{Refusal rate}_3 = \frac{\text{Respondent refusals} + \text{household refusals}}{\text{All potential respondents with known telephone numbers that were selected}}.$$

(Wiseman and McDonald, 1978:45)

Completion Rate

... the percentage of interviews in which the required information is given by the respondent. (Hauck and Steinkamp, 1964:13).

... the number of completed interviews divided by the number of completions plus partials, plus callbacks, plus refusals. (O'Neil, et al., 1979:254)

... the extent to which a task has been accomplished ... with the following meanings:

(a) $\text{C.R.} = \dfrac{\text{Number of completed interviews}}{\text{Number of contacts}}$

(b) $\text{C.R.} = \dfrac{\text{Number of respondents who answered all questions}}{\text{Number of respondents who started an interview}}$

(c) $\text{C.R.} = \dfrac{\text{Number of completed interviews}}{\text{Number of eligible units in sample}}$

(d) $\text{C.R.} = \dfrac{\text{Number of completed interviews}}{\text{Total units in sample (eligible plus ineligible)}}$

(e) $\text{C.R.} = \dfrac{\text{Number of completed interviews plus eligibles}}{\text{Total units in sample (eligible plus ineligible)}}$

(f) $\text{C.R.} = \dfrac{\text{Number of households that completed a census form}}{\text{Number of households that received a census form}}$

(g) $\text{C.R.} = \dfrac{\text{Number of telephone numbers that have been established to be residential, other working, or non-working}}{\text{Number of telephone numbers dialed}}$

(h) $\text{C.R.} = \dfrac{\text{Number of units for which eligibility status has been determined}}{\text{Number of units in sample}}$.

(CASRO Task Force on Completion Rates, 1982:2)

... the term or characteristic coverage rate, X_R/X_T, where X_R is the total of some variable, e.g., employment or acreage for eligible responding units in the sample, and X_T is the corresponding total for all eligible units in the population (sometimes for the sample). (Madow et al., 1983a:31)

BIAS ASSOCIATED WITH NONRESPONSE

Nonresponse Bias

... the bias resulting from confiding the survey analysis to the available (i.e., respondent) data. (Kalton, 1983:6)

MEASUREMENT ERROR DEFINITIONS

Survey Measurement

Definitions of measurement are presented in Chapter 9. Three specific definitions from the survey research literature are quoted below.

There exists a set {0} of N objects and a set of {Y} of numbers; as a special case, {Y} denotes the real numbers. Each object is assigned one and only one number; and two or more objects may be assigned the same number. (Dalenius, 1974:63)

(a) The value of the characteristic to be measured has been precisely defined for every population unit in a manner consonant with the uses to which the data are to be put.

(b) For any given population unit, this value exists and is unique. It will be called the *true value* of the characteristic for the population unit.

(c) There exist procedures to obtain information on the true value of the characteristic for every unit although these procedures may be costly and very difficult to use. (Sukhatme and Sukhatme, 1970:380)

TRUE VALUE / MEASUREMENT ERRORS

True Value Independent of Survey Procedures

The *individual true value* will be conceived of as a characteristic of the individual quite independent of the survey conditions .... Three criteria for definition of the true value are (the first two essential, the third useful but not essential):

(1) The true value must be uniquely defined.

(2) The true value must be defined in such manner that the purposes of the survey are met. For example, in a study of school children's intelligence, we would ordinarily not define the true value as the score assigned by the child's teacher on a given date although this might be perfectly satisfactory for some studies (if, for example, our purpose was to study intelligence as measured by teacher's ratings).

(3) Where it is possible to do so consistently with the first two criteria, the true value should be defined in terms of operations which can actually be carried through (even though it might be difficult or expensive to perform the operations). (Hansen et al., 1951:149)

Individual Response Error

...the difference between an individual observation and the true value for the individual. (Hansen et al., 1951:152)

Response Error

...nonsampling errors introduced during the course of data collection. (Hansen et al., 1951:147)

Individual Response

...the value obtained on a particular observation (e.g., the result obtained in a specific measurement or interview with a specific respondent at a given time). (Hansen et al., 1951:153)

Individual Response Bias

...the response error for a particular individual in a given survey will be though of as having an expected value (the individual response bias) and a random component of variation around that expected value. (Hansen et al., 1951:153)

Error in a Survey

...the difference between a survey estimate and the value which is estimated. (Hansen et al., 1951:147)

Related terms:

1. The *ideal goals* are the set of statistics that would be produced if all the survey requirements had been precisely defined and rigorously met.
2. The *defined goals* are a more operationally feasible set of statistics that could be achieved if the actual specifications of the survey are carried out precisely and rigorously.
3. The *expected values of the survey operations* are a set of statistics which would be conceived of as the expected value of a set of survey statistics y over a large number of independent replicates of the survey, all conducted under the same essential conditions. (Hansen et al., 1967:50)

[Sukhatme and Sukhatme give a definition that requires true values to be defined operationally—see the definition of *survey measurement* above.]

Observational or Response error

The discrepancy between the survey value and the corresponding true value. (Sukhatme and Sukhatme, 1970:381)

[Kish (1965) uses the Hansen et al. (1951) definition of true values and defines the following:]

Population Value and True Value

Both population value and true value refer to numerical expressions derived from the entire population. The difference between them arises from errors of observation. The true value would be obtained from all population elements, if the observations were not subject to error. The population value, also a function of all the observations, is subject to the same nonsampling errors as its sample estimate; it is a value that would be obtained if the entire population—rather than just a sample—were designated for observation under the actual survey conditions. Even that value is subject to fluctuation, since repeated measurements would obtain different values. (Kish, 1965:9)

TOTAL ERROR; ROOT-MEAN-SQUARE ERROR; BIAS

In sampling theory, a widely accepted model combines the variable error and the bias into the *total error*. This error factor, often called the *root mean square error* (RMSE), replaces the standard error (SE), and the mean square error (MSE) replaces the variance. This relationship between the variable sampling error and the bias is

$$E\left[\bar{y}_c - \bar{Y}_{\text{true}} \right]^2 = E\left[\bar{y}_c - E(\bar{y}_c) \right]^2 + \left[E(\bar{y}_c) - \bar{Y}_{\text{true}} \right]^2.$$

The expectation is taken over the distribution of all possible values of the estimator $\bar{y}_c$ (see 1.3). The mean square deviations of the possible sample results from the *true value* are analyzed into two components: (1) the mean square deviation of the variable errors around the average value $E(\bar{y}_c)$ of the survey design; plus (2) the square of the deviation of that average from the true value. Thus the total error, or the root mean square error, is

$$\text{Total Error} = \sqrt{\text{VE}^2 + \text{Bias}^2}.$$

When the variable errors, VE, are caused only be sampling errors, VE^2 equals the sampling variance; other components will be discussed in the next section. The deviation of the average survey value from the true population value is the *bias*; this is mostly caused by measurement biases. (Kish, 1965:510)

Bias

Bias refers to systematic errors that affect any sample taken under a specific survey design with the same constant error. (Kish, 1965:509)

[O'Muircheartaigh (1977b) uses the Hansen et al. (1951) definition of true value and defines the following:]

Response Variance

A measure of the effect of those variable errors whose absolute effects have expected value zero. (O'Muircheartaigh, 1977b:11)

[Murthy (1967:452) states that the true value "is to be conceived of as a characteristic of the unit independent of the survey conditions."]

Characteristics of True Value in Order to Be Useful in Practice

It should serve the purpose of the survey by being well defined and observable under reasonable conditions of survey relating to subject coverage, method of inquiry, survey period, reference period and method of tabulation. (Murthy, 1967:453)

DEFINITIONS BASED ON TRUE VALUES THAT DO NOT EXIST EXTERNAL FROM THE SURVEY PROCEDURES

[Zarkovich (1966) defines true values in the context of an *adopted system of work*: The adopted system of work is composed of] such factors as concepts and definitions, methods of collecting data, the units to be used in expressing the response, the tabulation program, the survey program, the wording of questions.... [It represents a fixed system of concepts, definitions, procedures, and operations that constitute the survey.] (Zarkovich, 1966:1)

[The true value is then defined as follows:] *True value*: the result that would be obtained in a particular survey if the adopted system of work is carried out correctly. It is an ideal result...obtained if the work is done in absolute conformity with the adopted system of work. (Zarkovich, 1966:1)

[Zarkovich (1966) says that there are two types of true values; individual true values for a given unit of a population and true values of certain aggregates, population totals, averages, coefficients of correlation, and so on. Zarkovich's definition of a measurement error uses the idea of *survey values*, "the results factually achieved." Error is then defined as follows:]

Error

The difference between the survey value and the corresponding true value. (Zarkovich, 1966:2)

[Deming (1960) does not accept the idea of a true value, and defines:]

Survey Technique

We may speak of the definitions, the methods of test, the questions, the methods of interviewing, the method of supervision, the treatment of nonresponse, etc. as the survey-technique. (Deming, 1960:62)

Preferred Survey Technique

There is no such things as a true value. ...we do have the liberty to define and accept a specific set of operations as preferred.... (Deming, 1960:62)

Working Survey Technique

Unfortunately, it often happens that the preferred technique...is too expensive to apply in a full scale survey, or it may be objectional otherwise. The experts must then supply also a working technique. (Deming, 1960:63)

Bias of the Working Technique

The difference of the 2 techniques applied to a complete coverage of the frame, is the bias of the working technique. A working technique is *accurate* if its bias is small. (Deming, 1960:63)

Equal Complete Coverage

The equal complete coverage is the result that would be obtained from examination of all the sampling units in the frame—by the same field workers or inspectors, using the same definitions and procedures, and exercising the same care as they exercised in the sample, at about the same period of time. (Deming, 1960:50)

RELATED CONCEPTS / DEFINITIONS

Accuracy and Precision

Precision generally refers to small variable errors; sometimes it denotes the inverse of the sampling variance; in any case it excludes the effects of bias. (Kish, 1965:510)

Accuracy refers to small total errors, and includes the effects of bias. ...an accurate design must be precise and have zero or small bias. (Kish, 1965:510)

Because of the difficulty of ensuring that no unsuspected bias enters into estimates, we will usually speak of the precision of an estimate instead of its accuracy. Accuracy refers to the size of the deviations from the true mean, whereas precision refers to the size of deviations from the mean *m* obtained by repeated application of the sampling procedure. (Cochran, 1977:16)

The precision, or a measure of the closeness of the sample estimates to the census count taken under identical conditions, is judged in sampling theory by the variance of the estimators concerned. On the other hand, accuracy refers to the closeness to the true values. (Raj, 1968:28)

Reliability and Validity

Fundamentally, reliability concerns the extent to which an experiment, test, or any measuring procedure yields the same results on repeated trials. The measurement of any phenomenon always contains a certain amount of chance error. ... Two sets of measurements of the same features of the same individuals will never exactly duplicate each other. ... The tendency toward consistency found in repeated measurements of the same phenomenon is referred to as reliability. (Carmines and Zeller, 1979:11, 12)

... any measuring device is valid if it does what it is intended to do. An indicator of some abstract concept is valid to the extent that it measures what it purports to measure. ... strictly speaking, one does not assess the validity of an indicator but rather the use to which it is being put. ... The distinction is central to validation because it is quite possible for a measuring instrument to be relatively valid for measuring one kind of phenomenon but entirely invalid for assessing other phenomena. Thus, one validates not the measuring instrument itself but the measuring instrument in relation to the purpose for which is being used. (Carmines and Zeller, 1979:12, 17)

Sensitivity and Specificity

Sensitivity is the extent to which patients who truly manifest a characteristic are so classified. ... Specificity is the extent to which patients who do not manifest a characteristic are correctly classified.... (MacMahon and Pugh, 1970:261–262)

RESPONSE VARIABILITY

The basic concepts necessary to the understanding of the mathematical model underlying the Response Variance Study were presented by Hansen, Hurwitz, and Madow. If it were possible to interview each individual repeatedly, a population of responses for each individual would be generated. In a survey, then, we get a sample of respondents and thus a sample of possible responses from each of the sample persons. Assuming that each survey is regarded as a trial and that the same general conditions hold for each trial, a random variable is defined.... (U.S. Bureau of the Census, 1968:11)

The basic assumption of the model is that the response from a given unit in the population is a random variable having some probability distribution.

Thus, if it were possible to record responses for each individual repeatedly, a population of responses for each unit would be generated. (U.S. Bureau of the Census, 1979:8)

Assumes the response for the jth unit by the ith enumerator has a random component e_{ijk} with $Ee_{ijk} = 0$ and $V(e_{ijk}) = \sigma_e^2$, i.e., constant for all i, j. (Sukhatme and Seth, 1952; Sukhatme and Sukhatme, 1970)

The conditions which determine the response of any individual may be regarded as divided into two groups:

(a) Those conditions which are "constant," "controlled," and predetermined for a given individual response, e.g., the questions to be asked, the type of interviewer. We have referred to these as the essential survey conditions.

(b) Those conditions which are adventitious and "unpredictable," e.g., the mood of the respondent, a momentary distraction which results in a question being misunderstood.

This division is similar to the division between "assignable (i.e., controllable)" causes and "residual" causes of variation in discussions of quality control. We have treated these two groups of factors in the same way as they are treated in the quality control field. Thus, we consider the "adventitious" factors as giving rise to a random variable, the response obtained for a given individual being one of the values of this variate. The "controlled" causes would determine the expected value of this random variable. They also affect its variance. (Hansen, et al., 1953b:308)

On any particular interview of a "unit" by an interviewer a response occurs. It is assumed that this response is a random variable for any interviewer and any respondent; i.e., a given response is considered to be only one of the possible responses which might be obtained from this respondent. Let

P_{hijk} = the probability that the response value X_{hijk} will be obtained if the ith interviewer and the hth group interviews the jth unit.

$P_{hijkuvw}$ = the probability that responses X_{hijk} and X_{huvw} are obtained (in the hth group) if the ith and uth interviewers interview the jth and vth units.

We have then

$$\sum_k P_{hijk} = 1$$

$$E_{hij} X_{hijk} = \bar{X}_{hij} = \sum_k P_{hijk} X_{hijk}$$

where the sum is taken over all possible responses of the jth unit to the ith interviewer. (Hansen et al., 1953b:309, 310)

The particular survey is regarded as one trial, i.e., one survey from among the possible conceived repetitions of the survey under the same general conditions. (Hansen et al., 1961:360)

...postulate a survey procedure—not attainable in the real world in which the process of recording a response for any element in the survey is (i) repeatable and (ii) gives rise to a random variable whose value at trial t is not correlated with its value on any other trial ($t \neq k$). (Hansen et al., 1964:114)

Defines a random variable for the ith population element Y_{itG} where t indexes a conceptual sequence of repeated trials while G specifies a set of general survey conditions. (Koch, 1973; Nathan, 1973)

The response error made is based on the concept of independent repetition. [In the case of a Census] suppose hypothetically, that many independent trials of the Census could be made under the same general conditions. ... A particular Census figure may then be considered to be a random observation from a distribution of all possible Census estimates. (Brackstone et al., 1975:145)

Conceptually we imagine a large number of independent repetitions of the measurement on the ith unit are possible; let $y_{i\alpha}$ be the value obtained on the αth repetition. Then $y_{i\alpha} = \mu_i + e_{i\alpha}$ where μ_i = correct value, $e_{i\alpha}$ = error of measurement. ... under repeated measurements for the same unit, the errors $e_{i\alpha}$ follows some frequency distribution. (Cochran, 1977:377)

...postulate that the survey is conceptually repeatable under identical conditions or, more precisely, that a measurement derived from the survey has a well defined, though quite likely unknown probability distribution. ... In practice it may not be possible to have a sample of more than one observation from the same distribution. (Fellegi, 1964:1016)

The response is a random variable for respondent i. (Madow, 1965)

The survey procedure is such that the process of recording a response for any unit is repeatable; it gives rise to a random variable whose value at trial t is not correlated with its value at any other trial.

[Bailar and Dalenius shows an incidence matrix as follows:

Element Number	Trial Number			
	1	2	$\cdots$	t
1	0	1		0
2	1	0		1
j	1	0		1
N	0	1		0

In this case there is the possibility of a different sample at a particular trial. (Bailar and Dalenius 1969:342)

The unique nature of each response must be stressed, and the fact that in statistical surveys the only thing involved in defining the quality check is the response obtained under some specific circumstances. As a consequence [one must] discard all other concepts that follows therefrom such as individual response bias, response variance, etc. (Zarkovich 1966:46)

The response obtained by the interviewer j on unit i is a random variable having a distribution with mean $\bar{\bar{Y}}_{ij}$ and variance S_{ij}^2. [Raj, 1968]

Does not discuss the concept of individual response variance; speaks instead of variability across two sets of randomization—selection of a sample of population elements and selection of a sample of survey personnel. (Murthy, 1967)

The errors have the structure $b_i + \delta b_s$ where b_i is an error contribution from the ith source common to all units affected by the jth source (all units interviewed by the ith interviewer while δb_s, sometimes referred to as the "elementary non-sampling error" varies randomly from unit to unit (D). (Hartley and Rao, 1978:36) [Biemer (1978, 1979) also uses this approach.]

Under the conditions G, let x_{it} be a (finite) variate value realized at the survey (or trial) t on unit u_i that is a member of sample s. ... On repeated trials on u_i, all of which are hypothetical except the one actually made, i.e., the survey, let $E(x_{it}|s) = X_{is}, V(x_{it}|s) = V_{is}$. ... (Koop, 1974:22, 23)

References

Alwin, Duane F. (1977). "Making Errors in Surveys. An Overview." *Sociological Methods and Research*, **6**, 131–150.

Andersen, Ronald, Judith Kasper, Martin R. Frankel, and associates (1979). *Total Survey Error; Applications to Improve Health Surveys*. Social and Behavioral Science Series, NORC Series in Social Research. San Francisco: Jossey-Bass.

Anderson, Harald (1979). "On Nonresponse Bias and Response Probabilities." *Scandinavian Journal of Statistics*, **6**, 107–112.

Anderson, Dallas W., Fred A. Bryan, Jr., Benjamin S. H. Harris III, Judith T. Lessler, and Jean-Paul Gagnon (1985). "A Survey Approach for Finding Cases of Epilepsy." *Public Health Reports*, **100**, 386–393.

Andrews, Frank M. (1984). "Construct Validity and Error Components of Survey Measure: A Structural Modeling Approach." *Public Opinion Quarterly*, **48**, 409–442.

Andrews, Frank M. and Rick Crandall (1976). "The Validity of Measures of Self-Reported Well-Being." *Social Indicators Research*, **3**, 1–19.

Armstrong, Barbara (1979). "Test of Multiple Frame Techniques for Agricultural Surveys: New Brunswick, 1978." *American Statistical Association 1979 Proceedings of the Section on Survey Research Methods*, 295–300.

Astin, Alexander W. and L. D. Molm (1972). *Correcting for Nonresponse Bias in Follow-up Surveys*. Unpublished report. Washington, DC: Office of Research of the American Council on Education.

Babbie, Earl R. (1973). *Survey Research Methods*. Belmont, CA: Wadsworth.

Backstrom, Charles H. and Gerald Hursh-César (1981). *Survey Research*, 2nd ed. New York: Wiley.

Bailar, Barbara A. (1968). "Recent Research in Reinterview Procedures." *Journal of the American Statistical Association*, **63**, 41–63.

——— (1975). "The Effects of Rotation Group Bias on Estimates from Panel Surveys." *Journal of the American Statistical Association*, **70**, 23–30.

——— (1976). "Some Sources of Error and Their Effect on Census Statistics." *Demography*, **13**, 273–286.

Bailar, John C., III (1978). "Nonresponse and Imputation Issues" (Discussion). *American Statistical Association 1978 Proceedings of the Section on Survey Research Methods*, 230–232.

Bailar, John C., III, and Barbara A. Bailar (1978). "Comparison of Two Procedures for Imputing Missing Survey Values." *American Statistical Association 1978 Proceedings of the Section on Survey Research Methods*, 462–467.

Bailar, Barbara A. and John C. Bailar III (1983). "Comparison of the Biases of the Hot-Deck Imputation Procedure with an 'Equal-Weights' Imputation Procedure." In William G. Madow and Ingram Olkin, eds., *Incomplete Data in Sample Surveys*, Vol. 3, *Proceedings of the Symposium*. New York: Academic, 299–311.

Bailar, Barbara A. and Tore Dalenius (1969). "Estimating the Response Variance Components of the U.S. Bureau of the Census Survey Model." *Sankhyā: The Indian Journal of Statistics*, **31B**, 341–360.

Bailar, Barbara A. and C. M. Lanphier (1978). *Development of Survey Methods to Assess Survey Practices*. Washington, DC: American Statistical Association.

Bailar, Barbara A., Leroy Bailey, and Joyce Stevens (1977). "Measures of Interviewer Bias and Variance." *Journal of Marketing Research*, **14**, 337–343.

Bailar, Barbara A., Leroy Bailey, and Carol Corby (1978). "A Comparison of Some Adjustment and Weighting Procedures for Survey Data." In N. Krishnan Namboodiri, ed., *Survey Sampling and Measurement*. New York: Academic, 175–198.

Bailey, Leroy, Thomas F. Moore, and Barbara A. Bailar (1978). "An Interviewer Variance Study of the Eight Impact Cities of the National Crime Survey Cities Sample." *Journal of the American Statistical Association*, **73**, 16–23.

Banks, Martha J., Ronald Andersen, and Martin R. Frankel (1983). "Total Survey Error." In William G. Madow and Ingram Olkin, eds., *Incomplete Data in Sample Surveys*, Vol. 1, *Report and Case Studies*. New York: Academic, 391–434.

Bartholomew, D. J. (1961). "A Method of Allowing for 'Not-at-Home' Bias in Sample Surveys." *Applied Statistics*, **10**, 52–59.

Beller, Norman D. (1979). *Error Profile for Multiple-Frame Surveys*. ESCS-63. Washington, DC: Economics, Statistics, and Cooperative Service, U.S. Department of Agriculture.

Benus, Jacob and Jean Coley Ackerman (1971). "The Problem of Nonresponse in Sample Surveys." In John B. Lansing, Stephen B. Withey, and Arthur C. Wolfe, eds., *Working Papers on Survey Research in Poverty Areas*. Ann Arbor, MI: Survey Research Center, Institute for Social Research, University of Michigan, 26–59.

Bershad, Max A. (1969). "The Index of Inconsistency for an L-fold Classification System $L \geq 2$." *Census Technical Notes*, no. 2, 1–30.

Biemer, Paul P. (1978). "The Estimation of Nonsampling Variance Components in Sample Surveys." Ph.D. dissertation, Texas A & M University.

_____ (1979). "An Improved Procedure for Estimating the Components of Response Variance in Complex Surveys." *American Statistician Association 1979 Proceedings of the Section on Social Research Methods*, 201–215.

Biemer, Paul P. and S. Lynne Stokes (1985). "Optimal Design of Interviewer Variance Experiments in Complex Surveys." *Journal of the American Statistical Association*, **80**, 158–166.

Birbaum, Z. W. and Monroe G. Sirken (1950). "Bias Due to Nonavailability in Sampling Surveys." *Journal of the American Statistical Association*, **45**, 98–111.

_____ (1965). "Design of Sample Surveys to Estimate the Prevalence of Rare Diseases: Three Unbiased Estimates." *Vital and Health Statistics*, PHS Publication 1000, Ser. 2, *Data Evaluation and Methods Research*, no. 11. Hyattsville, MD: National Center for Health Statistics, Public Health Service, U.S. Department of Health and Human Services.

Bishop, Yvonne M., Stephen Fienberg, and Paul W. Holland (1975). *Discrete Multivariate Analysis: Theory and Practice*. Cambridge, MA: MIT Press.

Blalock, Hubert M., Jr., and Ann B. Blalock, eds. (1968). *Methodology in Social Research*. McGraw-Hill Series in Sociology. New York: McGraw-Hill.

Borus, Michael E. (1966). "Response Errors in Survey Reports of Earnings Information." *Journal of the American Statistical Association*, **61**, 729–738.

Bosecker, Raymond R. and Barry L. Ford (1976). "Multiple Frame Estimation with Stratified Overlap Domain." *American Statistical Association Proceedings of the Social Statistics Section, 1976*, Part 1, 219–224.

Bowley, Sir Arthur Lyon (1926). "Measurement of the Precision Attained in Sampling." *Bulletin of the International Statistical Institute*, **22**, 6–62.

Bowley, Sir Arthur Lyon and A. R. Burnett-Hurst (1915). *Livelihood and Poverty: A Study in the Economic Conditions of Working-Class Households in Northampton, Warrington. Stanley, and Reading*. London: Bell. Reprint. New York: Garland, 1980.

Brackstone, G. J. and J. N. K. Rao (1981). "An Investigation of Rating Ratio Estimators." *Sankhyā: The Indian Journal of Statistics*, **41C**, 97–114.

Brackstone, G. J., J. F. Gosselin, and B. E. Garton (1975). "Measurement of Response Errors in Censuses and Sample Surveys." *Survey Methodology*, **1**, 144–157.

Bradburn, Norman M. (1983). "Response Effects." In Peter H. Rossi, James D. Wright, and Andy B. Anderson, eds., *Handbook of Survey Research*. New York: Academic, 289–328.

Bradburn, Norman M. and Seymour Sudman with Edward Blair, William Locander, Carrie Miles, Eleanor Singer, and Carol Stocking (1979). *Improving Interview Method and Questionnaire Design, Response Effects to Threatening Questions in Survey Research*. NORC Series in Social Research, Jossey-Bass Social and Behavioral Science Series. San Francisco: Jossey-Bass.

Brewer, K. R. W. (1963). "A Model of Systematic Sampling with Unequal Probabilities." *Australian Journal of Statistics*, **5**, 5–13.

Brooks, Camilla Anita (1982). "Probabilistic Survey Error Models to Correct for Nonresponse." Ph.D. dissertation, University of North Carolina at Chapel Hill.

Brooks, Camilla Anita and William D. Kalsbeek (1982). "The Double Sampling Scheme and Its E(MSE)." *American Statistical Association 1982 Proceedings of the Section on Survey Research Methods*, 235–239.

Bross, Irwin (1954). "Misclassification in 2×2 Tables." *Biometrics*, **10**, 478–486.

Campbell, Donald T. and Donald W. Fiske (1959). "Convergent and Discriminant Validity by the Multitrait–Multimethod Matrix." *Psychological Bulletin*, **56**, 81–105.

Cannell, Charles F. (1977). "Discussion of Response Rates." In *Proceedings of the Second Biennial Conference on Health Survey Research Methods*. DHEW Publication (PHS) 79-3207. Washington, DC: National Center for Health Services Research, U.S. Department of Health, Education and Welfare, 13–17.

Cannell, Charles F. and Ramon Hensen (1974). "Incentives, Motives, and Response Bias." *Annals of Economic and Social Measurement*, **3**, 307–317.

Carmines, Edward G. and Richard A. Zeller (1979). *Reliability and Validity Assessment*. Sage University Papers. Beverly Hills, CA: Sage.

Casady, Robert J. and Monroe G. Sirken (1980). "A Multiplicity Estimator for Multiple Frame Sampling." *American Statistical Association 1980 Proceedings of the Section on Survey Research Methods,* 601–605.

Casady, Robert J., Cecelia B. Snowden, and Monroe G. Sirken (1981). "A Study of the Dual Frame Estimators for the National Health Interview Survey." *American Statistical Association 1981 Proceedings of the Section on Survey Research Methods,* 444–447.

Cassel, Claes-Magnus, Carl-Erik Sarndal, and Jan Hakan Wretman (1977). *Foundations of Inference in Survey Sampling*, New York: Wiley.

CASRO Task Force on Completion Rates (1982). *On the Definition of Response Rates.* Special Report. New York: Council of American Survey Research Organizations.

Cassel, Claes-Magnus, Carl-Erik Sarndal, and Jan Hakan Wretman (1983). "Some Uses of Statistical Models in Connection with the Nonresponse Problem." In William G. Madow and Ingram Olkin, eds., *Incomplete Data in Sample Surveys*, Vol. 3, *Proceedings of the Symposium*. New York: Academic, 143–160.

Chai, John J. (1971). "Correlated Measurement Errors and the Least Squares Estimator of the Regression Coefficient." *Journal of the American Statistical Association*, **66**, 478–483.

Chapman, David W. (1976). "A Survey of Nonresponse Imputation Procedures." *American Statistical Association Proceedings of the Social Statistical Section, 1976,* Part 1, 245–251.

———— (1983). "The Impact of Substitution on Survey Estimates." In William G. Madow and Ingram Olkin, eds., *Incomplete Data in Sample Surveys*, Vol. 2, *Theory and Bibliographies*. New York: Academic, 45–61.

Chen, T. Timothy (1979a). "Analysis of Randomized Response as Purposively Misclassified Data." *American Statistical Association 1979 Proceedings of the Section on Survey Research Methods,* 158–163.

———— (1979b). "Log-Linear Models for Categorical Data with Misclassification and Double Sampling." *Journal of the American Statistical Association*, **74**, 481–488.

Chervy, Gabriel (1949). "Control of a General Census by Means of an Area Sampling Method." *Journal of the American Statistical Association*, **44**, 373–379.

Chromy, James R. (1979). "Sequential Sample Selection Methods." *American Statistical Association 1979 Proceedings of the Survey Research Methods Section,* 401–406.

Chromy, James R. and Daniel G. Horvitz (1978). The Use of Monetary Incentives in National Assessment Household Surveys." *Journal of the American Statistical Association*, **73**, 473–478.

Chua, Tin Chiu and Wayne A. Fuller (1987). "A Model for Multinomial Response Error Applied to Labor Flows." *Journal of the American Statistical Association*, **82**, 46–51.

Cochran, William G. (1963). *Sampling Techniques*, 2nd ed. New York: Wiley.

_____ (1968). "Errors of Measurement in Statistics." *Technometrics*, **10**, 637–666.

_____ (1977). *Sampling Techniques*, 3rd ed. New York: Wiley.

Cohen, R. (1955). "An Investigation of Modified Probability Sampling Procedures in Interview Surveys." Master's thesis, American University.

Cohen, Steven B. and William D. Kalsbeek (1981). *NCMES Estimation and Sampling Variances in the Household Survey. Instruments and Procedures 2*. National Health Care Expenditures Study. DHHS Publication (PHS) 81-3281. Washington, DC: National Center for Health Services Research; Office of Health Research, Statistics, and Technology; Public Health Service; U.S. Department of Health and Human Services.

Colledge, M. J., J. H. Johnson, R. M. Pare, and Innis G. Sande (1978). "Large Scale Imputation of Survey Data." *Survey Methodology*, **4**, 203–224.

Coombs, Clyde Hamilton (1952). *A Theory of Psychological Scaling*. Bulletin 34. Ann Arbor, MI: Engineering Research Institute, University of Michigan.

Cornfield, J. (1944). "On Samples from Finite Populations." *Journal of the American Statistical Association*, 236–239.

Cox, Brenda G. (1980). "The Weighted Sequential Hot Deck Imputation Procedure." *American Statistical Association 1980 Proceedings of the Section on Survey Research Methods*, 721–726.

Cox, Brenda G. and Gordon Scott Bonham (1983). "Sources and Solutions for Missing Data in the NMCUES." *American Statistical Association 1983 Proceedings of the Section on Survey Research Methods*, 444–449.

Cox, Brenda G. and Steven B. Cohen (1985). *Methodological Issues for Health Care Surveys*. New York: Dekker.

Cox, Brenda G. and Ralph E. Folsom, Jr. (1978). "An Empirical Investigation of Alternative Non-response Adjustments." *American Statistical Association 1978 Proceedings of the Section on Survey Research Methods*, 219–223.

_____ (1981). "An Evaluation of Weighted Hot Deck Imputation for Unreported Health Care Visits." *American Statistical Association 1981 Proceedings of the Section on Survey Research Methods*, 412–417.

Cox, Brenda G., A. Elaine Parker, Scott S. Sweetland, and Sara C. Wheeless (1982). *Imputation of Missing Item Data for the National Medical Care Utilization and Expenditure Survey*. RTI/1898/06-02F. Research Triangle Park, NC: Research Triangle Institute.

Craig, C. Samuel and John M. McCann (1978). "Item Nonresponse in Mail Surveys: Extent and Correlates." *Journal of Marketing Research*, **15**, 285–289.

Dalenius, Tore (1957). *Sampling in Sweden: Contributions to the Methods and Theories of Sample Survey Practice*. Stockholm: Almqvist & Wiksell.

_____ (1961). "Treatment of the Non-response Problem." *Journal of Advertising Research*, **1**(5), 1–7.

_____ (1962). "Recent Advances in Sample Survey Theory and Methods." *Annals of Mathematical Statistics*, **33**, 325–349.

_____ (1974). *The Ends and Means of Total Survey Design*. Stockholm: The University of Stockholm.

_____ (1983a). "Discussion" (Session on Taxonomy of Survey Errors). *American Statistical Association 1983 Proceedings of the Section on Survey Research Methods*, 61–62.

_____ (1983b). "Informed Consent or R.S.V.P." In William G. Madow and Ingram Olkin, eds., *Incomplete Data in Sample Surveys*, Vol. 3, *Proceedings of the Symposium*. New York: Academic, 85–106.

Daniel, Wayne W. (1975). "Nonresponse in Sociological Surveys: A Review of Some Methods for Handling the Problem." *Sociological Methods and Research*, **3**, 291–307.

David, Martin (1962). "The Validity of Income Reported by a Sample of Families Who Received Welfare Assistance During 1959." *Journal of the American Statistical Association*, **57**, 680–685.

David, Martin, Roderick J. A. Little, Michael E. Samuhel, and Robert K. Triest (1986). "Alternative Methods for CPS Income Imputation." *Journal of the American Statistical Association*, **81**, 29–41.

Deighton, Richard E., James R. Poland, Joel R. Stubbs, and Robert D. Tortora (1978). *Glossary of Nonsampling Error Terms: An Illustration of a Semantic Problem in Statistics*. Statistical Policy Working Paper 4. Washington, DC: U.S. Department of Commerce.

Deming, W. Edwards (1944). "On Errors in Surveys." *American Sociological Review*, **9**, 359–369.

_____ (1953). "On a Probability Mechanism to Attain an Economic Balance Between the Resultant Error of Response and the Bias of Nonresponse." *Journal of the American Statistical Association*, **48**, 743–772.

_____ (1960). *Sample Design in Business Research*. New York: Wiley.

Deming, W. Edwards and Frederick F. Stephan (1940). "On a Least Squares Adjustment of a Sampled Frequency Table When the Expected Marginal Tables Are Known." *Annals of Mathematical Statistics*, **11**, 427–444.

Dempster, A. P., N. M. Laird, and Donald B. Rubin (1977). "Maximum Likelihood from Incomplete Data via the EM Algorithm." *Journal of the Royal Statistical Society*, **39B**, 1–38.

Dillman, Don A. (1972). "Increasing Mail Questionnaire Response in Large Samples of the General Public." *Public Opinion Quarterly*, **36**, 254–257.

_____ (1978). *Mail and Telephone Surveys: The Total Design Method*. New York: Wiley.

Dohrenwend, Barbara Snell (1970–1971). "An Experimental Study of Payments to Respondents." *Public Opinion Quarterly*, **34**, 621–624.

Donald, Marjorie N. (1960). "Implications of Nonresponse for the Interpretation of Mail Questionnaire Data." *Public Opinion Quarterly*, **24**, 99–114.

Dowling, T. A. and Richard H. Shachtman (1975). "On the Relative Efficiency of Randomized Response Models." *Journal of the American Statistical Association*, **70**, 84–87.

Drew, J. H., and Wayne A. Fuller (1980). "Modeling Nonresponse in Surveys with Callbacks." *American Statistical Association 1980 Proceedings of the Section on Survey Research Methods*, 639–642.

_____ (1981). "Nonresponse in Complex Multiphase Surveys." *American Statistical Association 1981 Proceedings of the Section on Survey Research Methods*, 623–628.

Duncan, R. Paul (1979). "Survey Quality and Interviewer Attributes." *Southern Sociologist*, **10**(3), 16–21.

Dunkelberg, William C. and George S. Day (1973). "Nonresponse Bias and Callbacks in Sample Surveys." *Journal of Marketing Research*, **10**, 160–168.

Durbin, J. (1954). "Non-response and Call-Backs in Surveys." *Bulletin of the International Statistical Institute*, **34**(2), 72–86.

Eckler, Albert Ross and Leon Pritzker (1951). "Measuring the Accuracy of Enumerative Surveys." *Bulletin of the International Statistical Institute*, **33**(4), 7–24.

Eisenhart, Churchill (1963). "Realistic Evaluation of the Precision and Accuracy of Instrument Calibration Systems." *Journal of Research, National Bureau of Standards, Engineering and Instrumentation*, **67C**, 161–187.

El-Badry, M. A. (1956). "A Sampling Procedure for Mailed Questionnaires." *Journal of the American Statistical Association*, **51**, 209–227.

Ellis, Brian (1967). "Measurement." In Paul Edwards, ed., *The Encyclopedia of Philosophy*. New York: Macmillan and Free Press, Vol. 5, 241–247.

Elton, R. A. and S. W. Duffy (1983). "Correcting for the Effect of Misclassification Bias in a Case-Control Study Using Data from Two Different Questionnaires." *Biometrics*, **39**, 659–665.

Emrich, Lawrence (1983). "Randomized Response Techniques." In William G. Madow and Ingram Olkin, eds., *Incomplete Data in Sample Surveys*, Vol. 2, *Theory and Bibliographies.* New York: Academic, 73–80.

Ericson, W. A. (1967). "Optimal Sample Design with Nonresponse." *Journal of the American Statistical Association*, **62**, 63–78.

Ernst, Lawrence R. (1978). "Weighting to Adjust for Partial Nonresponse." *American Statistical Association 1978 Proceedings of the Section on Survey Research Methods,* 468–473.

———— (1980). "Variance of the Estimated Mean for Several Imputation Procedures." *American Statistical Association 1980 Proceedings of the Section on Survey Research Methods,* 716–720.

Fellegi, Ivan P. (1963). "Sampling with Varying Probabilities Without Replacement: Rotating and Non-rotating Samples." *Journal of the American Statistical Association*, **58**, 183–201.

———— (1964). "Response Variance and Its Estimation." *Journal of the American Statistical Association*, **59**, 1016–1041.

———— (1974). "An Improved Method of Estimating the Correlated Response Variance." *Journal of the American Statistical Association*, **69**.

Ferber, Robert (1948–1949). "The Problem of Bias in Mail Returns: A Solution." *Public Opinion Quarterly*, **12**, 669–676.

———— (1966). "Item Nonresponse in a Consumer Survey." *Public Opinion Quarterly*, **30**, 399–415.

Ferber, Robert and Seymour Sudman (1974). "Effects of Compensation in Consumer Expenditure Surveys." *Annals of Economic and Social Measurement*, **3**, 319–331.

Filion, F. L. (1976). "Exploring and Correcting for Nonresponse Bias Using Follow-ups of Nonrespondents." *Pacific Sociological Review*, **19**, 401–408.

Fleischer, Jack, Daniel G. Horvitz, J. Malcolm Airth, and A. L. Finkner (1958). "Measurement Errors Associated with Obtaining Acreage Estimates of Cotton Fields." *Biometrics*, **14**, 401–407.

Fleiss, Joseph L. (1973). *Statistical Methods for Rates and Proportions*. New York: Wiley.

Fleiss, Joseph L., Jacob Cohen, and B. S. Everitt (1969). "Large Sample Standard Errors of Kappa and Weighted Kappa." *Psychological Bulletin*, **72**, 323–337.

Folsom, Ralph E., Jr. (1980). "*U*-Statistics Estimation of Variance Components for Unequal Probability Samples with Nonadditive Interviewer and Respondent Errors." *American Statistical Association 1980 Proceedings of the Survey Research Methods Section,* 137–142.

Folsom, Ralph E., Jr., Bernard G. Greenberg, Daniel G. Horvitz, and James R. Abernathy (1973). "The Two Alternative Questions Randomized Response Model for Human Surveys." *Journal of the American Statistical Association*, **68**, 525–530.

Ford, Barry L. (1976). "Missing Data Procedures: A Comparative Study." *American Statistical Association: Proceedings of the Social Statistics Section 1976,* Part 1, 324–329.

—————— (1983). "An Overview of Hot-Deck Procedures." In William G. Madow and Ingram Olkin, eds., *Incomplete Data in Sample Surveys*, Vol. 2, *Theory and Bibliographies*. New York: Academic, 185–207.

Ford, Neil M. (1968). "Questionnaire Appearance and Response Rates in Mail Surveys." *Journal of Advertising Research*, **8**(3), 43–45.

Ford, Barry L., Douglas G. Kleweno, and Robert D. Tortora (1980). "The Effects of Procedures Which Impute for Missing Items: A Simulation Study Using an Agricultural Survey." *American Statistical Association 1980 Proceedings of the Section on Survey Research Methods,* 251–256.

Francis, Joe D. and Lawrence Busch (1975). "What We Know About 'I Don't Knows.'" *Public Opinion Quarterly*, **34**, 207–218.

Frankel, Martin R. (1979). "Models for the Use of Verification Information." In Ronald Andersen, Judith Kasper, Martin R. Frankel, and Associates, *Total Survey Error. Applications to Improve Health Surveys*. NORC Series in Social Research, Social and Behavioral Science Series. San Francisco: Jossey-Bass, 113–121.

Frankel, Lester R. and Solomon Dutka. (1983). "Survey Design in Anticipation of Nonresponse and Imputation." In William G. Madow and Ingram Olkin, eds., *Incomplete Data in Sample Surveys*, Vol. 3, *Proceedings of the Symposium*. New York: Academic, 69–83.

Gannon, Martin J., Joseph C. Nothern, and Stephen J. Carroll (1971). "Characteristics of Nonrespondents Among Workers." *Journal of Applied Psychology*, **55**, 586–588.

Giesbrect, Francis G. (1967). "Classification Errors and Measures of Association in Contingency Tables." *American Statistical Association Proceedings of the Social Statistics Section, 1967*, 271–273.

Gillo, Martin W. and Maynard W. Shelly (1974). "Predictive Modeling of Multivariable and Multivariate Data." *Journal of the American Statistical Association*, **69**, 646–653.

Goldberg, Judith D. (1975). "The Effects of Misclassification on the Bias in the Difference Between Two Proportions and the Relative Odds in the Fourfold Table." *Journal of the American Statistical Association*, **70**, 561–567.

Goodman, Leo A. (1949). "On the Estimation of the Number of Classes in a Population." *Annals of Mathematical Statistics*, **20**, 572–579.

_____ (1978). *Analyzing Qualitative / Categorical Data: Log-Linear Models and Latent-Structure Analysis*. London: Addison-Wesley.

Goudy, Willis J. (1976). "Nonresponse Effects on Relationships Between Variables." *Public Opinion Quarterly*, **40**, 360–369.

Gray, Gerald B. (1975). "Components of Variance Model in Multi-stage Stratified Samples." *Survey Methodology*, **1**, 27–43.

Greenberg, Bernard G., Abdel-Latif A. Abul-Ela, Walt R. Simmons, and Daniel G. Horvitz (1969). "The Unrelated Question Randomized Response Model: Theoretical Framework." *Journal of the American Statistical Association*, **64**, 520–539.

Greenlees, John S., William S. Reece, and Kimberly D. Zieschang (1982). "Imputation of Missing Values When the Probability of Response Depends on the Variable Being Imputed." *Journal of the American Statistical Association*, **77**, 251–261.

Groves, Robert M. and Robert Louis Kahn (1979). *Surveys by Telephone: A National Comparison with Personal Interviews*. New York: Academic.

Groves, Robert M. and James M. Lepkowski (1982). "Alternative Dual Frame Mixed Mode Survey Designs." *American Statistical Association 1982 Proceedings of the Section on Survey Research Methods,* 154–159.

Groves, Robert M. and Lou J. Magilavy (1980). "Estimates of Interviewer Variance in Telephone Surveys." *American Statistical Association 1980 Proceedings of the Section on Survey Research Methods*, 622–627.

Gunn, Walter J. and Isabelle N. Rhodes (1981). "Physician Response Rates to a Telephone Survey: Effects of Monetary Incentive Level." *Public Opinion Quarterly*, **45**, 109–115.

Gurney, Margaret and Maria E. Gonzalez (1972). "Estimates for Samples When Some Units Have Multiple Listings." *Proceedings of the Social Statistics Section of the American Statistical Association*, 283–288.

Hansen, Morris H., and William N. Hurwitz (1946). "The Problem of Nonresponse in Sample Surveys." *Journal of the American Statistical Association*, **41**, 516–529.

Hansen, Morris H. and William G. Madow (1976). "Some Important Events in the Historical Development of Sample Surveys." In Donald Bruce Owen, ed., *On the History of Statistics and Probability*. Statistics, Textbooks and Monographs, Vol. 17. New York: Dekker, 73–102.

Hansen, Morris H., William N. Hurwitz, Eli S. Marks, and W. Parker Mauldin (1951). "Response Errors in Surveys." *Journal of the American Statistical Association*, **46**, 147–190.

Hansen, Morris H., William N. Hurwitz, and William G. Madow (1953a). *Sample Survey Methods and Theory*, Vol. 1, *Methods and Applications*. New York: Wiley.

_____ (1953b). *Sample Survey Methods and Theory*, Vol. 2, *Theory*. New York: Wiley.

Hansen, Morris H., William N. Hurwitz, and Max A. Bershad (1961). "Measurement Errors in Censuses and Surveys." *Bulletin of the International Statistical Institute*, **38**, 359–374.

Hansen, Morris H., William N. Hurwitz, and Thomas B. Jabine (1963). "The Use of Imperfect Lists for Probability Sampling at the U.S. Bureau of the Census." *Bulletin of the International Statistical Institute*, **40**(1), 497–517.

Hansen, Morris H., William N. Hurwitz, and Leon Pritzker (1964). "The Estimation and Interpretation of Gross Differences and the Simple Response Variance." In C. R. Rao with D. B. Lahiri, K. R. Nair, P. Pant, and S. S. Shrikhande, eds., *Contributions to Statistics, Presented to Professor P. C. Mahalonibis on the Occasion of His 70th Birthday.* Oxford, England: Pergamon; Calcutta: Statistical Publishing Society, 111–136.

_____ (1967). "Standardization of Procedures for the Evaluation of Data: Measurement Errors and Statistical Standards in the Bureau of the Census." *Bulletin of the International Statistical Institute*, **42**(1): 49–66.

Hansen, Morris H., William G. Madow, and B. J. Tepping (1983). "An Evaluation of Model-Dependent and Probability-Sampling Inferences in Sample Surveys." *Journal of the American Statistical Association*, **78**, 776–807.

Hartigan, John A. (1975). *Clustering Algorithms*. New York: Wiley.

Hartley, H. O. (1946). "Discussion of 'A Review of Recent Statistical Developments in Sampling and Sampling Surveys' by F. Yates." *Journal of the Royal Statistical Society*, **109**, 37–38.

_____ (1962). "Multiple Frame Surveys." *Proceedings of the Social Statistics Section, American Statistical Association*, 203–206.

_____ (1974). "Multiple Frame Methodology and Selected Applications." *Sankhyā: The Indian Journal of Statistics*, **36C**, 99–118.

Hartley, H. O. and J. N. K. Rao (1978). "The Estimation of Nonsampling Variance Components in Sample Surveys." In N. Krishnan Namboodiri, ed., *Survey Sampling and Measurement*. New York: Academic.

Hauck, Matthew (1974). "Planning Field Operations." In Robert Ferber, ed., *Handbook of Marketing Research*. New York: McGraw-Hill, 2-147–2-159.

Hauck, Matthew and Stanley Steinkamp (1964). *Survey Reliability and Interviewer Competence*. Studies in Consumer Savings, no. 4. Urbana, IL: Bureau of Economic and Business Research, University of Illinois.

Hawkins, Darnell F. (1975). "Estimation of Nonresponse Bias." *Sociological Methods and Research*, **3**, 461–488.

Heckman, James J. (1976). "The Common Structure of Statistical Models of Truncation, Sample Selection and Limited Dependent Variables and a Sample Estimator for Such Models." *Annals of Economic and Social Measurement*, **5**, 475–492.

Heise, David R. and George W. Bohrnstedt (1970). "Validity, Invalidity, and Reliability." In Edgar F. Borgatta, ed., *Sociological Methodology 1970*. San Francisco: Jossey-Bass, 104–129.

Hendricks, Walter A. (1949). "Adjustment for Bias Caused by Nonresponse in Mailed Surveys." *Agricultural Economic Research*, **1**, 52–56.

Henley, James R., Jr. (1976). "Response Rate to Mail Questionnaires with a Return Deadline." *Public Opinion Quarterly*, **40**, 374–375.

Herzog, Thomas N. (1980). "Multiple Imputation of Individual Social Security Benefit Amounts—Part II." *American Statistical Association 1980 Proceedings of the Section on Survey Research Methods,* 404–407.

Herzog, Thomas N. and Clarise Lancaster (1980). "Multiple Imputation of Individual Social Security Benefit Amounts—Part I." *American Statistical Association 1980 Proceedings of the Section on Survey Research Methods,* 398–403.

Herzog, Thomas N. and Donald B. Rubin (1983). "Using Multiple Imputations to Handle Nonresponse in Sample Surveys." In William G. Madow and Ingram Olkin, eds., *Incomplete Data in Sample Surveys,* Vol. 2, *Theory and Bibliographies.* New York: Academic, 209–245.

Hindelang, Michael J., Michael R. Gottfredson, and Timothy J. Flanagan, eds. (1981). *Sourcebook of Criminal Justice Statistics—1980.* Washington, DC: Bureau of Justice Statistics, U.S. Department of Justice.

Hochberg, Yosef (1977). "On the Use of Double Sampling Schemes in Analyzing Categorical Data with Misclassification Errors." *Journal of the American Statistical Association,* **72,** 914–921.

Hochstim, Joseph R. and Demetrios A. Athanasopoulos (1970). "Personal Follow-up in a Mail Survey: Its Contributions and Its Cost." *Public Opinion Quarterly,* **34,** 69–81.

Hoeffding, Wassily (1948). "A Class of Statistics with Asymptotically Normal Distribution." *Annals of Mathematical Statistics,* **19,** 293–325.

Hoinville, Gerald and Robert Jowell in association with others (1978). *Survey Research Practice.* London: Heinemann.

Horvitz, Daniel G. (1978). "Some Design Issues in Sample Surveys." In N. Krishnan Namboodiri, ed., *Survey Sampling and Measurement.* New York: Academic, 3–11.

Horvitz, Daniel G. and D. J. Thompson (1952). "A Generalization of Sampling without Replacement from a Finite Universe." *Journal of the American Statistical Association,* **47,** 663–685.

Horvitz, Daniel G., B. V. Shah, and Walt R. Simmons (1967). "The Unrelated Question Randomized Response Model." *American Statistical Association Proceedings of the Social Statistics Section, 1967,* 65–72.

Jessen, Raymund J. (1978). *Statistical Survey Techniques.* New York: Wiley.

Jones, Roger G. (1983). "An Examination of Methods of Adjusting for Nonresponse to a Mail Survey: A Mail–Interview Comparison." In William G. Madow and Ingram Olkin, eds., *Incomplete Data in Sample Surveys,* Vol. 3, *Proceedings of the Symposium.* New York: Academic, 271–290.

Kahle, Lynn R. and Bruce Dennis Scales (1978). "Personalization of the Outside Envelope in Mail Surveys." *Public Opinion Quarterly,* **42,** 547–550.

Kalsbeek, William D. and Tyler D. Hartwell (1977). "Head and Spinal Cord Injuries: A Pilot Study of Morbidity Survey Procedures." *American Journal of Public Health,* **67,** 1051–1057.

Kalsbeek, William D. and Judith T. Lessler (1978). "Total Survey Design: Effect of Nonresponse Bias and Procedures for Controlling Measurement Error." *Proceedings of the Second Biennial Conference on Health Survey Research Methods.* DHEW Publication (PHS) 79-3207. Washington, DC: National Center for Health Services Research, U.S. Department of Health, Education and Welfare, 19–42.

Kalsbeek, William D., Ralph E. Folsom, Jr., and Anne F. Clemmer (1974). "The National Assessment No-Show Study: An Examination of Nonresponse Bias." *American Statistical Association Proceedings of the Social Statistics Section 1974*, 180–189.

Kalton, Graham (1983). *Compensating for Missing Survey Data*. Research Report Series. Ann Arbor, MI: Institute for Social Research, University of Michigan.

Kalton, Graham and Daniel Kasprzyk (1982). "Imputing for Missing Survey Response." *American Statistical Association 1982 Proceedings of the Section on Survey Research Methods*, 22–31.

Kalton, Graham and Leslie Kish (1984). "Some Efficient Random Imputation Methods." *Communications in Statistics. Theory and Methods*, **13**, 1919–1939.

Kanuk, Leslie and Conrad Berenson (1975). "Mail Surveys and Response Rates: A Literature Review." *Journal of Marketing Research*, **12**, 440–453.

Kaufman, G. M. and Benjamin King (1973). "A Bayesian Analysis of Nonresponse in Dichotomous Processes." *Journal of the Statistical Association*, **68**, 670–678.

Kendall, Maurice George and William R. Buckland (1960). *A Dictionary of Statistical Terms*, 2nd ed. London: Oliver & Boyd.

Kernan, Jerome B. (1971). "Are 'Bulk-Rate Occupants' Really Unresponsive?" *Public Opinion Quarterly*, **35**, 420–422.

King, Benjamin F. (1983). "Quota Sampling." In William G. Madow and Ingram Olkin, eds., *Incomplete Data in Sample Surveys*, Vol. 2, *Theory and Bibliographies*. New York: Academic, 63–71.

Kiranandana, Suchada Seridhoranakul (1976). "Imperfect Frames in Sample Surveys." Ph.D. thesis, Harvard University.

Kish, Leslie (1962). "Studies of Interviewer Variance for Attitudinal Variables." *Journal of the American Statistical Association*, **57**, 92–115.

_____ (1965). *Survey Sampling*. New York: Wiley.

_____ (1974). "Optimal and Proximal Multipurpose Allocation." *American Statistical Association Proceedings of the Social Statistics Section 1974*, 111–118.

_____ (1978). "Populations for Survey Sampling." *Survey Statistician* (International Statistical Institute, International Association of Survey Statisticians), **1**, 14–15.

Kish, Leslie and Dallas W. Anderson (1978). "Multivariate and Multipurpose Stratification." *Journal of the American Statistical Association*, **73**, 24–34.

Kish, Leslie and Irene Hess (1959). "A 'Replacement' Procedure for Reducing the Bias of Nonresponse." *American Statistician*, **13**(4), 17–19.

Kish, Leslie and John B. Lansing (1954). "Response Errors in Estimating the Value of Homes." *Journal of the American Statistical Association*, **49**, 520–538.

Kivlin, Joseph E. (1965). "Contributions to the Study of Mail-Back Bias." *Rural Sociology*, **30**, 322–326.

Koch, Gary G. (1969). "The Effect of Non-sampling Errors on Measurements of Association in 2×2 Contingency Tables." *Journal of the American Statistical Association*, **64**, 852–863.

_____ (1971). *A Response Error Model for a Simple Interviewer Structure Situation*. Technical Report 4, Project SU-618. Research Triangle Park, NC: Research Triangle Institute.

_____ (1973). "An Alternative Approach to Multivariate Response Error Models for Sample Survey Data with Applications to Estimators Involving Subclass Means." *Journal of the American Statistical Association*, **68**, 906–913.

Konijn, H. S. (1981). "Biases, Variances, and Covariances of Rating Ratio Estimators for Marginal and Cell Totals and Averages of Observed Characteristics." *Metrika*, **28**, 109–121.

Koop, J. C. (1974). "Notes for a Unified Theory of Estimation for Sample Surveys Taking into Account Response Errors." *Metrika*, **21**, 19–39.

Korn, Edward, L. (1982). "The Asymptotic Efficiency of Tests Using Misclassified Data in Contingency Tables." *Biometrics*, **38**, 445–450.

Kovar, Mary Grace and Gail Scott Poe (1985). "The National Health Interview Survey Design, 1973–84, and Procedures 1975–83." *Vital and Health Statistics*, Series 1, *Programs and Collection Procedures*, no. 18. DHHS Publication (PHS) 85-1320. Hyattsville, MD: National Center for Health Statistics, Public Health Service, U.S. Department of Health and Human Services.

Krewski, Daniel, Richard Platek, and J. N. K. Rao, eds. (1981). *Current Topics in Survey Sampling*. New York: Academic.

Kruskal, William and Frederick Mosteller (1980). "Representative Sampling, IV: The History of the Concept in Statistics, 1895–1939." *International Statistical Review*, **48**, 169–195.

Kviz, Frederick J. (1977). "Toward a Standard Definition of Response Rate." *Public Opinion Quarterly*, **41**, 265–267.

Landis, J. Richard and Gary G. Koch (1975a). "A Review of Statistical Methods in the Analysis of Data Arising from Observer Reliability Studies (Part I)." *Statistica Neerlandica*, **29**, 101–123.

_____ (1975b). "A Review of Statistical Methods in the Analysis of Data Arising from Observer Reliability Studies (Part II)." *Statistica Neerlandica*, **29**, 151–161.

Larson, Richard F. and William R. Catton, Jr. (1959). "Can the Mail-Back Bias Contribute to a Study's Validity?" *American Sociological Review*, **24**, 243–245.

Lehman, Edward C., Jr. (1963). "Tests of Significance and Partial Returns to Mailed Questionnaires." *Rural Sociology*, **28**, 284–289.

Lessler, Judith T. (1974). "A Double Sampling Scheme Model for Eliminating Measurement Process Bias and Estimating Measurement Errors in Surveys." Ph.D. dissertation, University of North Carolina at Chapel Hill.

_____ (1976). "Survey Designs Which Employ Double Sampling Schemes for Eliminating Measurement Process Bias." *American Statistical Association Proceedings of the Social Statistics Section, 1976*, Part 2, 520–525.

_____ (1977). "Use of Error Rates and Discrepancy Rates to Calculate the Mean Square Error of Survey Data." Paper presented at the annual meeting of the American Public Health Association, Washington, DC, October 30–November 3.

_____ (1981). "Multiplicity Estimators with Multiple Counting Rules for Multistage Sample Surveys." *American Statistical Association 1981 Proceedings of the Social Statistics Section*, 12–16.

_____ (1983). "An Expanded Survey Error Model." In William G. Madow and Ingram Olkin, eds., *Incomplete Data in Sample Surveys*, Vol. 3, *Proceedings of the Symposium*. New York: Academic, 259–270.

Lessler, Judith T., Ralph E. Folsom, Jr., and William D. Kalsbeek (1982). *A Taxonomy of Error Sources and Error Measures for Surveys.* Final report, Project RTI/1791/00-03F. Research Triangle Park, NC: Research Triangle Institute.

Levy, Paul S. (1977). "Optimum Allocation in Stratified Random Network Sampling for Estimating the Prevalence of Attributes in Rare Populations." *Journal of the American Statistical Association,* **72**, 758–763.

Linsky, Arnold S. (1975). "Stimulating Responses to Mailed Questionnaires: A Review." *Public Opinion Quarterly,* **39**, 82–101.

Little, Roderick J. A. (1982). "Models for Nonresponse in Sample Surveys." *Journal of the American Statistical Association,* **77** 237–250.

_____ (1983). "Superpopulation Models for Nonresponse. Part IV." In William G. Madow and Ingram Olkin, eds., *Incomplete Data in Sample Surveys,* Vol. 2, *Theory and Bibliographies.* New York: Academic, 337–413.

_____ (1988). "Missing Data Adjustments in Larger Surveys." *Journal of Business and Economic Statistics,* **6**, 287–296.

Little, Roderick J. A. and Donald B. Rubin (1987). *Statistical Analysis with Missing Data.* New York: Wiley.

Little, Roderick J. A. and Philip J. Smith (1983). "Multivariate Edit and Imputation for Economic Data." *American Statistical Association 1983 Proceedings of the Section on Survey Research Methods,* 518–522.

Love, Lawrence T. and Anthony G. Turner (1975). "The Census Bureau's Experience: Respondent Availability and Response Rates." *American Statistical Association 1975 Proceedings of the Business and Economics Statistics Section,* 76–85.

Lund, Richard E. (1968). "Estimators in Multiple Frame Surveys." *American Statistical Association Proceedings of the Social Statistics Section, 1968,* 282–288.

MacMahon, Brian and Thomas F. Pugh (1970). *Epidemiology: Principles and Methods.* Boston: Little, Brown.

Madow, William G. (1965). "On Some Aspects of Response Error Measurement." *American Statistical Association Proceedings of the Social Statistics Section, 1965,* 182–192.

_____ (1983a). "Annual Survey of Manufactures." In William G. Madow and Ingram Olkin, eds., *Incomplete Data in Sample Surveys,* Vol. 1, *Report and Case Studies.* New York: Academic, 237–268.

_____ (1983b). "Readership of Ten Major Magazines." In William G. Madow and Ingram Olkin, eds., *Incomplete Data in Sample Surveys,* Vol. 1, *Report and Case Studies.* New York: Academic, 367–389.

Madow, William G. and Ingram Olkin, eds. (1983). *Incomplete Data in Sample Surveys,* Vol. 3, *Proceedings of the Symposium.* New York: Academic.

Madow, William G., Harold Nisselson, and Ingram Olkin, eds. (1983a). *Incomplete Data in Sample Surveys,* Vol. 1. *Report and Case Studies.* New York: Academic.

Madow, William G., Ingram Olkin, and Donald B. Rubin, eds. (1983b). *Incomplete Data in Sample Surveys,* Vol. 2, *Theory and Bibliographies.* New York: Academic.

Mahalanobis, P. C. (1944). "On Large-Scale Sample Surveys." *Philosophical Transactions of the Royal Society of London,* **231B**, 329–451.

_____ (1946). "Recent Experiments in Statistical Sampling in the Indian Statistical Institute." *Journal of the Royal Statistical Society*, **109**, 327–378.

Marks, Eli S., Parker W. Maudlin, and Harold Nisselson (1953). "The Post-enumeration Survey of the 1950 Census: A Case History in Survey Design." *Journal of the American Statistical Association*, **48**, 220–243.

Marks, Eli S., William Seltzer, and Karol J. Krotki (1974). *Population Growth Estimation: A Handbook of Vital Statistics Measurement*. New York: The Population Council.

Mayer, Charles S. and Robert W. Pratt, Jr. (1966). "A Note on Nonresponse in a Mail Survey." *Public Opinion Quarterly*, **30**, 637–646.

Messmer, Donald J. and Daniel T. Seymour (1982). "The Effects of Branching on Item Nonresponse." *Public Opinion Quarterly*, **46**, 270–277.

Mood, Alexander McFarlane and Franklin A. Graybill (1963). *Introduction to the Theory of Statistics*, 2nd ed. New York: McGraw-Hill.

Moore, R. P. and B. L. Jones (1973). "Sampling 17-Year-Olds Not Enrolled in School." *American Statistical Association Proceedings of the Social Statistics Section, 1973*, 369–374.

Morgan, James N. and John A. Sonquist (1963). "Problems in the Analysis of Survey Data and a Proposal." *Journal of the American Statistical Association*, **58**, 415–435.

Moser, Claus Adolf and Graham Kalton (1972). *Survey Methods in Social Investigation*, 2nd ed. New York: Basic Books.

Mosteller, Frederick (1978). "Errors. I. Nonsampling Errors." In William H. Kruskal and Judith M. Tanur, eds., *International Encyclopedia of Statistics*, Vol. 1. New York: MacMillan, 208–229.

Mote, V. L. and R. L. Anderson (1965). "An Investigation of the Effect of Misclassification on the Properties of χ^2 Tests in the Analysis of Categorical Data." *Biometrika*, 52, 95–109.

Mulford, Charles L., Gerald E. Klonglan, Richard D. Warren, and David A. Hay (1974). "Influence of Attrition Rates of Responses to Mailed Questionnaire on Measurement and Modeling." Paper presented at the annual meeting of the American Sociological Association, Montreal.

_____ (1967). *Sampling Theory and Methods*. Calcutta: Statistical Publishing Society.

Namboodiri, N. Krishnan (1978). *Survey Sampling and Measurement*. New York: Academic.

Nargundkar, M. S. and G. B. Joshi (1975). "Non-response in Sample Surveys." *40th Session of the International Statistical Institute, Warsaw, 1975, Contributed Papers*, 626–628.

Nathan, Gad (1973). "Response Errors of Estimators Based on Different Samples." *Sankhyā: The Indian Journal of Statistics*, **35A**, 205–220.

_____ (1976). "An Empirical Study of Response and Sampling Errors for Multiplicity Estimates with Different Counting Rules." *Journal of the American Statistical Association*, **71**, 808–815.

National Research Council, Commission on Behavioral and Social Sciences and Education, Committee on National Statistics, Panel on Incomplete Data. (1983). Part 1. "Report." In William G. Madow, Harold Nisselson, and Ingram Olkin, eds., *Incomplete Data in Sample Surveys*, Vol. 1, *Report and Case Studies*. New York: Academic, 1–103.

National Research Council, Panel on Privacy and Confidentiality as Factors in Survey Response (1979). *Privacy and Confidentiality as Factors in Survey Response*. Washington, DC: National Academy of Sciences.

Nelson, Forrest D. (1977). "Censored Regression Models with Unobserved Stochastic Censoring Thresholds." *Journal of Econometrics*, **6**, 581–592.

Neter, John and Joseph Waksberg (1964). "A Study of Response Errors in Expenditures Data from Household Surveys." *Journal of the American Statistical Association*, **59**, 18–55.

Neter, John, E. Scott Maynes, and R. Ramanathan (1965). "The Effect of Mismatching on the Measurement of Response Errors." *Journal of the American Statistical Association*, **60**, 1005–1027.

Neyman, Jerzy (1934). "On the Two Different Aspects of the Representative Method: The Method of Stratified Sampling and the Method of Purposive Selection." *Journal of the Royal Statistical Society*, **97**, 558–625.

Nisselson, Harold (1983). "Overview." In William G. Madow and Ingram Olkin, eds., *Incomplete Data in Sample Surveys*, Vol. 2, *Report and Case Studies*. New York: Academic, 107–120.

Nordbotten, Svein (1963). *Automatic Editing of Individual Statistical Observations*. Conference of European Statisticians Statistical and Standards Studies, no. 2. ST/CES/2. New York: United Nations.

Norris, Carl N. (1983). "Nonresponse Issues in Public Policy Experiments, with Emphasis on the Health Insurance Study." In William G. Madow and Ingram Olkin, eds., *Incomplete Data in Sample Surveys*, Vol. 3, *Proceedings of the Symposium*. New York: Academic, 313–327.

Ognibene, Peter (1971). "Correcting Nonresponse Bias in Mail Questionnaires." *Journal of Marketing Research*, **8**, 233–235.

Oh, H. Lock and Frederick J. Scheuren (1978). "Multivariate Raking Ratio Estimation in the 1973 Exact Match Study." *American Statistical Association 1978 Proceedings of the Section on Survey Research Methods*, 716–722.

⸻ (1980). "Estimating the Variance Impact of Missing CPS Income Data." *American Statistical Association 1980 Proceedings of the Section on Survey Research Methods*, 408–415.

⸻ (1983). "Weighting Adjustment for Unit Nonresponse." In William G. Madow and Ingram Olkin, eds., *Incomplete Data in Sample Surveys*, Vol. 2, *Theory and Bibliographies*. New York: Academic, 143–184.

Oh, H. Lock, Frederick J. Scheuren, and Harold Nisselson (1980). "Differential Bias Impacts of Alternate Census Bureau Hot Deck Procedures for Imputing Missing CPS Income Data." *American Statistical Association 1980 Proceedings of the Section on Survey Research Methods*, 416–420.

O'Muircheartaigh, Colm A. (1977a). "Response Errors." In Colm A. O'Muircheartaigh and Clive Payne, eds., *The Analysis of Survey Data*, Vol. 2, *Model Fitting*. New York: Wiley.

_____ (1977b). "Statistical Analysis in the Context of Survey Research." In Colm A. O'Muircheartaigh and Clive Payne, eds., *The Analysis of Survey Data*, Vol. 1, *Exploring Data Structures*. New York: Wiley.

O'Neil, Michael J., Robert M. Groves, and Charles F. Cannell (1979). "Telephone Interview Introductions and Refusal Rates: Experiments in Increasing Respondent Cooperation." *American Statistical Association Proceedings of the Section on Survey Research Methods*, 252–255.

Ono, Mitsuo and Herman P. Miller (1969). "Income Nonresponses in the Current Population Survey." *American Statistical Association Proceedings of the Social Statistics Section*, 277–288.

Pare, R. M. (1978). *Evaluation of 1975 Methodology: Simulation Study of the Imputation System Developed by BSMD*. Ottawa: Business Survey Methods Division, Statistics, Canada.

Parten, Mildred Bernice (1966), *Survey, Polls, and Samples: Practical Procedures*. New York: Cooper Square.

Pavalko, Ronald M. and Kenneth G. Lutterman (1973). "Characteristics of Willing and Reluctant Respondents." *Pacific Sociological Review*, **16**, 463–476.

Pearce, Nancy D. (1981). "Data Systems of the National Center for Health Statistics." *Vital and Health Statistics*, Series 1, *Programs and Collection Procedures*, No. 16. DHHS Publication (PHS) 62-1318. Hyattsville, MD: National Center for Health Statistics, Public Health Service, U.S. Department of Health and Human Services.

Pearson, Karl (1902). "On the Mathematical Theory of Errors of Judgement, with Special Reference to the Personal Equation." *Philosophical Transactions of the Royal Society of London*, **198A**, 235–299.

Platek, Richard (1977). "Some Factors Affecting Non-response." *Survey Methodology*, **3**, 191–214.

Platek, Richard and Gerald B. Gray (1983). "Imputation Methodology: Total Survey Error, Part V." In William G. Madow and Ingram Olkin, eds., *Incomplete Data in Sample Surveys*, Vol. 2, *Theory and Bibliographies*. New York: Academic, 249–333.

Platek, Richard, M. P. Singh, and V. Tremblay (1977). "Adjustment for Non-response in Surveys." *Survey Methodology*, **3**, 1–24.

Politz, Alfred N. and Willard R. Simmons (1949). "An Attempt to Get 'Not-at-Homes' into the Sample Without Call-Backs." *Journal of the American Statistical Association*, **44**, 9–31.

Pucel, David J., Howard F. Nelson, and David N. Wheeler (1971). "Questionnaire Follow-up Returns as a Function of Incentives and Responder Characteristics." *Vocational Guidance Quarterly*, **19**, 188–193.

Raghavarao, Damarjv and Walter T. Federer (1979). "Block Total Response as an Alternative to the Randomized Response Method in Surveys." *Journal of the Royal Statistical Society*, **41B**, 40–45.

Raj, Des (1956). "Some Estimators in Sampling with Varying Probabilities without Replacement." *Journal of the American Statistical Association*, **51**, 269–284.

_____ (1968). *Sampling Theory*. New York: McGraw-Hill.

Rao, J. N. K. (1968). "Some Nonresponse Sampling Theory When the Frame Contains an Unknown Amount of Duplication." *Journal of the American Statistical Association*, **63**, 87–90.

———— (1973). "On Double Sampling for Stratification and Analytical Surveys." *Biometrika*, **60**, 125–133.

Rao, P. S. R. S. (1983a). "Callbacks, Follow-ups, and Repeated Telephone Calls." In William G. Madow and Ingram Olkin, eds., *Incomplete Data in Sample Surveys*, Vol. 2, *Theory and Bibliographies*. New York: Academic, 33–44.

———— (1983b). "Randomization Approach." In William G. Madow and Ingram Olkin, eds., *Incomplete Data in Sample Surveys*, Vol. 2, *Theory and Bibliographies*. New York: Academic, 97–105.

Rao, J. N. K. and P. D. Ghangurde (1972). "Bayesian Optimization in Sampling Finite Populations." *Journal of the American Statistical Association*, **67**, 439–443.

Rao, J. N. K. and Edward Hughes. (1983). "Comparison of Domains in the Presence of Nonresponse." In William G. Madow and Ingram Olkin, eds., *Incomplete Data in Sample Surveys*, Vol. 3, *Proceedings of the Symposium*. New York: Academic, 215–226.

Reeder, Leo G. (1960). "Mailed Questionnaires in Longitudinal Health Studies: The Problem of Maintaining and Maximizing Response." *Journal of Health and Human Behavior*, **1**, 123–129.

Refior, Wendell F. (1980). "Virginia Fourth Grade Sample: Preliminary Report on the Sample Design." Working Paper for RTI Project 255U-2056, RTI/2056/00-01. Research Triangle Park, NC: Research Triangle Institute.

Remington, Richard D. and M. Anthony Schork (1970). *Statistics with Applications to the Biological and Health Sciences*. Englewood Cliffs, NJ: Prentice Hall.

Rizvi, M. Haseeb (1983a). "An Empirical Investigation of Some Item Nonresponse Adjustment Procedures." In William G. Madow and Ingram Olkin, eds., *Incomplete Data in Sample Surveys*, Vol. 1, *Reports and Case Studies*. New York: Academic, 299–366.

———— (1983b). "Hot-Deck Procedures: Introduction." In William G. Madow and Ingram Olkin, eds., *Incomplete Data in Sample Surveys*, Vol. 3, *Proceedings of the Symposium*. New York: Academic, 351–352.

Roberts, Robert E., Owen F. McCrory, and Ronald N. Forthofer (1978). "Further Evidence on Using a Deadline to Stimulate Responses to a Mail Survey." *Public Opinion Quarterly*, **42**, 407–410.

Rogers, Theresa F. (1976). "Interviews by Telephone and in Person: Quality of Responses and Field Performance." *Public Opinion Quarterly*, **40**, 51–65.

Roshwalb, Alan (1982). "Respondent Selection Procedures Within Households." *American Statistical Association 1982 Proceedings of the Section on Survey Research Methods*, 93–98.

Royall, Richard M. (1971). "Linear Regression Models in Finite Population Sampling Theory." In V. P. Godambe and D. A. Sprott, eds., *Foundations of Statistical Inference*. Toronto: Holt, Reinhart, & Winston of Canada, 259–279.

Rubin, Donald B. (1977). "Formalizing Subjective Notions about the Effect of Nonrespondents in Sample Surveys." *Journal of the American Statistical Association*, **72**, 538–543.

_____ (1978). "Multiple Imputations in Sample Surveys—A Phenomenological Bayesian Approach to Non-response." *American Statistical Association 1978 Proceedings of the Section on Survey Research Methods*, 20–28.

_____ (1979). "Illustrating the Use of Multiple Imputations to Handle Nonresponse in Sample Surveys." Paper presented at the 42nd Session of the International Statistical Institute, Manila, December 4–14.

_____ (1983). "Conceptual Issues in the Presence of Nonresponse." In William G. Madow and Ingram Olkin, eds., *Incomplete Data in Sample Surveys*, Vol. 2, *Theory and Bibliographies*. New York: Academic, 123–142.

_____ (1987). *Multiple Imputation for Nonresponse in Surveys*. New York: Wiley.

Rubin, Donald. B. and Nathaniel Schenker (1986). "Multiple Imputation for Interval Estimation from Simple Random Samples with Ignorable Nonresponse." *Journal of the American Statistical Association*, **81**, 366–374.

Rubin, Donald B., Roderick J. A. Little, and Max A. Woodbury (1983). "Discussion." In William G. Madow and Ingram Olkin, eds., *Incomplete Data in Sample Surveys*, Vol. 3, *Proceedings of the Symposium*. New York: Academic, 203–212.

Sande, Gordon T. (1983). "Replacement for a Ten-Minute Gap." In Williams G. Madow and Ingram Olkin, eds., *Incomplete Data in Sample Surveys*, Vol. 3, *Proceedings of the Symposium*. New York: Academic, **338**.

Sande, Innis G. (1979). "A Personal View of Hot Deck Imputation Procedures." *Survey Methodology*, **5**, 238–258.

_____ (1982). "Imputation in Surveys: Coping with Reality." *The American Statistician*, **36**, 145–152.

_____ (1983). "Hot-Deck Imputation Procedures." In William G. Madow and Ingram Olkin, eds., *Incomplete Data in Sample Surveys*, Vol. 3, *Proceedings of the Symposium*. New York: Academic, 339–349.

Santos, Robert L. (1981). *Effects of Imputation on Complex Statistics*. Income Survey Development Program, Survey Development Research Center in Nonresponse and Imputation Report on Additional Task 2. Ann Arbor, MI: Survey Research Center, Institute for Social Research, University of Michigan.

Schaible, Wesley L. (1983). "Estimation of Finite Population Totals from Incomplete Sample Data: Prediction Approach." In William G. Madow and Ingram Olkin, eds., *Incomplete Data in Sample Surveys*, Vol. 3, *Proceedings of the Symposium*. New York: Academic, 131–141.

Scheaffer, Richard L., William Mendenhall, and Lynn Ott (1979). *Elementary Survey Sampling*, 2nd ed. North Scituate, MA: Duxbury Press.

Schieber, S. J. (1987). "A Comparison of Three Alternative Techniques for Allocating Unreported Social Security Income on the Survey of the Low-Income Aged and Disabled." *American Statistical Association 1978 Proceedings of the Section on Survey Research Methods*, 212–218.

Schuman, Howard and Stanley Presser (1981). *Questions and Answers in Attitude Surveys: Experiments on Question Form. Wording and Context*. New York: Academic.

Schwirian, Kent P. and Harry R. Blaine (1966–1967). "Questionnaire Return Bias in the Study of Blue Collar Workers." *Public Opinion Quarterly*, **30**, 656–663.

Scott, Christopher (1961). "Research on Mail Surveys." *Journal of the Royal Statistical Society*, **124A**, 143–205.

Sekar, C. Chandra and W. Edwards Deming (1949). "On a Method of Estimating Birth and Death Rates and the Extent of Registration." *Journal of the American Statistical Association*, **44**, 101–105.

Settergren, Susan K., Curtis S. Wilbur, Tyler D. Hartwell, and John H. Rassweiler (1983). "Comparison of Respondents and Nonrespondents to a Worksite Health Screen." *Journal of Occupational Medicine*, **25**, 475–480.

Shah, B. V. (1981). *SESUDAAN: Standard Errors Program for Computing of Standardized Rates from Sample Survey Data*. RTI/5250/00-01S. Research Triangle Park, NC: Research Triangle Institute.

Singh, Bahadur (1983). "Bayesian Approach." In William G. Madow and Ingram Olkin, eds., *Incomplete Data in Sample Surveys*, Vol. 2, *Theory and Bibliographies*. New York: Academic, 107–120.

Singh, Bahadur and Joseph H. Sedransk (1978a). "Sample Size Selection in Regression Analysis When There Is Nonresponse." *Journal of the American Statistical Association*, **73**, 362–365.

_____ (1978b). "A Two-Phase Sample Design for Estimating the Finite Population Mean Where There Is Nonresponse." In N. Krishnan Namboodiri, ed., *Survey Sampling and Measurement*. New York: Academic, 143–155.

_____ (1983). "Bayesian Procedures for Survey Design When There Is Nonresponse." In William G. Madow and Ingram Olkin, eds., *Incomplete Data in Sample Surveys*, Vol. 3, *Proceedings of the Symposium*. New York: Academic, 227–247.

Sirken, Monroe G. (1972a). "Household Surveys with Multiplicity." *Journal of the American Statistical Association*, **65**, 257–266.

_____ (1972b). "Stratified Sample Surveys with Multiplicity." *Journal of the American Statistical Association*, **65**, 224–227.

Sirken, Monroe G. and Robert J. Casady (1982). "Nonresponse in Dual Frame Surveys Based on Area/List and Telephone Frames." *American Statistical Association 1982 Proceedings of the Section on Survey Research Methods*, 151–153.

Sirken, Monroe G, and Paul S. Levy (1974). "Multiplicity Estimation of Proportions Based on Ratios of Random Variables." *Journal of the American Statistical Association*, **69**, 68–73.

Smith, T. M. F. (1976). "The Foundations of Survey Sampling: A Review." *Journal of the Royal Statistical Society*, **139A**, 183–195.

Smith, Linda L., Walter T. Federer, and Damaraju Raghavarao (1974). "A Comparison of Three Techniques for Eliciting Truthful Answers to Sensitive Questions." *American Statistical Association Proceedings of the Social Statistics Section, 1974*, 447–452.

Snedecor, George W. and William G. Cochran (1967). *Statistical Methods*, 6th ed. Ames, IA: Iowa State University Press.

Solomon, Herbert and S. Zacks (1970). "Optimal Design of Sampling from Finite Populations: A Critical Review and Indication of New Research Areas." *Journal of the American Statistical Association*, **65**, 653–677.

Sonquist, John A., Elizabeth Lauh Baker, and James N. Morgan (1971). *Searching for Structure. An Approach to Analysis of Substantial Bodies of Micro-Data Documentation for a Computer Program*. Ann Arbor, MI: Institute for Social Research, University of Michigan.

Srinath, K. P. (1971). "Multiphase Sampling in Nonresponse Problems." *Journal of the American Statistical Association*, **66**, 583–586.

Stephen, Frederick F. (1948). "History of the Uses of Modern Sampling Procedures." *Journal of the American Statistical Association*, **43**, 12–39.

Stevens, S. S. (1959). "Measurement, Psychophysics, and Utility." In C. West Churchman and Philburn Ratoosh, eds., *Measurement: Definitions and Theories*. New York: Wiley, 18–63.

Suchman, Edward A. (1962). "An Analysis of 'Bias' in Survey Research." *Public Opinion Quarterly*, **26**, 102–111.

Sudman, Seymour (1976). *Applied Sampling*. New York: Academic.

Sudman, Seymour and Norman M. Bradburn (1974). *Response Effects in Surveys: A Review and Synthesis*. Chicago: Aldine.

_____ (1982). *Asking Questions*. Jossey-Bass Series in Social and Behavioral Sciences. San Francisco: Jossey-Bass.

Sudman, Seymour, Norman M. Bradburn, Ed Blair, and Carol Stocking (1977). "Modest Expectation: The Effects of Interviewers' Prior Expectations on Responses." *Sociological Methods and Research*, **6**, 171–182.

Sukhatme, Pandurang Vasudeo and G. R. Seth (1952). "Non-sampling Errors in Surveys." *Journal of the Indian Society of Agricultural Statistics*, **4**, 5–41.

Sukhatume, Pandurang Vasudeo and Balkrisha V. Sukhatme (1970). *Sampling Theory of Surveys with Applications* 2nd rev. ed. Ames, IA: Iowa State University Press.

Szameitat, Klaus and Karl-August Schäffer (1963). "Imperfect Frames in Statistics and the Consequences for Their Use in Sampling."*Bulletin of the International Statistical Institute*, **40**, 517–544.

Tenebein, Aaron (1970). "A Double Sampling Scheme for Estimating from Binomial Data with Misclassifications." *Journal of the American Statistical Association*, **65**, 1350–1361.

Thomsen, Ib (1973). "A Note on the Efficiency of Weighting Subclass Means to Reduce the Effects of Non-response When Analyzing Survey Data." *Statistisk Tidskrift*, **4**, 278–283.

Thomsen, Ib and Erling Siring (1983). "On the Causes and Effects of Nonresponse: Norwegian Experiences." In William G. Madow and Ingram Olkin, eds., *Incomplete Data in Sample Surveys*, Vol. 3, *Proceedings of the Symposium*. New York: Academic, 25–59.

Thornberry, O. T. and J. T. Massey (1978). "Correcting for Undercoverage Bias in Random Digit Dialed National Health Surveys." *American Statistical Association 1978 Proceedings of the Section on Survey Research Methods*, 224–227.

Tourangeau, Roger (1984). "Cognitive Sciences and Survey Methods." In Thomas B. Jabine, Miron L. Straf, Judith M. Tanur, and Roger Tourangeau, eds., *Cognitive Aspects of Survey Methodology: Building a Bridge Between Disciplines*. Report of the Advanced Research Seminar on Cognitive Aspects of Survey Methodology. Washington, DC: National Academy Press, 73–100.

Tremblay, V., M. P. Singh, and L. Clavel (1976). "Methodology of the Labour Force Survey Re-interview Program." *Survey Methodology*, **2**, 43–62.

Turner, Charles F. and Elizabeth Martin, eds. (1984). *Surveying Subjective Phenomena*. New York: Russell Sage Foundation.

U.S. Bureau of the Census (1968) *Evaluation and Research Program of the U.S. Census of Population and Housing, 1960: Effects of Interviewers and Crew Leaders*. Series ER60, No 7. Washington, DC.

U.S. Bureau of the Census (1975). Course on Nonsampling Errors. Washington, DC: Training Branch, International Statistical Program Center.

_____ (1976). *Statistical Abstract of the United States, 1976,* 97th ed. Washington, DC: Bureau of the Census, U.S. Department of Commerce.

_____ (1978). *The Current Population Survey: Design and Methodology*. Technical Paper 40. Suitland, MD: Bureau of the Census, U.S. Department of Commerce.

_____ (1979). *Enumerator Variance in the 1970 Census: 1970 Census of Population and Housing. Evaluation and Research Program*. PHC(E)-13. Washington, DC: Bureau of the Census, U.S. Department of Commerce.

U.S. Bureau of the Census, Data Access and Use Laboratory (1970). *Census User Guide*, Part 1. Washington, DC: Bureau of the Census, U.S. Department of Commerce.

U.S. Bureau of the Census, Reinterview Section of the Statistical Methods Division (1978). *The Current Population Survey: Reinterview Program, January 1961 through December 1966*. Technical Paper 19. Washington, DC: Bureau of the Census, U.S. Department of Commerce.

U.S. Department of Health, Education, and Welfare (1977). *Advances in Health Survey Research Methods: Proceedings of a National Invitational Conference*. Research Proceeding Series, DHEW Publication (HRA) 77-3154. Washington, DC: U.S. Department of Health, Education and Welfare.

Vacek, Pamela M. and Takamaru Ashikaga (1980). "An Examination of the Nearest Neighbor Rule for Imputing Missing Values." *American Statistical Association 1980 Proceedings of the Statistical Computing Section*, 326–331.

Vogel, Frederic A. (1975). "Surveys with Overlapping Frames—Problems in Application." *American Statistical Association Proceedings of the Social Statistics Section*, 694–699.

Von Riesen, R. D., and T. J. Novotry (1979). "Regression Estimates of Nonresponse Bias in Mail Surveys." Contributed Paper, 24th International Meeting, Institute of Management Sciences.

Waksberg, Joseph (1978). "Sampling Methods for Random Digit Dialing." *Journal of the American Statistical Association*, **73**, 40–46.

Warner, Stanley L. (1965). "Randomized Response: A Survey Technique for Eliminating Evasive Answer Bias." *Journal of the American Statistical Association*, **60**, 63–69.

Warwick, Donald P. and Charles A. Lininger (1975). *The Sample Survey: Theory and Practice*. New York: McGraw-Hill.

Weeks, Michael F., Richard A. Kulka, Judith T. Lessler, and Roy W. Whitmore (1983). "Personal Versus Telephone Surveys for Collecting Household Health Data at the Local Level." *American Journal of Public Health*, **73**, 1389–1394.

Wells, H. Bradley (1971). "Dual Record Systems for Measurement of Fertility Change." Working Papers of the East–West Population Institute, no. 13. Honolulu: East–West Population Institute.

Welniak, Edward J. and John F. Coder (1980). "A Measure of the Bias in the March CPS Earnings Imputation System." *American Statistical Association 1980 Proceedings of the Section on Survey Research Methods*, 421–425.

Williams, W. H. (1968). "The Systematic Bias Effects of Incomplete Responses." *American Statistical Association Proceedings of the Social Statistics Section, 1968*, 308–312.

_____ (1978). "Selection Biases in Fixed Panel Surveys." In Herbert Aron David, ed., *Contributions to Survey Sampling and Applied Statistics. Papers in Honor of H. O. Hartley*. New York: Academic, 89–112.

Williams, S. R. and Ralph E. Folsom, Jr. (1977). *Bias Resulting from School Nonresponse: Methodology and Findings*. Research report. Research Triangle Park, NC: Research Triangle Institute.

Williams, W. H. and C. L. Mallows (1970). "Systematic Biases in Panel Surveys Due to Differential Nonresponse." *Journal of the American Statistical Association*, **65**, 1338–1349.

Wiseman, Frederick and Philip McDonald (1978). *The Nonresponse Problem in Consumer Telephone Surveys*. Report 78-116. Cambridge, MA: Marketing Science Institute.

_____ (1980). *Toward the Development of Industry Standards for Response and Nonresponse Rates*. Report 80-101. Cambridge, MA: Marketing Science Institute.

Wolter, Kirk M. (1983). "Coverage Error Models for Census and Survey Data." In *Booklet*, Vol. 1, *Invited and Contributed Papers, Madrid, September 12–22, 1983*. Paris: International Statistical Institute, 306–323.

Woltman, Henry F., Anthony G. Turner, and John M. Bushery (1980). "A Comparison of Three Mixed-Mode Interviewing Procedures in the National Crime Survey." *Journal of the American Statistical Association*, **75**, 534–543.

Wright, Tommy and How J. Tsao (1983). "A Frame on Frames: An Annotated Bibliography." In Tommy Wright, ed., *Statistical Methods and the Improvement of Data Quality*. Orlando, FL: Academic, 25–72.

Yaffe, Richard, Sam Shapiro, Robert R. Fuchsberg, Charles A. Rohde, and Helen C. Corpeño (1978). "Medical Economics Survey—Methods Study: Cost-Effectiveness of Alternative Survey Strategies." *Medical Care*, **16**, 641–659.

Zarkovich, S. S. (1966). *Quality of Statistical Data*. Rome: Food and Agricultural Organization of the United Nations.

Zeller, Richard A. and Edward G. Carmines (1980). *Measurement in the Social Sciences: The Link Between Theory and Data*. New York: Cambridge University Press.

Index

Accuracy, 238, 374
Adopted system of work, 243
Area sampling, 361
Associative measurement, 235

Bayesian methods, 133, 137, 180–181, 199
  application to estimation in the presence
    of nonresponse, 180–181
  nonrespondent subsampling, application
    to, 133
  prediction approach, 202–203
  relationship to non-Bayesian model-based
    methods, 200
  two-phase sampling, use of, 203
Bias(es):
  definition of, 372–373 ✓
  preferred versus working technique, 244
Bias due to nonresponse, 120, 140–148, 370.
  See also Nonresponse bias
Bias ratio, 251
Bias(es) in survey measurements:
  adjusting for:
    in categorical data measurement,
      273–274
    difference method, 272–273
    ratio method, 268, 269–270, 271–272
    regression method, 268, 270–271, 272
    substitution method, 268–269
  assessment of, 251, 253–254
  consistency studies for examining, 262
  differential bias, 261
  element, 248
  estimator of net bias, 249, 251
  individual response bias, 278
  model for numeric data, 248–249
  net bias due to misclassification, 258
  variance of element biases, 250, 252
Boundary problems, 363

Callbacks, multiple, 135–136
Capture-recapture methods to evaluate
    frame coverage, 63, 64–67
Chebyshev inequality, 9
Choosing among methods for dealing with
    nonresponse, 229–233
  comparison studies, 231–233
  cost considerations, 231
  criteria for making choices, 229–231
  statistical properties, 229–231
Coefficient of reliability, 259
Coefficient of variation of element biases,
    251, 252
Cold-deck method, 214
Complementary frame, 83
Completion rate, 368
Conditioning in panel surveys, 263, 264
Consistency studies, 262
Correlated measurement variance. *See also*
    Correlated response variance
  in general error model for total, 353
  under models with interviewers:
    Census Bureau/Cochran model, 304
    linear models, 313, 314
    Murthy model, 308, 310
Correlated response variance. *See also*
    Correlated measurement variance
  estimates from National Crime Survey,
    324, 326
  Hansen–Hurwitz–Bershad model, 283
  Koch model, 284
  Koop model, 287–288
  in models with interviewers:
    Census Bureau/Cochran model, 304
    Murthy model, 308
Cost models, 9
  in adjusting for bias, 268–273
  for use with multiple frame(s), 85–86

403